# Chemical Oceanography

VOLUME 5
2ND **EDITION**

# Chemical Oceanography

*Edited by*

## J. P. RILEY

*and*

## R. CHESTER

*Department of Oceanography,*
*The University of Liverpool, England*

VOLUME 5
2ND EDITION

1976

ACADEMIC PRESS

LONDON     NEW YORK     SAN FRANCISCO

*A Subsidiary of Harcourt Brace Jovanovich, Publishers*

ACADEMIC PRESS INC. (LONDON) LTD.
24/28 Oval Road,
London NW1

*United States Edition published by*
ACADEMIC PRESS INC.
111 Fifth Avenue
New York, New York 10003

Library of Congress Catalog Card Number: 74–5679
ISBN: 0–12–588605–5

Printed in Great Britain by
PAGE BROS (NORWICH) LTD
NORWICH

# Contributors to Volume 5

WOLFGANG H. BERGER, *Scripps Institution of Oceanography, University of California, San Diego, La Jolla, California 92037, U.S.A.*

D. S. CRONAN,* *Department of Geology, University of Ottawa, Ottawa 2, Canada*

THOMAS A. DAVIES,† *Deep Sea Drilling Project, Scripps Institution of Oceanography, San Diego, La Jolla, California 92037, U.S.A.*

H. ELDERFIELD, *Department of Earth Sciences, The University of Leeds, Leeds, England*

DONN S. GORSLINE, *Department of Geological Sciences, University of Southern California, Los Angeles, California, U.S.A.*

G. D. NICHOLLS, *Department of Geology, University of Manchester, Manchester, England*

HERBERT H. WINDOM, *Skidaway Institute of Oceanography, P.O. Box 13687, Savannah, Georgia 31406, U.S.A.*

* Present address: Geology Department, Imperial College of Science and Technology, London SW7, England.

† Present address: Department of Geology, Middleburg College, Middleburg, Vermont, U.S.A.

# Preface to the Second Edition

Rapid progress has occurred in all branches of Chemical Oceanography since the publication of the first edition of this book a decade ago. Particularly noteworthy has been the tendency to treat the subject in a much more quantitative fashion; this has become possible because of our much improved understanding of the physical chemistry of sea water systems in terms of ionic and molecular theories. For these reasons chapters dealing with sea water as an electrolyte system, with speciation and with aspects of colloid chemistry are now to be considered as essential in any up-to-date treatment of the subject. Fields of research which were little more than embryonic only ten years ago, for example sea surface chemistry, have now expanded so much that they merit separate consideration. Since the previous edition, there has arisen a general awareness of the potential threat to the sea caused by man's activities, in particular its use as a "rubbish bin" and a receptacle for toxic wastes. Although it was inevitable that there should be some over-reaction to this, there is real cause for concern. Clearly, it is desirable to have available reasoned discussions of this topic and also an examination of the role of the sea as a potential source of raw materials in view of the imminent exhaustion of many high grade ores; these subjects are treated in the second, third, fourth and seventh volumes.

Most branches of marine chemistry make use of analytical techniques; the number and range of these has increased dramatically over recent years. Consequently, it has been necessary to expand greatly and restructure the sections dealing with analytical methodology. These developments are extending increasingly into the very important and rapidly developing area of organic chemistry.

Many dramatic advances have taken place in all aspects of marine geochemistry during the last decade. Topics which have received increasing attention include the metalliferous sediments found at some active centres of sea floor spreading, the chemistry of interstitial waters, the formation of deep-sea carbonates and the chemistry and mineralogy of both atmospheric and sea water particulates. Many of the most important developments in the study of deep-sea sediments themselves have been linked to the Deep-Sea Drilling Project. This was initiated in 1968, and for the first time the entire length of the marine sedimentary column was sampled. The extent of these many advances has made it necessary to devote three volumes, i.e. the fifth, sixth and seventh, to various topics in marine geochemistry.

Both the range and accuracy of the physical constants available have increased since the first edition and a selection of tabulated values of these constants are to be found at the end of each of the first four volumes.

No attempt has been made to discuss Physical Oceanography except where a grasp of the physical concepts is necessary for a better understanding of the chemistry. For a treatment of the physical processes occurring in the sea the reader is referred to the numerous excellent texts now available on physical oceanography. Likewise, since the distribution of salinity in the sea is of greater relevance to the physical oceanographer and is well discussed in these texts, it will not be considered in the present volumes.

This series is not intended to serve as a practical handbook of Marine Chemistry, and if practical details are required the original references given in the text should be consulted. In passing, it should be mentioned that, although those practical aspects of sea water chemistry which are of interest to biologists are reasonably adequately covered in the "Manual of Sea Water Analysis" by Strickland and Parsons, there is an urgent need for a more general laboratory manual.

The editors are most grateful to the various authors for their helpful co-operation which has greatly facilitated the preparation of this book. They would particularly like to thank Messrs A. Dickson and M. Preston for their willing assistance with the arduous task of proof reading; without their aid many errors would have escaped detection. They would also like to acknowledge the courtesy of the various copyright holders, both authors and publishers, for permission to use tables, figures, and photographs. In conclusion, they wish to thank Academic Press, and in particular Mr. E. A. S. Cotton, for their efficiency and ready co-operation which has much lightened the task of preparing this book for publication.

*Liverpool*                                                          J. P. RILEY
*January*, 1976                                                      R. CHESTER

# CONTENTS

## Chapter 24 by THOMAS A. DAVIES and DONN S. GORSLINE

### Oceanic Sediments and Sedimentary Processes

## Chapter 25 by G. D. NICHOLLS

### Weathering of the Earth's Crust

## Chapter 26 by HERBERT L. WINDOM

### Lithogenous Material in Marine Sediments

## Chapter 27 by H. ELDERFIELD

### Hydrogenous Material in Marine Sediments; Excluding Manganese Nodules

## Chapter 28 by D. S. CRONAN

### Manganese Nodules and other Ferro-Manganese Oxide Deposits

## Chapter 29 by WOLFGANG H. BERGER

### Biogenous Deep Sea Sediments: Production, Preservation and Interpretation

# Contents of Volume 1

# Contents of Volume 2

# Contents of Volume 3

# Contents of Volume 4

# Contents of Volume 6

# Symbols and units used in the text

*Concentration.* There are several systems in common use for expressing concentration. The more important of these are the molarity scale (g molecules $l^{-1}$ of solution $= mol\ l^{-1}$) usually designated by $c_i$, the molality scale (g molecules $kg^{-1}$ of solvent* $= mol\ kg^{-1}$) designated by $m_i$ and the mole fraction scale usually denoted by $x_i$, which is of more fundamental significance in physical chemistry. In each instance the subscript $i$ indicates the solute species; when $i$ is an ion the charge is not included in the subscript unless confusion is likely to arise. Some other means of indicating the concentration are also to be found in the text, these include: g or mg $kg^{-1}$ of solution (for major components), $\mu$g or ng $l^{-1}$ or $kg^{-1}$ of solution (for trace elements and nutrients) and $\mu$g-at $l^{-1}$ of solution (for nutrients).

*Activity.* When an activity or activity coefficient is associated with a species, the symbols $a_i$ and $\gamma_i$ are used respectively regardless of the method of expressing concentration, where the subscript $i$ has the significance indicated above. Further qualifying symbols may be added as superscripts and/or subscripts as circumstances demand. It is important to realize that the numerical values of the activity and activity coefficient depend on the standard state chosen. It should also be noted that since activity is a relative quantity it is dimensionless.

## UNITS

Where practicable SI units (and the associated notations) have been adopted in the text except where their usage goes contrary to established oceanographic practice.

### LENGTH

| | | |
|---|---|---|
| Å | = Ångstrom | = $10^{-10}$ m |
| nm | = nanometre | = $10^{-9}$ m |
| μm | = micrometre | = $10^{-6}$ m |
| mm | = millimetre | = $10^{-3}$ m |
| cm | = centimetre | = $10^{-2}$ m |
| m | = metre | |
| km | = kilometre | = $10^3$ m |
| mi | = nautical mile (6080 ft) | = 1·85 km |

* A common practice is to regard sea water as the solvent for minor elements.

## WEIGHT

| | | |
|---|---|---|
| pg | = picogram | $= 10^{-12}\,\text{g}$ |
| ng | = nanogram | $= 10^{-9}\,\text{g}$ |
| µg | = microgram | $= 10^{-6}\,\text{g}$ |
| mg | = milligram | $= 10^{-3}\,\text{g}$ |
| g | = gram | |
| kg | = kilogram | $= 10^{3}\,\text{g}$ |
| ton | = metric ton | $= 10^{6}\,\text{g}$ |

## VOLUME

| | | |
|---|---|---|
| µl | = microlitre | $= 10^{-6}\,\text{l}$ |
| ml | = millilitre | $= 10^{-3}\,\text{l}$ |
| l | = litre | |
| dm$^3$ | = litre | |

## CONCENTRATION

| | |
|---|---|
| ppm | = parts per million ($\mu\text{g g}^{-1}$ or $\text{mg l}^{-1}$) |
| ppb | = parts per billion ($\text{ng g}^{-1}$ or $\mu\text{g l}^{-1}$) |
| µg-at l$^{-1}$ | = µg atoms l$^{-1}$ = (µg/atomic weight) l$^{-1}$ |

## TIME

| | | | |
|---|---|---|---|
| s | = second | h | = hour |
| ms | = millisecond | d | = day |
| min | = minute | yr | = year |

## GENERAL SYMBOLS

| | |
|---|---|
| $a_x$ | activity of component $x$ in solution |
| $K$ | equilibrium constant |
| $M_x$ | molarity of component $x$ |
| $m_x$ | molality of component $x$ |
| $P_G$ | partial pressure of gas $G$ in solution |
| $T$ | temperature in K |
| $t$ | temperature in °C |

## ENERGY AND FORCE

| | | |
|---|---|---|
| J | = Joule | $= 0\cdot2390\,\text{cal}$ |
| N | = Newton | $= 10^{5}\,\text{dynes}$ |
| W | = Watt | |

Chapter 24

# Oceanic Sediments and Sedimentary Processes

THOMAS A. DAVIES*

*Deep Sea Drilling Project,
Scripps Institution of Oceanography,
La Jolla, California, U.S.A.*

and

DONN S. GORSLINE

*Department of Geological Sciences,
University of Southern California,
Los Angeles, California, U.S.A.*

## 24.1. INTRODUCTION

### 24.1.1. HISTORICAL BACKGROUND

The oceans occupy 71% of the Earth's surface (Figure 24.1) and exert a correspondingly large influence, directly and indirectly, over Man's activities. The sediments found beneath the oceans contain a record of the history

* Present address: Department of Geology, Middlebury College, Middlebury, Vermont, U.S.A.

1

THOMAS A. DAVIES AND DONN S. GORSLINE

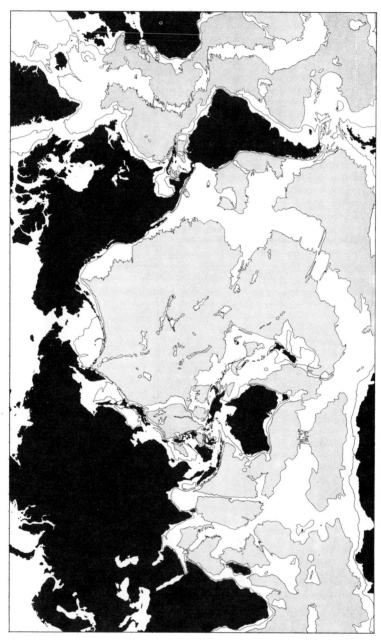

FIG. 24.1. Principal bathymetric features of the World Ocean, as defined by the 1000 m and 4000 m contours. Shaded areas are regions deeper than 4000 m. (Based on Chase, 1975.)

of the ocean basins and their waters. They are, consequently, of considerable interest. Nevertheless, because of the practical difficulties of working in the deep-sea, most of our knowledge of the ocean basins, and their sediments, in areas more than a short distance from land has been acquired surprisingly recently.

The first systematic attempt to gather information about the sediments beneath the deep ocean was made during the famous *Challenger* expedition of 1872–76, although samples had been collected as early as the 1850s by Maury (1856). Murray and Renard's account of their studies of the sediment samples collected by the *Challenger* remains a classic among the literature of marine geology (Murray and Renard, 1891). After H.M.S. *Challenger* came a series of distinguished expeditions, among them the *Albatross* expeditions (Hedgpeth, 1941), the *Meteor* expeditions (Defant, 1933), the *Carnegie* expeditions (Revelle, 1944), and the Swedish Deep Sea Expedition (Pettersson, 1957; Arrhenius, 1952), each of which added significant new information to that gathered as a result of the *Challenger* expedition. Following World War II there was a flurry of deep-sea oceanographic activity involving ships of many nations (see e.g. Lisitzin, 1972). However, until the advent of the operational phase of the Deep Sea Drilling Project in 1968 our knowledge of deep-sea sediments remained restricted to the surficial layer (sampled by gravity and piston coring) and to some older rocks (sampled by dredge hauls) (Menard, 1964; van Andel, 1968), and information about sediment thicknesses and distributions had to be inferred from seismic reflection records. Since mid-1968, however, the drilling vessel of the Deep Sea Drilling Project, the *Glomar Challenger*, has recovered cores representing significant stratigraphical sections from all the major ocean basins except the Arctic. A clearer picture of the distribution of ocean sediments in space and time is thus beginning to develop. For earlier information the chapters by Arrhenius, Griffin and Goldberg, and other authors in Hill (1963) should be consulted.

There is a wealth of data on near-shore sediments and a good review can be found in the recent text by Reineck and Singh (1973) (see also Chapter 33). A number of classic papers concerning shallow marine sediment types and their distributions deserve specific mention. Shepard (1932) examined a large number of shelf sediments and was the first to recognize that they do not show a regularly decreasing grain size away from the coast, as had been postulated by Johnson (1919). Emery (1952) expanded upon this hypothesis and defined a series of sediment types present in the surficial materials of all shelves, and showed later (1968) that there is a broad latitudinal arrangement of the major types. Modifications of these groupings and their applications have come recently from a number of workers and have been reviewed by Swift (1974) and by Swift *et al.* (1971). Continental margin sedimentation has been examined *in toto* in the compilations by Stanley, (1969; 1975) and processes have been discussed in a symposium volume edited by Swift *et al.* (1972). The reports of a recent symposium on estuarine and shelf sedimentation (Allen, 1974) are a source of data on shelf sedimentary materials, principally those of the Atlantic.

Knowledge about the distribution of oceanic sediments, coupled with new information from physical and chemical oceanography and marine biology, is leading to a new understanding of the marine sedimentary system (see, e.g. Berger, 1970; Hays, 1970; Swift *et al.*, 1972; Broecker, 1974; Hay, 1974; Hsü and Jenkyns, 1974). These advances will be illustrated here in terms of the sediments of the northern North Atlantic, the eastern equatorial Pacific and the Indian oceans.

### 24.1.2. PHILOSOPHY OF SEDIMENT CLASSIFICATION

The aim of any system of classification is to provide a tool for the understanding and interpretation of data, and the ideal system uses all the information regarding the objects to be classified, and groups it in a way which is meaningful to the observer. Consequently, the name assigned to an object should convey a great deal more information than simply an indication of its place in the classification scheme. Unfortunately, we do not know enough about the oceanic environment to be able always to decide which information is the most important. For this reason most sediment classification schemes fall somewhere along a spectrum ranging from the purely descriptive through schemes which become increasingly more interpretative (Klein, 1963). The matter is further complicated by the very confusing terminology used for sediments which includes the indiscriminate use of archaic, common and modern scientific terms; an example of this is the use of the term *greywacke* (Crook, 1974). Briefly, early geologists simply appropriated the terms used by miners, sailors or farmers and defined them more precisely. Increasingly sophisticated equipment for sampling and analysis, and the development of new methods for the reduction and analysis of data have made possible the evolution of more complex descriptive terms and these have now largely supplanted or modified the older meanings.

The complexity and ephemeral nature of the oceanic environment, and the practical difficulties of collecting sediment samples from beneath the sea, have had a considerable impact on both the way in which oceanic sediments are studied and the extent to which our observations can be interpreted. Collections made on the early expeditions were limited to small surficial samples from relatively imprecisely defined locations. These permitted little more than simple descriptions of texture, and occasionally colour and composition, to be made. Seabed notations on modern navigational charts preserve this terminology by the use of terms such as *grey fine sand, shell* and *green mud*. This type of descriptive scheme serves to define broad patterns of sediment distribution, but is of little value in interpretation, particularly in the deep-sea, where many factors influence the nature of the sedimentary deposits. Consequently, little can be gained from such a simplistic approach. Even so, textural aspects of marine sediments have remained a strong element in their definition and description. Composition has been a secondary feature and has been approached from both chemical and mineralogical viewpoints. Griffiths (1967) has suggested that a sediment description must include at least the following five parameters: size, shape, orientation, packing and compositions of the grains in an aggregate of grains. Not all of these parameters can be accurately determined without producing changes in one or several of the others. For this reason the discussion by Griffiths highlights the attention which must be paid to sample collection and analysis, and also shows the inherent weaknesses of classifications that use only one or two of these parameters.

The purely textural approach has been employed by many workers; examples being the use of textural diagrams (see, e.g. Picard, 1971), the application of grain size distribution parameters (see, e.g. Folk and Ward, 1957; Inman, 1952; Tanner, 1958) and the use of size distribution curves (see, e.g. Doeglas, 1946; Visher, 1969). Davis and Ehrlich (1970) have noted some of the pitfalls in the application of textural parameters to the identification of the original sedimentary environment (see, e.g. Friedman 1958, 1962). The failure of these simple approaches has been reviewed recently by Reed *et al.* (1975).

In a complex system, such as the ocean, many parameters in addition to those of the sediments themselves must be considered before a significant insight into depositional environments can be gained. The development of seismic reflection profiling techniques, satellite navigation and deep-sea drilling has enabled marine geologists to take into account the form, location and thickness of the sediment bodies, and some of their temporal variations. This permits a more sophisticated approach to the study and classification of oceanic sediments, and is leading to a better understanding of sedimentary processes which formerly could only be discussed speculatively. An example of this approach is the identification of bottom current deposits in the North Atlantic, the mechanism of formation of which was not apparent until the form of the deposits, their location and the physical oceanography of the region were taken into account (Heezen *et al.*, 1966; Schneider *et al.*, 1967; Jones *et al.*, 1970).

Genetic terms have been used to describe sediment types and thereby assist in revealing processes, histories and sequences in sediment deposition. A good example of this is the classification of shelf sediments devised by

Emery (1952) in which they are grouped into five types: detrital, biological, authigenic, residual and relict. The identification of each type depends on both qualitative and quantitative factors such as size, composition, degree of staining or structure of the sediment particles. Swift *et al.* (1971) have argued that relict sediments, defined as remnants of previous sedimentary episodes, may be subject to reworking and have suggested that the term *palimpsest* be applied to them; thus indicating that although they retain much of their original character they are, in fact, undergoing some redistribution by present day transport mechanisms. Similar genetic terminologies have been developed for deep-sea sediments. All of these genetic schemes of classification, are, however, subject to the criticism that an unacceptable amount of subjective interpretation is necessary before a sediment can be correctly classified.

Sedimentologists now tend to view sedimentary deposits as dynamic products of the interaction of such factors as sediment supply rates, dissolution rates, structural activity, geometry of the depositional site and energy levels of the transporting or distributing agents. This approach has led to the development of mathematical or analogue models which can be used to generate deposit geometries, sedimentary structure assemblages and textural and compositional gradients which can then be checked against field observations (see, e.g. Sloss, 1962; Allen, 1964; Wright and Coleman, 1973). Contemporary analytical studies are much more likely to illustrate sediment data in the form of trend surfaces rather than as maps of arbitrarily defined sediment types (see Booth (1973) for a recent application of trend surfaces, and Imbrie and Purdy (1962) for an example of the sediment type form).

### 24.1.3. OUTLINE OF DISCUSSION AND GENERAL APPROACH

The aim of this chapter is to provide a broad background for the specialized discussions of sediment types and processes contained in the rest of the book. Following a section discussing the basic elements of the marine sedimentary system, present day knowledge and ideas concerning ocean sediments and their development will be illustrated by considering the sedimentation in three regions. These are: (i) the North Atlantic, a relatively young ocean surrounded by landmasses which have contributed substantial amounts of weathering products; (ii) the equatorial Pacific, an oceanic region relatively isolated from land, in which biological productivity and sea water chemistry are the dominant influences; and (iii) the Indian Ocean which may be regarded as being intermediate between them, and provides an opportunity for the examination of temporal effects on sedimentation in an ocean basin. In each oceanic system the same fundamental elements of the marine sedimentary system can be recognized, but in order to describe and understand the processes operating in each instance different facets of the system must be highlighted. The problems of developing global models for oceanic sedimentation, which depend on the assumption that the world ocean is a closed physical and chemical system, will then be briefly reviewed. For a more general treatment of geological, oceanographical and sedimentary concepts the reader is referred to Blatt *et al.* (1972), Sverdrup *et al.* (1941) and Garrels and Mackenzie (1971).

In the following discussions the processes occurring on the continental margins have been generally separated from those taking place in the deep ocean provinces. A brief study of the literature reveals that marine sedimentological studies can usually be classed as shallow water or deep water ones. which reflects factors which are both physical, e.g. rates of circulation, the distances from source and temperature, and chemical, e.g. salinity and dissolved gas content of the waters. Such a division, however, does not imply that there is a total isolation of these two environments. For example, the movement of sediments from a land source follows the gravity gradient from mountain top to deep ocean trench, although the movement may be episodic and may involve more than one agent of transport. This simple division can, however, be related to the general two-layered structure of the ocean, i.e. the *surface ocean*, with its wind driven circulation, which is separated by the permanent thermocline from the *deep ocean* water mass system, which is characterized by a thermohaline circulation. This is a useful distinction when the problems of modelling the marine sedimentary system are considered. A number of writers (see, e.g. Poldervaart, 1955; Menard, 1961; Berger, 1974) have noted that the great bulk of contemporary sediment is deposited at the continental margins, particularly on the continental rise. Probably no more than 20% of the total volume of material sedimented is deposited in the deep ocean, and an even smaller percentage, perhaps half, of this forms the pelagic biogenic oozes and clays. The sediments of the continental margins and abyssal plans are produced largely by the physical transport of terrigenous material. However, geochemical processes play a much more important part in the formation of pelagic sediments.

Basic assumptions which underlie our discussion of oceanic sedimentation is that the theory of plate tectonics is valid. Seismic reflection records indicate that the ocean basins contain much less sediment than might be anticipated on the basis of known rates of erosion of the continents (Kuenen, 1965). Even before the first cores were recovered by the *Glomar Challenger* almost unassailable geophysical evidence had been assembled to demonstrate that the continents were moving with respect to one another by a process of sea floor spreading (Dietz, 1961; Hess, 1962; Vine, 1966; Cox, 1973). Indeed, one of the first results of the Deep Sea Drilling Project was the verification of this hypothesis (Maxwell *et al.*, 1970). From the concept of sea floor spreading has grown the theory of plate tectonics, in which the surface of the Earth is considered to be made up of many rigid plates, all of which are in motion with respect both to one another and to the geographical coordinates of latitude and longitude (Le Pichon, 1968). Although plate tectonics will not be discussed here, two important consequences of this concept must be pointed out. Firstly, the oceans are geologically young. Thus, rocks older than $166 \times 10^6$ years (Jurassic) have not yet been recovered from beneath the deep oceans (Davies and Edgar, 1974), and present evidence makes it appear unlikely that they ever will be. Secondly, the ocean basins have changed in size, shape and geographical location through time. This has, naturally, influenced the nature and distribution of oceanic sediments. For these reasons when geologists consider oceanic sedimentation in the wider context of the Earth's history of $4.5 \times 10^9$ years they are presented with only a small number of fragments from the last few scenes of a constantly changing drama.

## 24.2. ELEMENTS OF THE MARINE SEDIMENTARY SYSTEM

### 24.2.1. GENERAL STATEMENT

Marine sediments are a subset of all sediments, and any analysis of their origins, movements and deposition should be viewed in this general framework. The basic elements of all sedimentary systems are: (a) source, or provenance; (b) transport, or distribution; (c) deposition, or accumulation; and (d) post-depositional processes. For example, the products of the weathering of mountain ranges are transported by rivers and deposited at the river mouths, eventually to become deltaic sandstones. The tests of dead planktonic foraminifera are transported from the photic zone by ocean currents and gravity and eventually come to rest on the ocean floor as deposits of calcareous ooze, which may subsequently be lithified to chalk. Each of these elements is influenced by several factors which can be broadly termed "environmental constraints". Thus, climate, relief, regional geology, or oceanographical conditions may influence the amount and nature of the sediment supply. The characteristics of a sedimentary deposit reflect the actions and interactions of these elements and their associated subsets of environmental constraints. The observed horizontal and vertical distributions of ocean sediments are the result of the sum total of all these actions and interactions integrated over the surface of the Earth and over geological time. This general approach has long been recognized as the framework within which to describe sedimentary geology (see Twenhofel, 1932).

## 24.2.2. ELEMENTS

### 24.2.2.1. Source (Provenance)

The components of sediments deposited on the floor of the oceans may be classified as being of terrigenous, biogenous, volcanic, authigenic or cosmic origin.

Terrigenous sediments have components which are largely derived through the weathering and erosion of the land masses by water, ice or wind, and which have been delivered to the ocean in particulate form. The annual addition of materials to the oceans from land is estimated to be $\sim 250 \times 10^{14}$ g (Garrels and Mackenzie, 1971) of which $\sim 206 \times 10^{14}$ g is solid matter. The way in which this material reaches the oceans, and the amount introduced via the different paths, are summarized in Fig. 24.2.

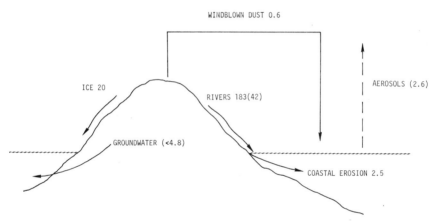

WINDBLOWN DUST 0.6

ICE 20

RIVERS 183(42)

AEROSOLS (2.6)

GROUNDWATER (<4.8)

COASTAL EROSION 2.5

FIG. 24.2. Annual transfer of erosion products to the ocean. Figures are in units of $10^{14}$ g. Figures in parentheses refer to dissolved products. (Based on Garrels and Mackenzie, 1971.)

Climate, relief and geology of the source areas are the principal factors which control the supply of sediment to the oceans. Rivers are by far the most important dispersal agent, transporting $\sim 183 \times 10^{14}$ g yr$^{-1}$ of particulate matter (Holeman, 1968), and $\sim 42 \times 10^{14}$ g yr$^{-1}$ of dissolved substances (Garrels and Mackenzie, 1971). Most of this material is derived from the erosion of mountainous semi-arid regions in which the rainfall is sufficient to cause substantial run-off, but insufficient, or too erratic, to permit the development of an erosion-resisting cover of vegetation (Fig. 24.3). Langbein and Schumm (1958) have shown that the maximum sediment transport in the United States occurs with rivers draining regions having an annual precipitation of 25–35 cm, but this generalization may or may not apply on a world-wide basis. It should be noted that small variations

Fig. 24.3. Drainage basins of the principal rivers of the world. Figures refer to rivers listed in Table 24.1. (Based on Gross, 1972.)

in climatic factors can cause large changes in the sediment supply to the oceans. For example, the average annual rainfall on the coastal mountain slopes of California is ~ 70 cm per year of which perhaps 50 cm is lost by evapo-transpiration processes leaving about 10 cm each for ground-water recharge and runoff (California Department of Water Resources, 1967). If the average annual rainfall were to increase by only 5 cm, without a change in the seasonality of the rainfall, then as the evapo-transpiration losses are likely to remain essentially constant, the net runoff may increase by 50%. Such minor changes are probably the causes of the marked mid-latitude changes in the sediment contribution seen in the sedimentary records of the California Continental Borderland (Gorsline and Prensky, 1975) in which the terrigenous contribution increased during glacial periods by from 3 to 10 times over that during interglacial periods. Variations in the proportion of dissolved to particulate load, and in the specific sources of the various types of detritus, have been examined for the Amazon River drainage basin by Gibbs (1967), who noted that the high relief areas govern the delivery of particulates and that the low lying floodplains and valley floors are the prime contributors to the dissolved load. Effects of climate produce varying degrees of weathering, and these in turn control the proportions of clay mineral species. These clay

TABLE 24.1

*Rivers discharging more than $10 \times 10^{13}$ g yr$^{-1}$ of sediment annually**

| Name | Location | Total drainage Area (km$^2 \times 10^3$) | Average annual suspended load (g $\times 10^{13}$) | Average discharge at mouth (km$^3$ yr$^{-1}$) |
|---|---|---|---|---|
| 1 Yellow | China | 673 | 189 | 47 |
| 2 Ganges | India | 956 | 145 | 369 |
| 3 Brahmaputra | East Pakistan | 666 | 73 | 383 |
| 4 Yangtze | China | 1943 | 50 | 685 |
| 5 Indus | West Pakistan | 969 | 44 | 174 |
| 6 Amazon | Brazil | 5776 | 36 | 5696 |
| 7 Mississippi | U.S.A. | 3222 | 31 | 561 |
| 8 Irrawaddy | Burma | 430 | 30 | 426 |
| 9 Mekong | Thailand | 795 | 17 | 347 |
| 10 Colorado | U.S.A. | 637 | 14 | 5 |
| 11 Red | North Vietnam | 119 | 13 | 123 |
| 12 Nile | Egypt | 2979 | 11 | 89 |

* Based on data from Holeman (1968). Data are incomplete since data for many rivers are lacking.

mineral provinces are reflected in deep ocean clay mineralogy as illustrated by Biscaye (1965) and by Griffin et al. (1968).

Four rivers, the Hwang Ho (Yellow), Ganges, Brahmaputra and Yangtze, account for 25% of the river-transported sediment delivered to the ocean annually, and the first nine rivers listed in Table 24.1 account for more than one third of the total. The bulk of this material is deposited in near-shore environments, such as estuaries, deltas or off-shore basins, and in only a few instances, such as the Congo and Magdelena (Colombia) rivers, does it find its way directly beyond the continental shelf into deep water. If should also be noted that very little sediment is discharged directly into open coastal-ocean areas and that the bulk of the discharge is into marginal seas. Of the 12 rivers listed in Table 24.1, only the Amazon, and the Ganges-Brahmaputra drain into open ocean areas.

Ice transport, as might be expected, is now of significance only in the polar regions, principally Antarctica, from which $19 \times 10^{14}$ g of erosion products are transported to the ocean each year (Garrels and Mackenzie, 1971). However, during the glacial periods which have occurred from time to time throughout the past $40 \times 10^6$ yr (Kennett et al., 1975), particularly in the Pliocene and Pleistocene (last $5 \times 10^6$ yr), ice transport was of much greater significance, as is shown by the extent to which ice-rafted debris is spread over the ocean floor. Lisitzin (1972) has documented the very large supply of suspended sediments from glacial action in the Antarctic continent, and has suggested an even larger role for this area than is generally conceded by Western European and U.S. workers.

Coastal erosion and wind blown transport are quantitatively insignificant as agents in the supply and transportation of terrigenous material to the ocean. Wind transport is of significance downwind of the arid, low latitude regions of the world, and quartz dusts of fine silt and clay sizes are excellent indicators of global wind patterns (Rex and Goldberg, 1958). Heath et al. (in press) have deduced from shifts in the zones of wind-contributed particulates in ocean sediments that the wind belts moved towards the equator during glacial episodes. Such effects are quantitatively discernable only in deep ocean sediments having extremely slow accumulation rates.

Biogenous sediments result from the fixation of mineral phases by marine organisms. In the near-shore region, molluscs and coral fix significant amounts of calcium carbonate in the form of aragonite. In contrast, in deep ocean areas zooplankton (principally foraminifera and radiolarians) and phytoplankton (coccoliths and diatoms) are responsible for fixing large amounts of calcium carbonate (as calcite) and silica (as opal). Fish and sponges are among the other groups of marine organisms which are capable of fixing various minerals (Bathurst, 1971). Ginsburg and James (1974) in a comprehensive and authoritative review of carbonate sedimentation on continental margins have noted that carbonates are significant contributors to sediments in mid-latitudes as well as in tropical areas. Calcium carbonate is usually a minor component in those environments which are dominated by terrigenous sediments; continental margin carbonates nevertheless contribute sizeable volumes to the marine sediment budget, even though these contributions are often ignored in geochemical summaries.

Shallow water carbonate sediments are typically best developed in the tropics (see, e.g. Emery, 1963b) and in areas where terrigenous sedimentation is limited by barriers, climate or circulation. In inner shelf areas such sediments are composed mainly of molluscan fragments with algal and foraminiferal debris farther offshore (Ginsburg and James, 1974). Carbonate muds and partially cemented aggregates are common in the quiet waters of lagoons and back reef areas (Imbrie and Purdy, 1962). In agitated waters, for example those of the Persian Gulf and the Bahama banks, oolites are a typical form (Newell and Rigby, 1957; Newell et al., 1960).

The coral reefs of the tropical Pacific reflect a purely biological response to oceanic conditions, and as first suggested by Darwin, their form and development are indicators of ocean floor dynamics. These deposits, and their associated talus slopes, are aggregates of algal and coral debris with a large admixture of mollusc debris and foraminifera. Fine sediment is relatively rare because of the turbulence of these well mixed and aerated systems (see Emery et al., 1954; Wells, 1957a,b).

Faecal pellets are generated by many types of organisms living in the surface waters in deep ocean areas; these remove the very fine suspended load and aggregate it to produce silt-sized particles. I. M. McCave (personal communication) has called these organisms "particle generators". Operation of this mechanism in the high productivity areas along the eastern sides of the ocean basins, at the eastern end of the equatorial circulation, and at the major oceanic convergences, effectively removes much of the detrital suspended matter which escapes flocculation in estuaries and coastal waters. This process explains the discrepancy between the theoretical settling rates of dispersed fine clays and the concentration of most land-derived clays in abyssal plains and rises adjacent to the continents. In addition, it helps to maintain the well-defined latitudinal pattern of clay mineral suites found in deep ocean sediments (see above).

The ocean contains $\sim 470 \times 10^{20}$ g of dissolved substances (Garrels and Mackenzie, 1971). Most of these are present in low concentrations and, although they may be temporarily removed by marine organisms, they are recycled back into solution in the ocean immediately after the death of the organism. Silica and calcium carbonate are exceptions to this as they are biologically fixed in such a way that dissolution takes a significant length of time. The production of biogenous sediment is thus determined by the biological productivity of the ocean; this, in turn, is controlled by the available nutrient supply (Berger, 1974). In regions where the biological productivity is sufficiently high, deposits of biogenous sediment accumulate. The factors controlling the accumulation of biogenous sediment are complex, and will be outlined in a later section of this chapter. If it is assumed that all of the calcium carbonate and silica which are permanently removed from the ocean as biogenous sediment are replaced by dissolved substances entering from rivers and groundwater, it can be concluded that $\sim 1.40 \times 10^{15}\,\mathrm{g\,yr^{-1}}$ of calcium carbonate and $\sim 0.49 \times 10^{15}\,\mathrm{g\,yr^{-1}}$ of silica are extracted as biogenous sediment (data from Garrels and Mackenzie, 1971). However, Broecker (1974), and others, have presented data which suggest that the amount of calcium carbonate precipitated annually may be only about one third of that given above. The calculations on which these conclusions are based assume that the oceans are in a steady state condition, and thus that the composition of sea water is constant. This is probably a valid assumption over periods extending to several tens of millions of years, but may not be so for longer periods (see Chapter 5 and Garrels and Mackenzie, 1971; Hay, 1974). The annual production of biogenous sediment is thus only about 10% of that of the terrigenous sediment introduced into the oceans by erosion of the continents – see Berger (1974) for a more comprehensive discussion of this subject.

Volcanism is a common process on the deep ocean floor and may make a significant contribution to marine sedimentation (Butze, 1955; Ninkovich and Heezen, 1965). Altered volcanic ash is a major component of Pacific red clays, and may also be a source of manganese and iron precipitates. Terrestrial volcanic eruptions may also contribute to marine sedimentation, via ash falls, rafting of pumice (see, e.g. Richards, 1958) and pyroclastic flows.

Extra-terrestrial material constitutes an insignificant but interesting fraction of deep-sea sediments, being most abundant in the slowly accumulating deposits of deep Pacific areas far from continental sources. Such material includes micrometeorites (cosmic spherules) and tektites (Glass, 1967).

Authigenic materials (chemical precipitates) may be in part biochemically produced. Such materials include manganese, iron and other metal oxides, barium sulphate, aluminosilicates and, in shallow water, phosphates (see Chapter 33). The most important of these are ferromanganese nodules (see Chapter 28).

Man's activities in dumping waste materials into the ocean, although not yet of geological significance on a global scale, are locally of considerable ecological and geological importance near densely populated, highly industrialized, regions such as northern Europe and the northeastern U.S.A. (Pearson, 1960). Gross (1970) has pointed out that New York City alone supplies more particulate matter to the ocean annually than does any Atlantic Seaboard river.

Sediment sources for sands can sometimes be identified by the presence of rare tracer minerals, and those for clays can be defined in a much more general way by the clay mineral suites present in them (Biscaye, 1965; Ross, 1970). Reworked fossil materials may be used to estimate the age of source materials and to identify specific sources for some exotic sediment types (Merriam and Bandy, 1968). Textural gradients in sedimentary deposits may also be used to indicate the sediment source, although they are primarily indicators of the point of entry into deep water areas and may give misleading information about the original source (Kuenen, 1964; 1970).

### 24.2.2.2. Transport

Transport mechanisms in the marine sedimentary system are all primarily responses to gravity gradients. Direct transport agents include currents which may be generated by waves, thermohaline anomalies, pressure gradients, wind, or density differences arising from suspensate loading. Other processes include mass movements and biological transport. Flotation by ice or surface tension are special cases of transport by surface currents.

Sediment movements in shallow water areas are generally dominated by wave generated flows and surface circulation, whereas, deep water areas are influenced by density currents, mass movements, thermohaline (geostrophic) currents and slope currents. Reworking and transport by biological agencies are usually local, although the transport of kelp holdfasts and log-carried debris may be long distance processes (Emery, 1963a).

Processes of sediment motion may be grouped into those processes which influence the suspended load, and those which move the bedload (see Gorsline and Swift, 1974; McCave, 1972; Swift et al., 1972). Most terrigenous sediment is processed by coastal wave action although in some environments this may be only a minor factor (e.g. on some deltaic coasts; but see Wright and Coleman, 1973). It therefore usually undergoes some initial sorting into suspended and bedload fractions which subsequently tend to follow separate paths or energy levels. In the simplest system, the bedload moves laterally along the coast from the source (streams, cliff erosion, bottom erosion) in the direction of wave motion (Inman and Brush, 1973). The suspended load generally moves in an off-shore direction along gravity gradients, although the process may include several periods of deposition and resuspension (McCave, 1972); internal waves are probably important in the movement of suspended loads (Southard and Cacchione, 1972).

In the deep ocean the principal influence of surface currents (slope currents) is on the distribution of fine terrigenous and biogenous material. In addition, surface currents in high latitude regions are responsible for the distribution of ice-rafted continental debris. With the exception of ice-rafted material and volcanic debris most of the sedimentary particles with settling velocities low enough for them to be carried by surface currents are so fine that they are carried vast distances. These fine particles consequently have a much more generalized distribution than do the coarser ones, and the imprint of the surface current patterns on their distribution in deep-sea sediments is largely lost. Movement of fine suspended material from both terrigenous and biogenous sources occurs in a stepwise manner, concentration occurring at all density discontinuities in the ocean water column along which lateral movement can take place. The most important of these zones of accumulation are the seasonal thermoclines in the surface layer and the permanent thermocline at the base of this layer. Another major concentration zone is typically located just above the sediment-water interface in those deep ocean areas which are open to lateral and downslope movement of suspensates from the continents (see Drake, 1972; Drake and Gorsline, 1973; Eittreim et al., 1969; Ewing and Connary, 1970).

Bottom transport in the deep-sea occurs by mass movement (slumping, creep), density currents (turbidity currents) and thermohaline (geostrophic) currents (see, e.g. Emery and Uchupi, 1972). All these processes transport material from shallow to deep water, although geostrophic currents may simply rework or erode existing deep water deposits.

The existence of deep-seas and layers interbedded with normal fine-grained pelagic sediments, the discovery of submarine canyons cutting many of the continental margins of the world, and the occurrence of sudden breaks in submarine cables, led to the realization that turbidity currents are significant agents in deep-sea sedimentation (see, e.g. Kuenen, 1937, 1951, 1964, 1966; Heezen, 1959, 1963; Heezen and Laughton, 1963; Kuenen and Migliorini, 1950; Middleton, 1966a, b, 1967). Turbidity currents are generally thought of as being occasional catastrophic events which spread large volumes of sediment over wide areas and cause massive extinctions of benthic fauna. However, it must be pointed out that the concept also embraces much gentler, and probably much more frequent, bottom-hugging density currents (nepheloid layers) which are perhaps related to storm run-off from arid regions and are responsible for the spread of thin sediment laminae over vast areas of the abyssal plains. The causes of most turbidity currents large, or small, are not known with any certainty (Walker, 1973).

The normal deep oceanic circulation results from density differences between water masses of differing salinity and/or temperature. The existence of bottom currents resulting from thermohaline circulation has been known for a long time, but the fact that these currents have any geological significance was only recognized comparatively recently (Heezen, 1959; Heezen and Hollister, 1964). Further studies have led to the realization that bottom currents in fact have a marked influence on the distribution of sediments on the ocean floor. This was first demonstrated by studies of the continental rise off the eastern United States which showed that the rise has been modified by deep, contour-following, geostrophic currents (Heezen et al., 1966; Hollister and Heezen, 1967; Schneider et al., 1967). It is now apparent that in this and many other continental rise areas of the World Ocean, the deep water and bottom currents are significant agents for the transport of fine sediment. Nepheloid flows are apparently major contributors of fine sediments to the abyssal plains adjacent to the continents, and probably arise from the reworking of bottom sediments by bottom currents, or are the tails of turbidity currents (Eittreim et al., 1969). On a local scale, where bottom currents encounter major solid obstacles, such as seamounts, they accelerate. This reduces the sedimentation rate, or even results in erosion of the sea bed, producing a characteristic moat (Heezen and Johnson, 1963; Johnson et al., 1971; Weser, 1970), or marginal channel (Fig. 24.4). On a regional scale, it should be noted that over very large areas of the Atlantic, Indian and S.W. Pacific Oceans sediments representing long periods of time during the Cenozoic are absent (Berggren and Hollister, 1974; Davies et al,. 1975; Kennett et al., 1972, 1974; Moore et al., in press). These hiatuses in sediment accumulation must be attributed to the activity of deep ocean bottom currents. Deposition from bottom currents has been elegantly demonstrated on the Blake-Bahama Outer Ridge, which Bryan (1970) and Markl et al. (1970) have suggested is the consequence of the interaction of the Gulf Stream and the Western Boundary Undercurrent. Other examples of such bottom current deposits are known from elsewhere in the oceans, particularly in the North Atlantic, and their formation is discussed in the following section.

### 24.2.2.3. Deposition

The sources of sediment, the processes of sediment transport within the ocean and the physico-chemical conditions within the ocean basins interact and result in the depostion and accumulation of sediment bodies of various types. Deposition occurs when the available energy decreases below that necessary to carry the available load. For current transported sediments this may occur under a number of conditions: (i) when the base of the topographical slope is reached; (ii) if the flow cross-section increases; (iii) if sufficient energy is lost through frictional effects; or, (iv) if the density of the transporting fluid decreases as the result of the sedimentation of the suspended load. Thus, for example, continental rise deposition occurs primarily because of the marked decrease in slope between the continental slope and the ocean floor. Deposition may also occur as a result of the death of an organism and the consequent cessation of swimming or, for very small

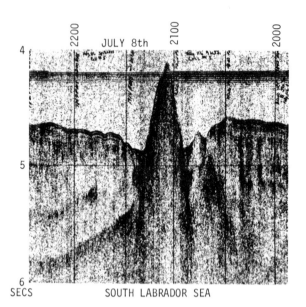

A

SOUTH LABRADOR SEA

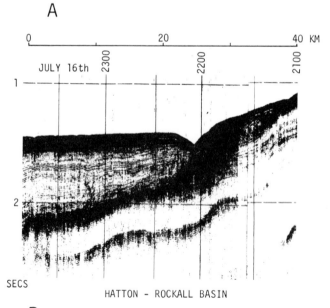

B

HATTON - ROCKALL BASIN

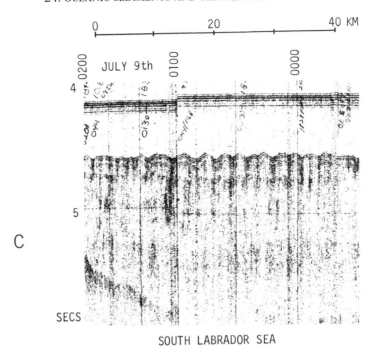

FIG. 24.4. Bottom features associated with the activities of deep ocean bottom currents: (A) moats around a seamount, (B) marginal channel, (C) sediment waves. These examples are from *Glomar Challenger* profiles from the North Atlantic.

life forms, by a decrease in water viscosity and the more rapid sinking consequent upon it. Chemical precipitation (deposition) can be the product of changes in chemical boundary conditions (e.g. the change from reducing to oxidizing conditions, or vice versa, when solubility limits are exceeded). A more detailed discussion of specific examples of sediment accumulation is given in subsequent sections and at this stage attention will be confined merely to some general comments on the accumulation of sediments in the deep ocean.

Turbidity currents are responsible for the spreading of sheets of sediment over wide areas. Because they are driven by gravity, the resultant deposits (turbidites) are found as sub-horizontal layers in basins and hollows in the seabed. When these hollows are filled, broad flat abyssal plans are formed, and the relationship between turbidites and such plains is now widely accepted. It is now known that turbidity current activity has not been confined to the Pleistocene as was once thought, but probably occurred to a greater or lesser extent throughout the history of the oceans (Whitaker, 1974). Deep-sea drilling has produced samples of turbidites ranging in age from Cretaceous to Recent, and has given evidence that this type of sediment can no longer be regarded as being an insignificant sediment type in pre-Plio-Pleistocene times.

Pleistocene sea level changes brought about by glaciation not only increased the supply of sediment, but also lengthened (high sea level stand) or shortened (low sea level stand) the canyon systems cutting the continental shelf and slope. These changes influenced the pattern of sediment deposition on the continental rise (submarine fans) and on the basin floors (abyssal plains). When the discharge increased and the gradient steepened the distributaries on the fan (rise) extended and so built this feature outwards over the basin floors of abyssal plains. For this reason, upper fan, middle fan, lower fan (see Haner, 1971) and suprafan (Normark, 1970), facies shift seawards from canyon mouths. When the sea level rose the canyon discharge generally decreased and the gradient therefore also decreased with the consequences that the fan (rise facies) either transgressed shoreward, or became inactive, and that normal bottom currents (contour currents) became the dominant influence over the fan surfaces. The characteristics (e.g. texture, thickness and sedimentary structure; see Bouma, 1962) of abyssal plain and rise turbidites in a given section provide evidence for the relative proximity of the source submarine canyon (Walker, 1970).

Turbidites can be recognized from the characteristics listed below:

(a) They form discrete beds with a well-defined base and have a grain size which decreases towards the top where it grades into the overlying strata. They often contrast strongly in colour and other physical properties, with sediments deposited at the same locality in other ways.

(b) The beds are usually graded from coarse at the base to fine at the top. In some instances the beds, especially in their upper parts, exhibit lamination (parallel and convolute). Evidence for burrowing is found only in the uppermost part of the beds.

(c) Shallow water and/or terrigenous materials are abundant in the beds, but are absent from, or rare in, the intervening layers of sediment.

(d) On echo-sounder records and seismic profiles turbidites are characterized by:

(i) high sea bed reflectivity (giving rise to sea bed multiples);

(ii) strong stratification;

(iii) flat beds and absence of waves and ridges, giving the appearance that the sediments have been ponded in regional depressions.

Sediments deposited by the action of deep-ocean bottom currents are found wherever the currents are slowed down and so lose their competence to carry sediment; this may occur either by interaction with another water mass, or after encountering a solid obstacle. Bottom current deposits usually form long narrow ridges parallel to the current direction at the margins of the flow where the velocity gradient is greatest rather than under the centre of the current. Such deposits are frequently found around the margins of the ocean basins (the best known examples being in the North Atlantic); these deposits are often referred to as contourites (see Hollister and Heezen, 1972) because the ocean bottom currents from which they are deposited are often orographically controlled and flow parallel to the bathymetric contours. However, because other forces (e.g. potential density differences and reactions with other moving water masses) may cause the path of the current to deviate from a contour, the use of the terms "contour current sediments" and "contourites" should be avoided. A more appropriate term would be bottom current deposits.

The principal characteristics by which sediment deposits formed under the influence of bottom currents can be identified are as follows:

(a) The sediments are relatively acoustically transparent and lack the strong internal layering which often characterizes turbidite deposits.

(b) The sediments are not uniformly distributed in relation to the basement topography (cf. pelagic sediments discussed later), but are often heaped into piles and ridges which are elongated in the direction of the bottom current and lie under the margins of the current.

(c) Where the sediment body encounters a basement high there is typically a marginal channel, or moat, and the sediments thin out and dip towards the obstacle (Fig. 24.4).

(d) The upper surface of the sediment body is commonly wavy with a typical wavelength of 2 km and an amplitude of 50 m (Fox et al., 1968; see Fig. 24.4).

The characteristic features of turbidites and bottom current deposits have been summarized and compared by Bouma and Hollister (1973).

In areas in which bottom current activity is slight, particles transported by surface currents settle out from the surface layer and cover the sea floor in a uniform blanket of sediment. Such sediments, which are referred to as pelagic, contain a number of components of terrigenous and biogenous origin. The terrigenous components, which are deposited in all areas of the world ocean, include river-transported solids (mainly clays), volcanic debris, wind-blown dust and, at high latitudes, ice-rafted debris. In contrast (see above) the biogenous component can only accumulate in regions where productivity exceeds dissolution.

The surface waters of the oceans are undersaturated with respect to silica, and for this reason dissolution of biogenous silica commences immediately after the death of the organism. It has been estimated that for this reason only $\sim 4\%$ of the skeletal silica formed in the ocean survives to reach the sea floor (Heath, 1974) and half of this is dissolved after deposition with the result that accumulations of siliceous sediment are not common. They are found only in those regions having high primary production (i.e. in sub-polar and equatorial latitudes and at some continental margins (Berger, 1974; see Fig. 24.5). Diatomaceous sediments are typical of the high latitudes and the ocean margin regions, and radiolarian oozes are characteristic of the equatorial areas. The correlation between the accumulation of siliceous sediments and areas of high productivity is enhanced not only by the fact that the precipitation of biogenous silica is increased, but also because the preservation of siliceous shells appears to be better in sediments rich in organic matter (Berger, 1970). The relationship between state of preservation and relative abundance of siliceous fossils has been studied by Riedel (1959) and by Goll and Bjørkland (1971). The oceanic silica cycle and the factors controlling the preservation and accumulation of siliceous sediments have been more extensively reviewed by Berger (1970, 1974). In summary, the cycle has two loops (Fig. 24.6). The major loop which runs parallel to phosphate, accounts for 95% of the silica transported. The second loop, which is not related to phosphate, involves post-oxidative dissolution. It should be noted that dissolution of silica at the sediment-water interface supplies silica to the ocean at a rate which may be ten times as great as the river-input. The continental shelves are especially important in the silica cycle since they act as both sinks and sources for dissolved silica (Calvert, 1966; Heath, 1974).

The factors governing the accumulation of calcareous sediments are in many ways similar to those controlling the formation of siliceous sediments. However, the surface waters of the ocean are supersaturated with respect to calcium carbonate, and hence calcareous sediments accumulate in shallow seas the world over. At depths greater than a few hundred metres, however, the oceans are undersaturated with respect to carbonate, and so dissolution commences. The process is particularly rapid at depths greater than about 4000 m, and below 5000 m virtually no carbonate is present in the sediments (see Chapter 29). Results from box-coring and plankton tows suggest that most of the dissolution of carbonate does not occur in the water column, but takes place at the sediment/water interface where it resides for much longer periods (Berger, 1974). The depth at which the rates of supply and dissolution of carbonate are equal is termed the carbonate compensation depth; below this depth there is no net accumulation of calcareous sediment (see Berger and Winterer, 1974, Fig. 1). The carbonate compensation depth tends to be depressed under the equatorial high productivity belt (Arrhenius, 1952; Berger, 1974; Chapter 29), but is much shallower in highly productive ocean margin regions. This is believed to be the result of the abundance of organic matter produced in these marginal regions which leads to the

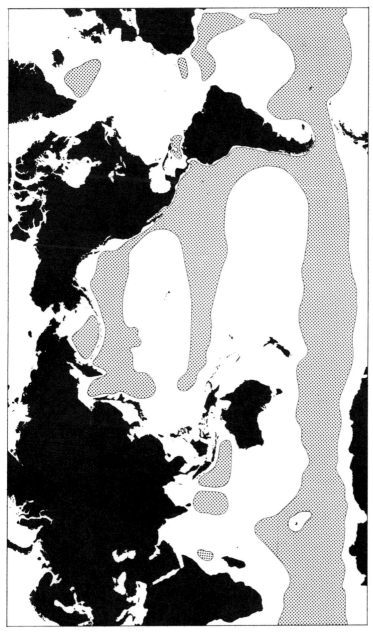

FIG. 24.5. Distribution of siliceous fossils (stippled) in ocean sediments. (Based largely on Berger, 1974, and Luyendyk and Davies, 1974).

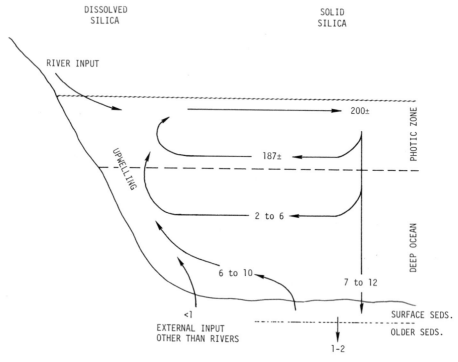

Fig. 24.6. The silica cycle in the oceans. Figures are in g m$^{-2}$ yr$^{-1}$. (Based on Berger, 1974, and Heath, 1974.)

development of carbon dioxide-rich, and hence corrosive, interstitial waters. In contrast, in the deep ocean the supply of organic matter to the seabed is relatively small, and its effects on dissolution are masked by the increased supply of carbonate shells.

The level at which the rate of carbonate dissolution increases abruptly, which is referred to as the lysocline, is marked by the relative enrichment of the faunas with solution-resistant forms (Berger, 1968). It varies in depth over the ocean basins, being at a level of ca. 3600 m in the tropical Pacific (Peterson, 1966). In the open ocean it usually coincides with the top of the Antarctic Bottom Water, especially in the southern hemisphere, and generally lies deeper in the Atlantic than it does in the Pacific as a result of the net flow of deep water from the Atlantic into the Pacific and Indian Oceans (Berger, 1970; Broecker, 1974).

The factors controlling the dissolution of calcium carbonate, and thus the accumulation of calcareous sediments, are extremely complex and will not be further discussed here; for a more detailed treatment see Chapter 29.

#### 24.2.2.4. *Post depositional processes*

The processes which occur in sediments following deposition are known as diagenesis. Chemical processes involved in diagenesis include; bacterial reactions (Kaplan, 1963); the formation of authigenic minerals by both precipitation and submarine weathering; and, precipitation associated with hydrothermal activity (see e.g. Boström and Peterson, 1966, 1969; Böstrom, 1970). Physico-chemical changes are brought about by compaction and loading as the sediments accumulate into thick deposits. For example, the transformation of nannofossil ooze into soft limestone has been described by Davies and Supko (1973) and by Schlanger *et al.* (1973).

The mobilization of opaline silica to form chert is an important post-depositional process. The discovery of extensive chert horizons in deep-sea sediments has been one of the more surprising findings of the deep-sea drilling program. Deep-sea cherts occur in several distinct forms (i.e. nodular or bedded) and in a variety of compositions ranging from pure to ones which are contaminated with varying amounts of volcanic or terrigenous debris. The conditions controlling their formation are not yet clearly understood, although factors such as the silica content of the bottom waters, the rate of sediment accumulation, together with an abundant supply of opal, are almost certainly important. For a detailed discussion of the origin of deep-sea cherts the reader should consult the papers by Heath and Moberly (1971), Lancelot (1973), Calvert, (1974), and von Rad and Rösch (1974).

The ocean floor is not the permanent abode of sediment. Strong erosion of deep ocean surfaces can occur as the result of major structural changes in ocean floor configuration and also as a result of tectonic movement of sediment masses, either to subduction zones or to depths at which chemical conditions can lead to the dissolution of the sediment.

### 24.3. REGIONAL PATTERNS OF SEDIMENT DISTRIBUTION

The general pattern of sediment distribution in the World Ocean is shown in Fig. 24.7. In order to illustrate the operation of the marine sedimentary system the pattern of sediment distribution, and its controlling factors, in three different regions will be described briefly.

#### 24.3.1. THE NORTH ATLANTIC

##### 24.3.1.1. *Introduction*

The sediments of the North Atlantic north of 45° N do not seem to have attracted widespread attention prior to 1940, when Bramlette and Bradley

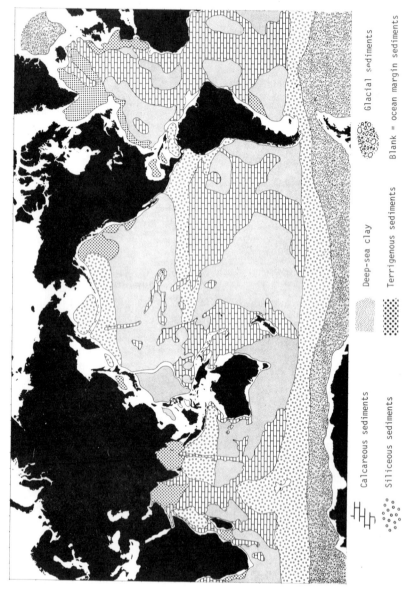

FIG. 24.7. Distribution of the principal types of sediment on the floors of the oceans. (Based on data from many sources.)

published the results of their studies of cores obtained between Ireland and Newfoundland by the cable ship *Lord Kelvin* (Bramlette and Bradley, 1940). Since that time, however, a wealth of information concerning the sediments of this region has been gathered through the activities of ships sent out by institutions on both sides of the Atlantic. The following summary is based in large part on a review presented by Davies and Laughton (1972).

Nearly all the sediments between the Gibbs Fracture Zone and the latitude of Iceland have been transported long distances by various mechanisms before being deposited, and the source of the sediment is often difficult to determine. In such polygenetic sediments, individual sources of sedimentary material exert a less significant influence on the nature of the resultant sedimentary deposit than do the agents of sediment transportation and deposition. In many areas even the traditional distinction between terrigenous and pelagic sediments is of little value in any attempts to understand the pattern of sediment distribution. It is immediately apparent from reports of dredging, coring and drilling operations that the details of the nature and composition of the sediments of the northern North Atlantic are complex, as would be expected. The North Atlantic is surrounded by land masses contributing sediments derived from virtually every known rock type, and ranging in age from Precambrian to Recent. Because of its geometry and geographical location this region of the ocean is oceanographically and biologically complex, and this leads to a corresponding heterogeneity in the distribution of pelagic material. Thus, it follows that the key to understanding sedimentation in the northern North Atlantic lies in viewing the sediments in terms of the processes of transportation and deposition. Once a working model has been developed it may be possible, through further detailed study, to develop the type of comprehensive model which can now be constructed for less subtle and less complex sedimentary systems.

### 24.3.12. *Sedimentary processes operating in the North Atlantic*

(a) *Near-shore sediments.* The bulk of the work on Atlantic near-shore sediments and processes has been carried out in the United States and Western Europe (see Allen, 1974; Swift, 1974). The major work by Emery and Uchupi (1972) is the best current source of data.

All of the areas studied are wide shelves over which wind drift and tidal currents are the prevailing agents for sediment transport; however, storm waves are the most effective agent, but occur only rarely. Emery (1968) has described broad areas of relict sediments on the shelves. Although Swift (1974) has agreed that the dominant bathymetric features of the shelves are relicts of lower sea level stands, he has concluded that they are affected by contemporary processes. The cape head massifs and detached bar systems described by Swift (1974) are to be attributed to the result of rising sea levels over the past 12000 years.

Most of these continental margins are relatively inactive at the present time; sediments from the rivers are trapped in the large estuaries (Meade, 1974), and little reaches the slopes and rises. Rates of sedimentation were much higher in glacial times and mass movements on kilometre scales were major factors in the movement of sediments from the slopes. During this period many shelf edge canyons were active, and turbidite deposition probably dominated those rises which are now the sites of bottom current dominance (Emery and Uchupi, 1972).

Around Great Britain, strong tidal streams generate extensive fields of bedforms that have been produced by the reworking of older Pleistocene glacial and fluvial deposits (Stride, 1963). As tide stream velocity and supply of sediment increase, these large scale bedforms shift from sand sheets and ribbons to large sand waves and submarine dune fields (Stride, 1963). The large tidal ranges in the North Sea and the large estuaries of the European North Sea coast have produced complexes of tidal flat-deltaic sediments and prograding marsh deposits that trap large quantities of sediment (see, e.g. Reineck et al., 1968; Seibold et al., 1971; van Straaten, 1950). Storms are also major factors in shelf sediment distribution in these areas, as are major floods with their associated large contributions to the near-shore zone.

The dominant Atlantic margin deposits are those of the massive continental rises and their associated abyssal plains. At the present time the largest volume of sediments entering the World Ocean is deposited in the marginal seas of southeast Asia. However, in times of lowered sea level large quantities of sediment entered the North Atlantic from the North American coastal plain rivers (Emery and Uchupi, 1972).

(b) *Turbidity currents.* In the Labrador Sea (Fig. 24.8) stratified sediments lie around the outside of a semicircle of basement ridges which bounds a large body of bottom current deposited sediment which will be discussed later. The stratified sediments lap onto, and over, this sediment body. The boundary between these sediments and the bottom current deposited sediments can easily be seen on both seismic profile (Fig. 24.9) and echo-sounding records. The profiles of the sediments show distributions and appearances which are typical of those which might be expected for sediments deposited by turbidity currents derived from the neighbouring continental margins, of both Greenland and Labrador.

Drilling at DSDP Site 113 confirmed this interpretation (Laughton et al., 1972), The sediments were found to be terrigenous clays, silts and sands containing rock fragments and heavy minerals. Turbidite beds were present, and there was also evidence of mud flows that carried Eocene clasts in a

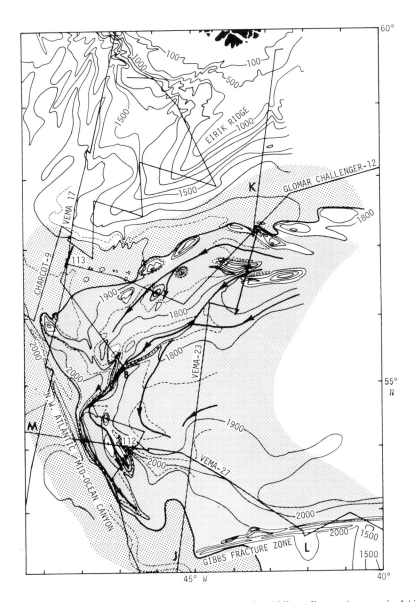

FIG. 24.8. Bottom current sediment body (fine stipple) and turbidite sediments (coarse stipple) in the southern Labrador Sea. Heavy lines indicate crests of sediment ridges. Straight lines are some of the more important ship tracks on which the map is based. Depths are in uncorrected fathoms. (From Davies and Laughton, 1972.)

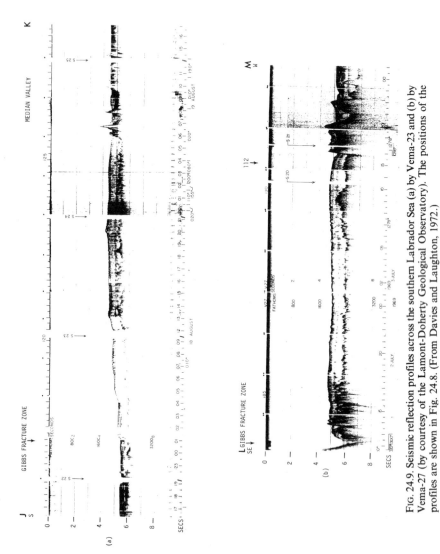

FIG. 24.9. Seismic reflection profiles across the southern Labrador Sea (a) by Vema-23 and (b) by Vema-27 (by courtesy of the Lamont-Doherty Geological Observatory). The positions of the profiles are shown in Fig. 24.8. (From Davies and Laughton, 1972.)

Pliocene matrix. The extremely high sedimentation rates (10–20 cm/1000 years) found reflect the large source area for terrigenous sediment. Although the Eirik Ridge does not reach as far south as Site 113, small channels run southwards from its southern end, and undoubtedly bring additional sediments from east of Greenland into the area of Site 113, initially by bottom currents and subsequently by local turbidity currents which contribute to the high sedimentation rate.

An unusual type of turbidite was encountered during drilling in the deep basin west of Rockall (Site 115) where hard volcanogenic sandstone bands (a few metres in thickness) interbedded with soft sediments were cored. These sandstone bands are composed of the eruptive products of underwater, or sub-ice, volcanoes. These volcanic materials are thought to have been transported by turbidity currents, probably from Iceland where flash floods, or *jokulhlaups*, carrying large quantities of hyaloclastites from sub-ice volcanism, were common during the Quaternary. The cores often show graded-bedding, horizontal and cross lamination and slump structures, and contain some shallow water benthonic foraminifera and molluscs. They were sampled to a sub-bottom depth of 228 m.

The areal extent of the sandstone layers can be inferred from geophysical data. Seismic profiles obtained on cruises of *Vema* and *Discovery* suggest that the layers extend over 100 km west from the base of Hatton Bank. The Maury Mid-Ocean Canyon, running southwards from the Iceland-Faroes Ridge, widens considerably where it has cut down to the top of the first sandstone layer. However, it narrows again 150 km south of the site where, perhaps, the sandstone stops. To the east of the site the layers abut against the foot of the west scarp of Hatton Bank, which suggests that it is not the source. A natural course for the eruptive products of Icelandic volcanoes would be through the canyons south of Iceland to the deepest part of the Iceland Basin between Hatton Bank and the Reykjanes Ridge.

Older turbidites, in addition to those of Pleistocene age, were sampled in the Bay of Biscay, at Sites 118 and 119. Site 119 yielded samples of Palaeocene turbidites composed of (?) shallow water carbonate debris with reworked Cretaceous coccoliths, and Site 118 yielded Miocene and Pliocene turbidites, consisting of detrital carbonate, and Pleistocene turbidites composed of mixed terrigenous debris. The relationship between the older turbidites at these sites and tectonic events in the Bay of Biscay region has been discussed by both Sibuet *et al.* (1971) and Davies and Laughton (1972). The Pleistocene turbidites are typical detrital sands like those which have been described for abyssal plains from all over the North Atlantic (Heezen and Laughton, 1963).

On the basis that accumulation of turbidites forms abyssal plains, the data from deep-sea drilling can be combined with that of earlier investigators (Ericson *et al.*, 1961; Heezen and Laughton, 1963) to define those areas of the North Atlantic in which turbidity currents have had a dominant influence on sedimentation throughout the Pleistocene These areas are delineated in Fig. 24.10.

(c) *Deep ocean and bottom Currents.* The geological significance of deep ocean and bottom currents has been discussed in general terms in preceding sections of this chapter. A major advance in the recognition of the significance of bottom currents in the distribution of sediments in the North Atlantic came when it was noted from seismic reflection records that in several places there were, in fact, long ridges composed of thick piles of sediment (Johnson and Schneider, 1969; Jones *et al.*, 1970). For example, the Feni Ridge lies along the east side of the Rockall Bank, the Gardar Ridge runs down the east side of the Reykjanes Ridge and the Eirik Ridge extends southwest from the southern tip of Greenland (see Fig 24.11).

The present day deep-water circulation of the North Atlantic has been widely studied by European and American oceanographers, and both Johnson and Schneider (1969) and Jones *et al.* (1970) were able to relate the sediment ridges to the deep water circulation pattern. Thus, the high salinity overflow water from the Iceland-Faroes Ridge (Worthington, 1970) travels southwest along both the eastern side of the Rockall Bank (producing the Feni Ridge) and the eastern side of the Reykjanes Ridge below about 1 500 m (producing the Gardar Ridge). Much of this water passes through the Gibbs Fracture Zone (Worthington and Volkmann, 1965) to drift northwards up the western side of the Reykjanes Ridge and then southwards along the eastern continental margin of Greenland (producing the Eirik Ridge). The currents have a very pronounced tendency to lie along the west side of the basins through which they flow because of the effect of the Coriolis force which guides the currents along bathymetric contours. A more complete discussion of this system has been given by Davies and Laughton (1972).

In the southern Labrador Sea there is a thick body of sediments which is acoustically relatively transparent, and over the whole area there is only one prominent but rather diffuse mid-sediment reflector. Neither the reflector, nor the upper surface, conform with the basement topography, and the sediments show all the characteristics of having been deposited under the influence of bottom currents. The results obtained from drilling have shown that these sediments accumulated at a rate varying between 1·5 and 4·0 cm/1000 years (Davies and Laughton, 1972). These rates are lower than those for other bottom current ridges, but because of the considerable area over which the bottom currents can wander a mean rate similar to that of pelagic sedimentation might have been expected. Sedimentation in the southern Labrador Sea

FIG. 24.10. Regions in the northern North Atlantic where sediment distribution by turbidity currents is significant. (From Davies and Laughton, 1972.)

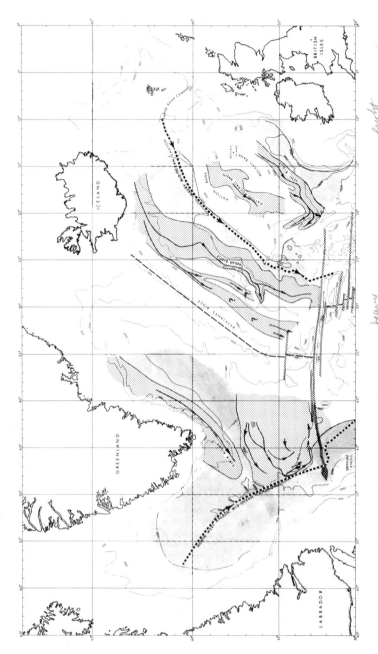

Fig. 24.11. Bottom current sediment deposits (light stipple) and turbidite sediments (heavy stipple) in the North Atlantic. Solid lines = axes of sediment ridges; dashed lines = axis of mid-ocean ridge; diagonal lines = fracture zones; dotted lines = axes of mid-ocean canyons. Depths in uncorrected fathoms. (From Davies and Laughton, 1972.)

has been discussed in detail by Davies and Laughton (1972) and by Egloff and Johnson (in press). The sediment body appears to be more or less confined by a semicircle of basement ridges, but overflows and thins to the west and south where it dips under the stratified terrigenous sediments derived from the continental margins. The upper surface of the sediment body is commonly wavy, but on a larger scale it is formed into long ridges. Johnson *et al.* (1969) have suggested that there is a similar anti-clockwise bottom current system in the Northern Labrador Sea, fed in part by Atlantic water, which moulds the underlying sediments. It is difficult, however, to distinguish these sediments from the turbidites derived from the continental margins.

Jones *et al.* (1970) have shown that in the eastern North Atlantic the Norwegian Sea overflow water spills from the Faroes-Iceland Ridge down the eastern side of the Rockall Plateau and gives rise to the Feni Ridge (see above). It is believed that sedimentation in the Hatton-Rockall Basin, on the top of the Rockall Plateau, is also under the control of bottom currents (Davies and Laughton, 1972). The sedimentation rate is high (about 3 cm/1 000 years), and the form of the sediment body, with its wavy surface and characteristic marginal channels, clearly indicates the influence of bottom currents. The simplest interpretation of these observations is that Norwegian Sea water cascading over the Faroes-Iceland Ridge, not only sinks into the basins on either side of Rockall Plateau, but also spills along the top of the plateau between the Hatton and the Rockall Banks (see Fig. 24.11).

(d) *Surface currents.* The principal influence of surface currents on marine sedimentation is on the distribution of fine terrigenous and biogenic material (*vide supra*). However at high latitudes surface currents also bring about the distribution of ice-rafted continental debris.

(i) *Ice-rafting.* The influence of the Plio-Pleistocene glaciation is seen at many locations in the North Atlantic by the presence of exotic pebbles and relatively poorly sorted heterogeneous detrital sediments. Figure 24.12 shows the approximate locations of deep-sea drilling sites at which the sediment is known to have been influenced by ice-rafting; the figure also shows the present limits of drifting ice and the thicknesses of sediment which accumulated at each site during the glaciation. Abnormally high sedimentation rates prevailed at Sites 113, 114 and 118, and at Sites 113 and 118 this was due largely to an increased turbidity current activity. The apparently high sedimentation rate at Site 114 was caused by local action of bottom currents which piled up sediments from a large area into one place. One interesting fact to emerge from an examination of Fig. 24.12 is that ice-rafting was apparently a major influence on deposition at Site 111, one of the more southerly sites drilled, but only a comparatively minor one at Site 116 on the Rockall Bank, one of the more northerly sites. The present day distributions of drift ice and pre-

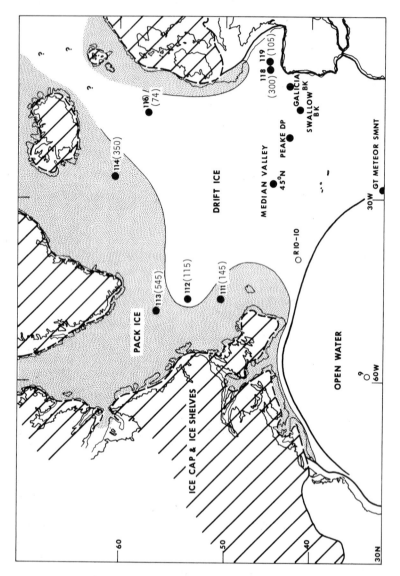

FIG. 24.12. Suggested ice limits at maximum extent of glaciation in North Atlantic. Solid dots = localities where evidence of ice rafting has been found, with thickness of glacial section in metres at drill-sites in parentheses. (From Davies and Laughton, 1972.)

vailing surface currents suggest that this is the result of a surface-current pattern not very different from the present day system.

Evidence for ice-rafting has been obtained from sediments cored at Lamont-Doherty Station R 10–10, and in the region of the Peake Deep. For station R 10–10 Ericson *et al.* (1961) reported the presence of "glacial marine layers" containing fragments of limestone and igneous rock up to 3 cm in diameter. In the Peake Deep region, Davies and Jones (1971) were able to establish variations in the composition and abundance of the detrital mineral component of the sediments which they attributed to ice-rafting during the cooler glacial episodes. Glacial erratics resulting from ice-rafting have been dredged from the Peake Deep region (Cann and Funnell, 1967), from the Galicia Bank (Black *et al.*, 1964), the Swallow Bank on the Iberian abyssal plain (Matthews, 1961) and the mid-Atlantic Ridge at 45°N (Paterson, in press). Erratics from the Galicia Bank contain a high proportion of recognizable European rocks (Matthews *in* Black *et al.*, 1964), whereas those from the Peake Deep region contain no recognizable European types and probably originated in Greenland or Labrador (J. R. Cann, personal communication; *in* Davies, 1967). Pratt (1961) has reported the presence of glacial erratics on the Great Meteor Seamount. The southerly limit of glacial erratics in the eastern part of the North Atlantic is generally considered to be 30°N (Matthews *in* Black *et al.*, 1964). However, Peterson *et al.* (1970) make no mention of ice-rafting in their report of sediments sampled at DSDP Site 9 in the western North Atlantic (Bermuda rise), even though a section of Pleistocene foraminiferal oozes more than 80 m in thickness was sampled at this site. This is not surprising in view of the fact that the predominant surface currents in the western North Atlantic south of Newfoundland are now from the south, and probably were from this same direction even at times of maximum glaciation.

On the basis of the above evidence it is possible to construct a map showing the ice conditions in the North Atlantic region at the time of maximum glaciation (Fig. 24.12). In this figure the limits of the icecaps and coastlines have been put at approximately the present day 100 m contour. The pack-ice/drift-ice boundary is largely speculative, and has been drawn on the assumption that the surface current system in the North Atlantic at the time of the glacial advance was substantially the same as it is now, but was somewhat depressed to the south.

(ii) *Pelagic sedimentation.* In contrast to bottom current sediments, pelagic ones are those which are laid down in deep water under quiet current conditions. They can be recognized partly by the nature of the sediment, and partly by the aspect of the beds as seen on seismic profiles. It should be noted that this usage refers to a process rather than to a sediment composition. Thus, sediments of the Hatton-Rockall Basic (foraminiferal nannofossil oozes) are essentially pelagic in composition, but have clearly been laid down under the dominant influence of southward flowing bottom currents; hence, they will be grouped with bottom current sediments. Pelagic sediments, defined in this way, are probably comparatively rare in the northern part of the North Atlantic, except perhaps on the summits of seamounts and on the mid-Atlantic Ridge.

### 24.3.1.3. *Interaction of sedimentary processes*

The basin south of Iceland provides an excellent example of the combined effects of the sedimentary processes just described. Four quite distinct mechanisms of sedimentation have been active in the basin, and have been responsible for the topography (Fig. 24.13). These are bottom currents, turbidity currents, low velocity density currents, and pelagic sedimentation.

(a) *Bottom currents.* The Gardar Ridge is the largest of several sediment

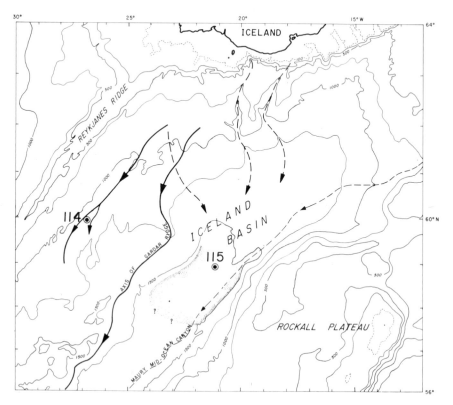

FIG. 24.13. Sediment distribution processes in the Iceland Basin. Solid arrows = bottom current ridges; broken arrows = turbidity current paths. (From Davies and Laughton, 1972.)

ridges found on the west side of the Iceland Basin. The sediments in these depositional ridges cover a total of about half of the Basin area and they have been shaped by southwesterly travelling Norwegian Sea overflow water.

Seismic profiles of the continental slope south of Iceland show a considerable thickness of layered sediment which may also have been deposited by bottom currents. These sediments are draped over the topography associated with two large canyons implying that the latter were cut prior to the deposition of these sediments. However, the canyons still appear to be used as paths by turbidity currents which have eroded a little of the draped sediment.

(b) *Turbidity currents.* Turbidity currents generated from Icelandic *jokulhlaups,* or from metastable accumulations of sediment on the shelf edge, travel down-slope and are controlled by gravity and by the Coriolis force which tends to swing their paths to the right. However, gravity is the dominant force and hence a turbidity current path may well cross over a bottom current path and sweep away any sediments deposited there. The ultimate destination of the material transported by turbidity currents will be the deepest topographical depression where the currents will die and the sediments will be deposited in graded beds. The turbidity currents will be episodic, the turbidite beds being interbedded with pelagic and other sediments.

(c) *Low velocity density currents.* The Maury Mid-Ocean Canyon (Ballard *et al.,* 1971; Johnson *et al.,* 1971) is in many respects similar to the Northwest Atlantic Mid-Ocean Canyon in the Labrador Sea (Heezen *et al.,* 1969). It is over 1000 km long stretching from the Faroe Bank Channel to a point southwest of the Rockall Plateau. Under the name of the Viking Seachannel, it has recently been traced as far as the Iberian Basin (Cherkis *et al.,* 1971). In most places it is not more than 10 km wide and 200 m deep. It lies along the axis of greatest depth in the basin, and steadily increases in depth to the southwest. Unlike the Northwest Atlantic Mid-Ocean Canyon, the Maury Canyon does not have prominent levees although there is a slight indication of one on the west side between 58° and 59°N.

Possible mechanisms for the production of mid-ocean canyons have been discussed by Heezen *et al.* (1969); however, they were unable to explain their origin satisfactorily. There can be no doubt that a flow of bottom water controlled by density difference was involved. This must have been of limited volume and low velocity for it to have been retained within the channel. The connection of the Maury Mid-Ocean Canyon with the Faroe Bank Channel at the northeast end suggests that the water originated in the Norwegian Sea. However, the flow may have been sporadic. The velocity, although low, must have been sufficient to either prevent (or retard) sedimentation along the axis, or to erode pre-existing sediments. However, whether the density difference arose from physical and chemical characteristics or from the suspended sediment load, or both, is not yet known.

(d) *Pelagic sedimentation.* When no other influences are at work, and the currents are small, pelagic sediments can accumulate. These will be partly biogenic and partly fine suspended terrigenous clays and authigenic minerals. The pelagic contribution cannot easily be distinguished from that resulting from bottom current deposition, and may be interbedded with the turbidites.

In the Iceland Basin the net result of these four sedimentary processes, which vary with time and in space, is a very complex three dimensional accumulation of sediments. Although the sediments of a particular area may be labelled "bottom current sediments" or "turbidites"· the beds interfinger and are often hard to distinguish from one another.

### 24.3.2. THE CENTRAL PACIFIC

#### 24.3.2.1. *Introduction*

In comparison with the North Atlantic, with its abundant supply of terrigenous sediment and vigorous bottom water circulation, the central equatorial Pacific is a relatively quiet region starved of external sediment inputs. Indeed, only in the northeast Pacific is there free access to the deep ocean for material derived from the erosion of the landmasses. Elsewhere, terrigenous debris is trapped in the major ocean trenches or, off California, in the deep basins of the continental borderland (Menard, 1964). Access for terrigenous debris to the equatorial Pacific is barred by the Middle America Trench and the Southern California Borderland. Any material that does reach the region from central America is prevented from being transported westwards by the East Pacific Rise. If the pattern of sediment distribution observed in the central equatorial Pacific is to be understood it is necessary to pay far more attention to the factors controlling pelagic sedimentation than to the physical processes concerned with the transportation and deposition of clastic sediments. Factors which are of primary importance include: biological productivity, surface circulation and winds, and ocean chemistry. Sediment distribution in the central Pacific has been discussed in these terms by Arrhenius (1952, 1963), Menard (1964), Berger (1973). Winterer (1973) and in a comprehensive monograph by van Andel *et al.* (in press). The following short summary is based largely on these sources.

#### 24.3.2.2. *Pacific margin sediments*

The margins of the North Pacific are the scene of the active transport of sediment over the shelves and shelf edges as dictated by the energy input and sediment supply; thin and patchy deposits of coarse sediments, gravels and sands are the typical sediments on these shelves. The continental borderland off California and Baja California is an exception to this and will be discussed later.

Sedimentation is active in marginal troughs and trenches. In some north central Pacific locales they are filled, and local rises have been built (Menard, 1955), e.g. that running from Point Conception north to the Gulf of Alaska. The remainder of the central and north Pacific basin is bounded by active trenches which effectively trap the bulk of the terrigenous clastic contribution.

Shelf and bank sediments are typically detrital and have a carbonate content which increases seawards (see Emery, 1960). Carbonates also comprise a large fraction of the sediments of the tropical margins, but never to the extent which they do in the sediments of the Caribbean and Atlantic margins at the same latitude (Emery et al., 1956). Phosphorites are a sea-bed feature of the central eastern Pacific zones of upwelling, e.g. on the banks of the California Borderland and the Chile-Peru shelf (see Pasho, 1973). These phosphorites are in part relicts and, indeed, the California deposits may all be of middle or late Miocene age. The presence of these older phosphorites is indicative of the slow rates of sediment accumulation reflecting the high energies associated with these narrow margins.

Pacific slopes are active sedimentation sites but, because of their tectonic activity and steepness, much of the sediment cover slips or creeps to the slope base and is trapped in the marginal basins or trenches. Trenches and marginal basins (borderland basins) are turbidity current provinces, and in the trenches turbidites overlie oceanic sediments seawards of the continental slope. Landward of the axial trench channels and landward slopes, the various sedimentary facies are sheared under the slope, or raised over the advancing scarp, by the complex processes of subduction (see Scholl and Marlowe, 1974).

The continental borderland off southern California and Baja California has a unique morphology produced by the interaction of the over-riding North American Plate and the East Pacific Rise (Atwater, 1970). This has produced a province some 700 km in length and up to 200 km wide that is folded and faulted into a checkerboard of basins and banks (see Shepard and Emery, 1941; Moore, 1970). The latter provide an unusual set of sediment traps that bridges the gap in sedimentation rates between those of the shelf and the deep ocean. Topography and winds combine to generate a major region of upwelling over the northern borderland, the area of which has fluctuated with the Holocene and Pleistocene climatic and sea level changes, and has left a record in the basin sediments in the varying proportions of biogenic carbonates, terrigenous muds, and turbidites (Gorsline et al., 1968; Gorsline and Barnes, 1972). These variations parallel the sedimentary record of the deep ocean (Shackleton and Opdyke, 1973), but in greater detail; cycles with frequencies of a few tens or hundreds of years can be discerned. As sea level fell, the frequency of turbidite formation and the terrigenous sedimentation rate both increased; biogenic contributions also increased, but at rates slower than those of terrigenous contributions. The rise of sea level led to a decreasing terrigenous contribution and to a lowering of the turbidity current activity with a corresponding decline in the biogenic sedimentation rate. Clay mineral suites also provide evidence for the decrease of the local terrigenous contribution which occurred as sea level rose (Fleischer, 1970).

### 24.3.2.3. *Equatorial Pacific sediments*

(a) *Biogenous sediments.* The present pattern of surface circulation in the equatorial Pacific is shown in Fig. 24.14, the dominant feature of this is the westward flowing equatorial current system, fed from north and south by the subtropical gyres. Upwelling occurs along the equator bringing cold nutrient-rich water to the surface and making the eastern equatorial Pacific one of the most biologically productive regions in the world, with a primary production of 200–500 mg $Cm^{-2}$ day$^{-1}$ (Koblentz-Mishke *et al.*, 1970), which supports a correspondingly high standing crop of zooplankton (see the maps by Knox, 1970). Biogenous material produced in this high productivity region dominates the equatorial sedimentation pattern and results in a zone of thick sediment accumulation. The presence of this broad zone of highly fossiliferous sediments was first noted during the *Challenger* expedition (Murray and Renard, 1891). Near the Equator itself, the sediments are highly calcareous, but give way north and south to belts of siliceous sediment. Beyond the boundaries of the high productivity zone, non-fossiliferous pelagic clays are accumulating. In the eastern part of the region the zone of calcareous sediment accumulation extends north and south over the shallow regions of the East Pacific Rise and its associated ridges.

Since the oceans are undersaturated with respect to silica (*vide supra*) the silica of siliceous organisms begins to dissolve immediately on the death of the organism. For this reason siliceous sediments can only accumulate in regions where the rate of supply exceeds the rate of dissolution, and therefore their distribution mirrors that of surface productivity. The distribution of calcareous sediments is complicated by the fact that the rate of dissolution of calcium carbonate varies with depth and increases rapidly below a depth of about 3600 m in the equatorial Pacific (Peterson, 1966). From this depth, i.e. the lysocline (see Section 24.2.2.3), down to the carbonate compensation depth there is a rapid dissolution of calcareous material with a correspondingly abrupt change in the faunal composition of the accumulating sediment (Berger, 1970; see also Chapter 29). The carbonate compensation depth is depressed beneath regions of high productivity (see Fig. 1 of Berger and Winterer, 1974). This phenomenon was first recognized by Arrhenius (1952),

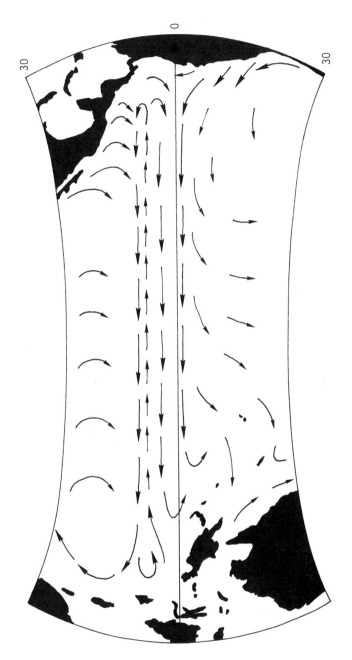

FIG. 24.14. Surface circulation in the tropical Pacific. (Based on Neumann and Pierson, 1966.)

and the factors controlling it have been elucidated by Peterson (1966) and by Berger (1967, 1970, 1974), among others. The occurrence of belts of siliceous sediments north and south of the carbonate-rich equatorial sediments merely reflects the fact that under the equatorial high productivity belt, where the carbonate compensation depth is depressed, the siliceous component of the sediments is masked by the much greater abundance of calcareous material. The factors controlling the accumulation of this equatorial belt of biogenous sediments are summarized in Fig. 24.15.

The accumulation rate of deep-sea pelagic clay in the Pacific is about 2 mm/1000 years. In contrast, the rate of calcareous material ranges from 10 to 20 mm/1000 years, and that for siliceous material from 4–5 mm/1000 years (van Andel et al., in press). It would be expected, therefore, that if the present sedimentation regime were maintained for periods of a few million years, a thick belt of biogenous sediment would accumulate under the equatorial current system. Since the Pacific Ocean floor is gradually moving northwards under the equatorial high productivity zone the bulge of thick equatorial sediments should be displaced northwards. Evidence for this displacement can be seen from seismic reflection records (Winterer, 1973), although the record is complicated by the occurrence of Eocene chert horizons, and great care must be taken in interpreting the data (van Andel et al., in press). Deep-sea drilling may provide information on the ages and present latitudes of ancient equatorial sediments, and so provide a powerful tool for the examination of past movements of the Pacific plate and, incidentally, an independent check on the validity of the predictions of the sea floor spreading hypothesis. This approach has been explored in detail by Winterer (1973), Berger and Winterer (1974), and van Andel et al. (in press).

By comparison with the bottom circulation of the northern North Atlantic that of the Pacific is not especially vigorous; the latter is, however, far from being a static region. Pre-Quaternary sediments have frequently been sampled a few degrees north and south of the equator, and increasingly older outcropping deposits are encountered with increasing distance from the equator (Riedel and Funnell, 1964). These occurrences have been explained in terms of deep-sea erosional processes of regional extent (Johnson and Johnson, 1970; Moore et al., in press) which are probably related to the northward movement of Antarctic Bottom Water.

(b) *Non-biogenous sediments.* With the exception of volcanic ashes and detrital (clastic) sediments, derived through the erosion of volcanic islands, which accumulate as archipelagic aprons (Menard, 1964), the predominant non-biogenous sediments accumulating in the central Pacific are composed of clay minerals and authigenic materials.

The dominant clay minerals are illites and montmorillonites (Fig. 24.16).

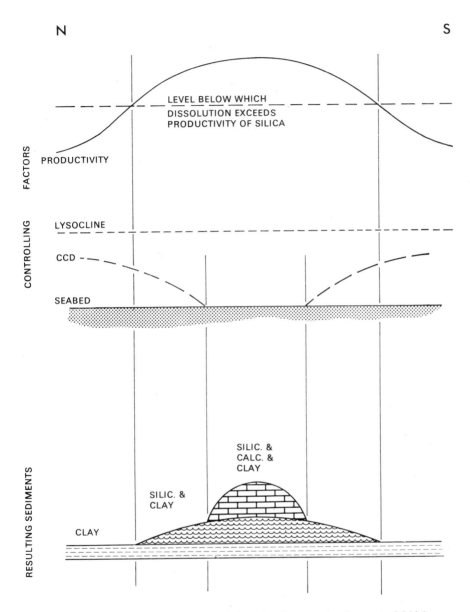

FIG. 24.15. Factors controlling the accumulation of sediment under the equatorial high productivity belt. (CCD = carbonate compensation depth.)

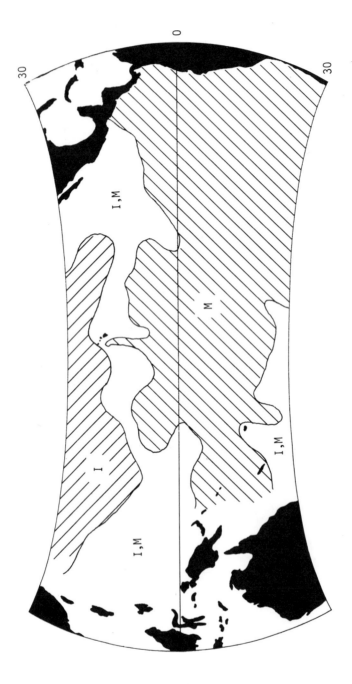

FIG. 24.16. Dominant clay minerals in the tropical Pacific. (Based on Griffin et al., 1968.)
I = illite; M = montmorillonite.

The illites, or detrital micas, are generally considered to be the products of the erosion of major landmasses, rather than to be of a marine origin (Griffin *et al.*, 1968). In the Pacific Ocean, illite is the dominant clay mineral and is found in a wide band extending between 20° and 40°N, which also contains high concentrations of finely divided quartz (Griffin and Goldberg, 1963). This band results from aeolian transport of lithogenous material from the European-Asian arid land areas. In contrast, montmorillonite is commonly characterized as a clay mineral indicative of volcanic regimes, although montmorillonite of continental origin is also known. Montmorillonite covers a broad region in the southeast Pacific, extending over the East Pacific Rise and westwards south of the equator through the volcanic islands of the South Pacific.

Phillipsite, a potassium-rich zeolite, is one of the most common authigenic minerals in the Pacific and may constitute more than half the sediment in some areas. Characteristically, phillipsite is associated with regions of high volcanic activity which are also rich in montmorillonite. Ferro-manganese nodules and concretions are other authigenic phases which are typically found in the sediments of the central Pacific. There is a broad belt rich in nodules extending east-west south of the Hawaiian Islands (Fig. 24.17), although such concretions are ubiquitous on the deep sea floor. These nodules have been discussed in detail in Chapter 28.

### 24.3.3. THE INDIAN OCEAN

#### 24.3.3.1. *Introduction*

The Indian Ocean is the smallest of the three major ocean basins, but is nevertheless geologically complex. Its major physiographical features are shown in Fig. 24.18, and they have been extensively described by Laughton *et al.* (1971). The dominant feature is the active mid-ocean ridge system which has the form of an inverted Y, the southeast branch of which joins the Pacific-Antarctic Ridge. Since the Eocene, sea floor spreading from this ridge has separated Australia from Antarctica. The southwest branch is less well known and appears to be much more complex, being cut by many north-northeast trending fracture zones.

The changing geography of the Indian Ocean during the past $65 \times 10^6$ years is fairly well known as a result of extensive geophysical investigations (see e.g. McKenzie and Sclater, 1971; Sclater *et al.*, 1974; Sclater and Fisher, 1974). By combining the results of geophysical studies on the changing position of the continents with the data from deep-sea drilling it is possible to deduce the changing patterns of sedimentation in the Indian Ocean through time.

Information about the sediments of the Indian Ocean has only been

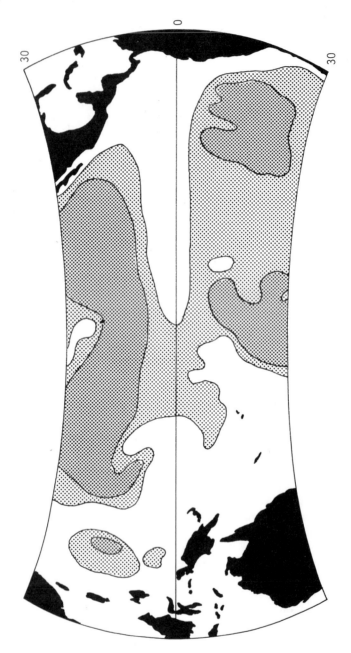

FIG. 24.17. Distribution of manganese nodules in the tropical Pacific. Heavy shading indicates regions of greatest abundance. (Based on Turekian, 1968.)

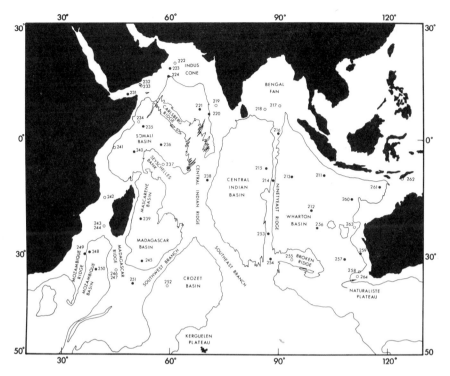

FIG. 24.18. Physiographic features of the Indian Ocean and locations of Deep Sea Drilling Project sites. (From Luyendyk and Davies, 1974.)

acquired relatively recently and is derived largely from the results of the Deep Sea Drilling Project (von der Borch *et al.*, 1974; Whitmarsh *et al.*, 1974; Fisher *et al.*, 1974; Simpson *et al.*, 1974; Davies *et al.*, 1974; Veevers *et al.*, 1974). However, important observations were also made during the International Indian Ocean Expedition of 1959–66. These various data are being summarized in the IIOE Atlas of Geology and Geophysics (chief editor, G. B. Udintsev). The distribution of sediment thickness in the Indian Ocean as determined by seismic reflection has been described by Ewing *et al.* (1969), and their results which agree with the findings from deep-sea drilling, are shown in Fig. 24.19. The greatest thicknesses of sediments are found in the Bengal and Indus fans and in the basins off the east coast of Africa. The controlling influence of physiography on sediment distribution is clearly evident in the figure. For example, the southern part of the Central Indian Basin is virtually devoid of sediment because of its isolated position, and similar conditions are to be found on the western side of the Wharton Basin.

The present day pattern of sedimentation is shown in Figure 24.20. Thick

FIG. 24.19. Sediment thickness in the Indian Ocean. as defined by seismic reflection records (modified from Ewing *et al.,* 1969). Blank = less than 0.1 secs of sediment; light stipple = 0.1–0.5 secs; coarse stipple = 0.5–1.0 secs, heavy stipple = greater than 1.0 secs. (From Luyendyk and Davies, 1974.)

terrigenous sediments are accumulating on the Indus Cone, the Bengal Fan, and the Zambezi Fan. Carbonate sediments are accumulating in the shallow areas along the Australian and African coasts, on the shallow ridges and platforms, and in the shallower parts of the Somali Basin. Siliceous sediments are found in significant amounts in the subpolar regions of the Crozet Basin and in the equatorial regions of the deep basins. Elsewhere, deep-sea clays are accumulating, except in the central Wharton Basin and southern Mascarene Basin, where sedimentation appears to have stopped. The proportion of terrigenous detrital debris accumulating in the Mozambique and Mascarene Basins appears to be much higher than it is in the central Indian and Wharton Basins. This is to be expected because of the comparative remoteness of the latter regions, and also because there is very little run-off from western Australia.

Strong currents sweep the narrow South African shelf and move sediment

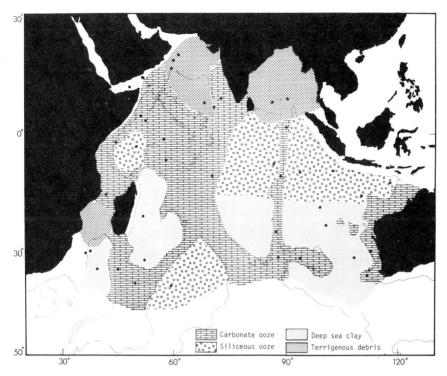

FIG. 24.20. Present-day sedimentation in the Indian Ocean. Solid circles represent DSDP sites. (From Luyendyk and Davies, 1974.)

to the deep sea (Vincent, in press). An interesting transport mechanism for suspended sediment, operating in the northeastern Indian Ocean margins, has been described by Rodolfo (1969). During the monsoons, the water drift from the major deltas of the Bay of Bengal and the Irrawaddy system in the northern Andaman Sea is towards the west. This period coincides with the time of the largest river discharge, and the result is a fast accumulation of fine deltaic muds on the western and northern areas of the Andaman shelf and in the Bay of Bengal. Thus, sediment from the Irrawaddy adds to the Ganges discharge and contributes to the accumulation in the Bengal Fan (see Curray and Moore, 1974). The net result of the circulation patterns, climatic characteristics and relief thus produces a starved basin (the Andaman Basin) downslope from a major river delta. This is a marked exception to the rule that starved basins are produced by tectonic barriers, and illustrates that they can also be produced by oceanographic barriers.

The Bengal Fan, the largest submarine fan in the World Ocean, is interesting in that it has prograded into the deep Indian Ocean as a result of the

C

stretching of the headward distributaries as the Indian subcontinent moved northwards relative to the Indian Ocean floor during the past several tens of millions of years (Curray and Moore, 1974). Thus, the oldest sediments are at the toe of the fan rather than at its head.

### 24.3.3.2. Patterns of sedimentation through time

In the following discussion it will only be possible to give a general description of the broad patterns of sediment distribution. For comprehensive and detailed reports of drilling at each site the reader is referred to the relevant Reports of the Deep-Sea Drilling Project.

(a) *Pre-Tertiary sedimentation*. The geography of the Indian Ocean about $75 \times 10^6$ years ago, as deduced by McKenzie and Sclater (1971), is shown in Fig. 24.21. At that time India lay south of the equator, and in fact its tip extended south of 30° S. Africa was situated more or less at its present latitude,

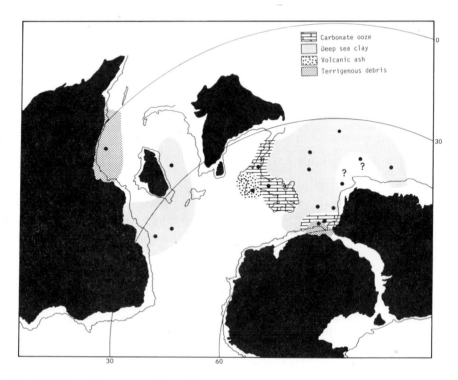

FIG. 24.21. Late Cretaceous sedimentation plotted on a palaeogeographical reconstruction for $75 \times 10^6$ yr B.P. (From McKenzie and Sclater, 1971). Solid circles represent DSDP sites which sampled this time interval. (From Luyendyk and Davies, 1974).

whereas Antarctica and Australia were still joined as one major landmass, asymetrically about the south pole. In the figure, the results obtained during recent deep-sea drilling have been superimposed on the inferred ancient geography.

Little is known of pre-Tertiary sedimentation in the western Indian Ocean, although it is known from drilling that the eastern basin was a region of quite deep water clay accumulation. There was probably volcanic activity along the line of incipient rifting marked by the Broken Ridge and the Naturaliste Plateau. Volcanic rocks are found in the Perth Basin (Brown et al., 1968), and on the Naturaliste Plateau detrital sediments composed of the weathering products of basaltic volcanic rocks were accumulating (Davies et al., 1974).

In the western basin, terrigenous sediments were accumulating off the northern coast of east Africa and terrigenous-volcanic ones on the Mozambique Ridge. However, elsewhere the only sediments observed are the thin non-fossiliferous clays of the Mascarene and Mozambique Basins. It seems likely that carbonate sediments were accumulating along the shallow western margin of India and around Madagascar; however, there is no direct evidence for this. In immediately pre-Tertiary times the carbonate compensation depth in both the western and the eastern basins appears to have been significantly shallower than it is at present (Luyendyk and Davies, 1974).

During the immediate pre-Tertiary period India, the Ninetyeast Ridge, the Broken Ridge and the Naturaliste Plateau formed a topographical barrier which effectively divided the newly forming Indian Ocean into two parts; the eastern basin being closed, and the western one having open connections with the South Atlantic to the west and with the Tethyan region to the north. Although this barrier must have had a profound effect on ocean circulation patterns, the sediments deposited in the eastern and western basins do not appear to be markedly different. The south end of the Ninetyeast Ridge was the site of accumulation of shallow water sediments and volcanic ashes, presumably reflecting the volcanic activity which was occurring at the southern end of the ridge. The northern part of the Naturaliste Plateau, the Broken Ridge and the northern part of the Ninetyeast Ridge formed shallow water platforms not more than a few hundred metres deep on which pelagic carbonates were accumulating under quiet conditions. In the deeper waters of the Wharton Basin sedimentation was either very slow or non-existent, and for this reason the sedimentary record at all the Wharton Basin sites consists only of very thin non-fossiliferous clays.

(b) *Middle Eocene sedimentation.* By middle Eocene times India had moved farther north to lie astride the equator, Africa was relatively farther south than it is at present, and Australia and Antarctica were still one landmass, markedly asymmetrical with respect to the South Pole (Fig. 24.22).

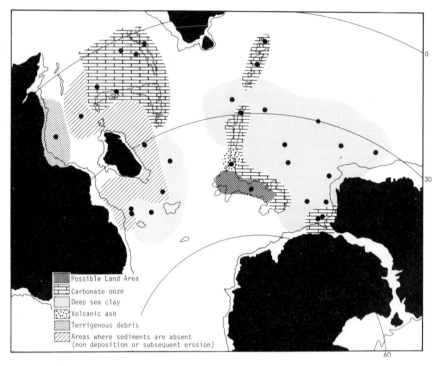

FIG. 24.22. Middle Eocene sedimentation plotted on a paleogeographical reconstruction for $45 \times 10^6$ yr B.P. (From McKenzie and Sclater, 1971). Solid circles represent DSDP sites which sampled this time interval. (From Luyendyk and Davies, 1974.)

The water circulation between the eastern and western basins was still restricted by the Ninetyeast Ridge and the Broken Ridge. At this time it must be assumed that the Broken Ridge was exposed above sea level since the oldest sediments found overlying the Cretaceous limestones on the Broken Ridge are late Eocene littoral gravels (Davies *et al.*, 1974). Volcanic activity was proceeding at the southern end of the Ninetyeast Ridge and a great thickness of volcanic ash was accumulating in very shallow water. Uninterrupted pelagic carbonate sedimentation was proceeding on the Naturaliste Plateau (the northern quiescent end of the Ninetyeast Ridge) and around the submerged margins of the Broken Ridge. Cores obtained from drilling on the Naturaliste Plateau provided evidence that Eocene sediments may have been deposited over much of the plateau, but must have been later removed by erosion, since only isolated pockets of them are found.

Carbonate sediments were also accumulating at this time on the Chagos-Laccadive and Mascarene plateaus and in the area to the northwest of them, and terrigenous debris continued to be deposited off the northern coast of

East Africa. The Wharton, Central Indian, Madagascar and Mozambique basins were areas in which there was very slow accumulation of deep-sea clays or no deposition at all. In the Somali Basin, the northern part of the Mascarene Basin, the western Mozambique Basin, and on the Mozambique and Madagascar ridges, the middle Eocene sediments were very thin and were subsequently removed by strong bottom currents at the time of the formation of the Oligocene unconformity (see below). Frakes and Kemp (1972) have suggested that at this time there was a large counterclockwise surface gyre which transported warm water from low latitudes down what is now the east coast of India to the southern end of Africa, with a return flow along the Australian-Antarctic coasts. Palaeontological results from drilling in the Indian Ocean provide no evidence either for or against the existence of this circulation pattern.

(c) *Early Oligocene sedimentation.* By the Early Oligocene the geography of the Indian Ocean region more closely resembled that of today (Fig. 24.23). There was a clear, though narrow, circumpolar passage around the Antarctic continent, and there was communication around and through the Ninetyeast Ridge between the Wharton Basin and the central Indian Basin. In the south, the Broken Ridge and the Kerguelen Plateau had begun to separate. In the northwest, the passage between India and Arabia, which connected Tethys to the western basin, was becoming restricted. The mid-ocean ridge system in its present configuration was becoming active.

   Although the Oligocene sedimentary record is poorly represented, some observations can be made. Volcanic activity, probably subaerial, was proceeding at the southern end of the Ninetyeast Ridge and resulted in the accumulation of detrital volcanic sands and silts in an essentially lagoonal or littoral environment. Elsewhere on the Ninetyeast Ridge and also probably on the Broken Ridge and the Kerguelen Plateau· pelagic carbonate sediments were accumulating sporadically in water depths of a few hundred metres. At all these locations the sedimentary record is incomplete. Pelagic carbonate sediments were also accumulating in the strait between Madagascar and the mainland of Africa, and on the Chagos-Laccadive and Mascarene plateaus, and in the region to the north of them. Terrigenous sediment continued to pile up off the northeast coast of Africa and to the south of Arabia. Oligocene sediments are absent from the Naturaliste Plateau, from large areas of the Wharton Basin, and from the entire deep water region west of the central Indian Ridge and to the north of its southwest branch.

24.3.3.3. *Regional unconformities*
One of the more unexpected results of deep-sea drilling has been the discovery

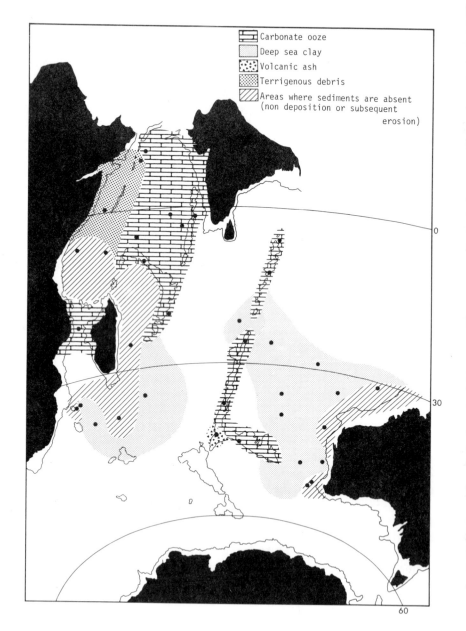

FIG. 24.23. Early Oligocene sedimentation plotted on a palaeogeographical reconstruction for $36 \times 10^6$ yr B.P. (From McKenzie and Sclater, 1971). Solid circles represent DSDP sites which sampled this time interval. (From Luyendyk and Davies. 1974.)

of large gaps in the sedimentary record in many regions. On a global scale these have been summarized by Moore *et al.* (in press). In the Indian Ocean such gaps have been encountered in a broad spectrum of terrigenous, biogenous and volcanogenic sediments encompassing late Mesozoic and Cenozoic time (Davies *et al.*, 1975). The hiatuses were encountered at well over half the 48 sites drilled, and were found to occur on a wide variety of topographical features over a broad range of depths. They appear to show temporal groupings centred on the Oligocene, early Tertiary, and late Cretaceous. At many sites true unconformities can be clearly recognized. In other cases, undateable intervals were encountered which were barren of microfossils, and which usually had the lithology of detrital or pelagic clay accompanied by zeolites and other indicators of extremely slow sedimentation. These intervals, which may be considered to be dissolution facies, have been produced by the slow accumulation of sediment beneath the carbonate compensation depth, and may or may not contain unconformities. In many instances there is indirect evidence that these intervals do encompass unconformities.

The stratigraphical intervals represented at sites in the western and eastern Indian Ocean are shown in Figs. 24.24 and 24.25 respectively. In the west, unconformities and/or undated intervals can be seen centred on the Oligocene and the early Tertiary (or the latest Cretaceous). In the east, hiatuses centred on the Oligocene, early Tertiary and/or late Cretaceous are shown. The limits of the earlier hiatuses are not well defined, but the upper limit of the early Tertiary hiatus in the east may lie in the middle of the Eocene, and the limit of the Cretaceous hiatus may be as young as late Palaeocene (Davies *et al.*, 1975).

Drill sites which sampled the Oligocene, and older time periods, are shown in Fig. 24.26 in which the extent of the Oligocene hiatus is indicated by the shaded regions. This hiatus is present at both shallow and deep sites, whereas complete Oligocene sections were recovered in the northwest Indian Ocean at shallower sites near the ancient ridge crest and nearer to the palaeo-equator. The figure also shows that a similar hiatus was recorded in the Western Pacific (Kennett *et al.*, 1972). The west Pacific hiatus apparently ends in the middle Miocene, at about the same time as that in the western Indian Ocean. An Oligocene hiatus is not generally found in the central Pacific, indeed drilling results have shown the Oligocene is represented by calcareous oozes which had very high sedimentation rates (Winterer *et al.*, 1973). High rates of calcareous sediment accumulation have also been found in the South Atlantic (Maxwell *et al.*, 1970).

Indian Ocean and southwest Pacific drilling sites plotted on an early Eocene reconstruction are shown in Fig. 24.27. The early Tertiary hiatus is more difficult to regionalize because it is often obscured by the overlying Oligocene hiatus, and at many sites the oldest sediments recovered were

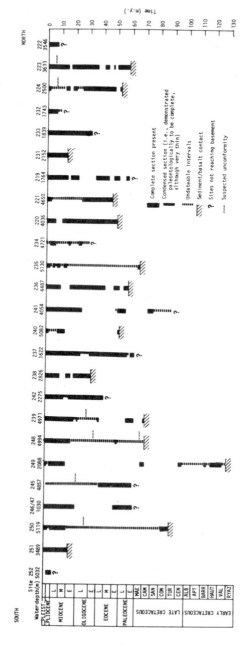

FIG. 24.24. Stratigraphical section sampled at sites in the western Indian Ocean. (From Davies et al., 1975; by courtesy of Macmillan Journals Limited.)

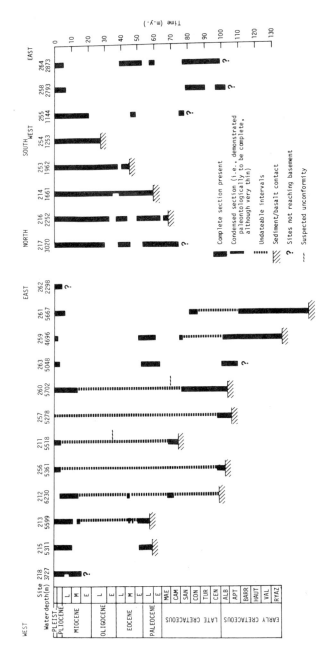

FIG. 24.25. Stratigraphical interval sample at sites in the eastern Indian Ocean. Ridge and plateau sites (right) are separated from basin sites (left). (From Davies *et al.*, 1975; by courtesy of Macmillan Journals Limited.)

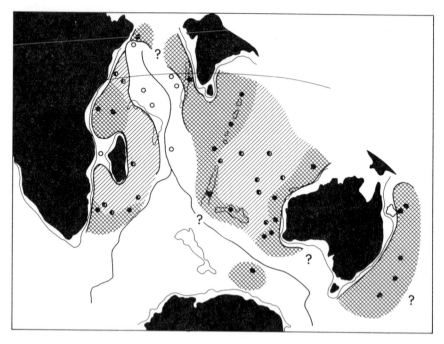

FIG. 24.26. DSDP sites plotted on an early Oligocene reconstruction of the Indian Ocean (from McKenzie and Sclater, 1971). Open circles represent sites where the sedimentary section is complete; solid circles where there is a proven unconformity, and half circles where there is an undated interval or an inferred unconformity. The shading indicates the extent of the Oligocene unconformity, cross-hatching showing the extent of the proven unconformity. (From Davies *et al.*, 1975; by courtesy of Macmillan Journals Limited.)

younger than Eocene. However, certain patterns, such as uninterrupted sedimentation in the northwest, unconformities on shallow structures like the Ninetyeast Ridge,and a well-developed hiatus in southerly latitudes appear to be similar to those of the Oligocene hiatus. Such similarity in the regional patterns of the early Tertiary and Oligocene hiatuses implies that they may have arisen from similar causes. Early Tertiary hiatuses have been identified in the central Pacific (Douglas *et al.*, 1973) and in the southwest Pacific, (Kennett *et al.*, 1972). However, in the South Atlantic the early Tertiary sequence is evidently complete (Maxwell *et al.*, 1970). It seems, therefore, that regional hiatuses in the Cretaceous and early Tertiary in the Indian Ocean are closely allied with contemporaneous ones in the central and southwest Pacific. However the Oligocene hiatus in the Indian Ocean and southwest Pacific occurred at the same time as rapid calcareous sedimentation in the central Pacific and South Atlantic.

The possible origins of these hiatuses have been discussed by Davies *et al.*

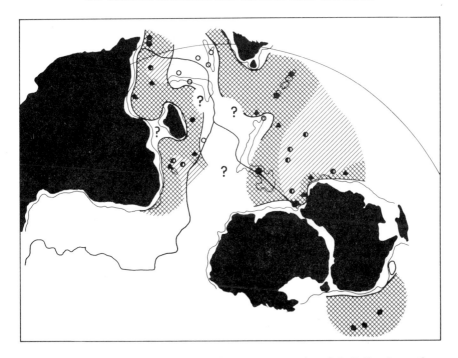

FIG. 24.27. DSDP sites plotted on a middle Eocene reconstruction of the Indian Ocean (from McKenzie and Sclater, 1971). Open circles represent sites where the sedimentary section is complete; solid circles where there is a proven early Tertiary unconformity; half circles where there is an undated interval or an inferred unconformity; and triangles where an early Tertiary unconformity cannot be separated from the overlying Oligocene unconformity. The shading indicates the probable extent of the early Tertiary unconformity, cross-hatching showing the extent of the proven unconformity. (From Davies *et al.*, 1975; by courtesy of Macmillan Journals Limited.)

(1975), and their discussion is summarized here. The dangers of extrapolating the observed hiatuses to a regional scale cannot be over-emphasized in view of our imperfect knowledge of the history of the local sedimentation regime at each site. and the relatively poor coverage of sites compared to the total size of the Ocean. When the cause or causes of these hiatuses are considered a number of variables must be taken into account. These include tectonism and continental drift, world climate, oceanic productivity, oceanic circulation and oceanic chemistry, all of which are interrelated in a complex fashion. True unconformities are the result of erosion or nondeposition caused by vigorous surface and/or deep ocean currents. The fact that the hiatuses observed in the Indian Ocean cover such a wide depth range indicates that both deep and surface circulation have been involved. Dissolution facies are formed below the carbonate compensation depth because of the dissolution of calcareous material. The compensation depth may become shallower as a result of a number of factors which produce an increase in the degree of undersaturation of bottom waters (Berger and Winterer, 1974). These include a decrease in the regional productivity, the removal of carbonate from the water by precipitation on shelf areas and an increase in the amount and rate of supply of cold bottom waters.

The bottom water of the southern hemisphere forms in the Antarctic shelf regions, notably in the Ross and Weddell Seas (Heezen and Hollister, 1971). Part of the Weddell Sea water flows eastwards into the southwestern regions of the Indian Ocean where it behaves as a western boundary undercurrent. Some of the bottom water also drifts northwards through fracture zones in the southeast Indian Ridge, and may be joined by Ross Sea water flowing west between Australia and Antarctica to pass through the gap between the Naturaliste Plateau and the Broken Ridge into the Wharton Basin (Heezen and Hollister, 1971). This pattern of circulation appears to mirror the regional development of the Oligocene hiatus, and possibly also the early Tertiary one, and provides a possible explanation of why the hiatus is more clearly defined in the western ocean regions. However, if the occurrence of regional unconformities is to be explained in terms of the circulation of aggressive bottom water, it is necessary to explain why these unconformities appear to be most pronounced only at certain times in the sedimentary record. A possible answer lies in the tectonic and climatic events which occurred in the Antarctic region during the Tertiary.

An important event in the late Cretaceous was the rifting of New Zealand from Antarctica and the opening of the Ross Sea (Hayes and Ringis, 1973). Shortly after this, perhaps in the early Eocene, climatic deterioration and the onset of glacial conditions may have occurred in Antarctica (Margolis and Kennett, 1970). This would have led to the formation of substantial amounts of aggressive Antarctic Bottom Water and this would have resulted in the formation of unconformities and dissolution facies in the middle Eocene sediments. Further climatic deterioration occurred in Antarctica in Oligocene times and this led to late Oligocene glaciation (Margolis and Kennett, 1970) which would have resulted in the formation of the Oligocene unconformity. Between the late Eocene and the middle of the Oligocene, climatic conditions in Antarctica were probably a little warmer. This may have been sufficient to reduce the supply of Antarctic Bottom Water, to encourage increased surface productivity, and to diminish the intensity of current activity to the point at which sedimentation could resume. The Oligocene hiatus was brought to a close by the gradual establishment of the present pattern of circum-Antarctic flow towards the end of the Oligocene (Kennett *et al.*, 1974). This again reduced the inhibiting effect of the Antarctic Bottom Water on sedimentation in the ocean basins farther north, because a substantial proportion of it was diverted into a circum-polar flow and, thus proportionately less of it found its way into the basins.

Although the above explanation is largely speculative it does illustrate the complexity of the interactions between the marine sedimentary system and tectonics. Thus, before the sedimentary record can be adequately interpreted the geologist must take account not only of those factors controlling the marine sedimentary system today, but also of the dynamics of the earth's crust and of the changing sizes, shapes and geographical positions of the ocean basins.

## 24.4. MODELS FOR OCEANIC SEDIMENTATION

It was stated above that in studying the oceanic sedimentary system it is necessary to consider the following factors: source, transportation, accumulation and post-depositional changes. Each of these is influenced by environmental constraints. The interactions of these have been illustrated using three regions of the World Ocean as examples. It became evident that the observed vertical and horizontal distribution of ocean sediments is the result of the actions and interactions of the four factors integrated over the surface of the earth and over geological time. This generalization will form the basis of a discussion of the problem of formulating models of oceanic sedimentation.

### 24.4.1. A STATIC MODEL

An attempt will be made to develop a static model depicting the conditions on the surface of the earth today. In this model the ocean will be considered as a two layered medium consisting of a surface ocean, extending from the surface to a depth of 100–1000 m and an underlying deep ocean. Transfer between the two layers takes place in regions of upwelling (divergence) and downwelling (convergence). The circulation of the surface ocean is under the control of the atmospheric circulation patterns. Circulation in the deep ocean is thermohaline. The northern part of an elliptical ocean extending from pole to pole is depicted in Fig. 24.28A together with the atmospheric circulation. The anticipated surface circulation patterns and environmental effects are also shown. It should be noted that the regions of up-welling occur along the eastern margins of the ocean particularly in the tropics, and that the regions of convergence are more pronounced in the west where the subpolar and subtropical gyres converge. Another important feature is that the atmospheric circulation pattern causes a displacement of the isotherms, with the result that warmer waters extend farther north on the western side of the

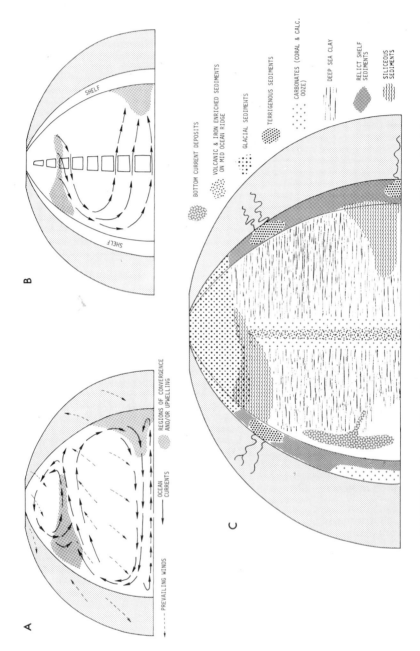

FIG. 24.28. A static model for ocean sedimentation: (A) surface circulation; (B) deep circulation; (C) the predicted pattern of sediment distribution (modified from Gross, 1972).

basin and cooler waters reach further south on the eastern side. The circulation patterns of the deep ocean are shown in Fig. 24.28B. Cold water which is formed in the subpolar gyre sinks along the convergence and travels towards the equator along the western side of the deep ocean in the form of a western boundary current. Deep water migrates eastwards as it approaches the subtropical and tropical regions and subsequently returns to the surface in the regions of upwelling.

The pattern of sediment distribution in this ideal ocean can be predicted using as a basis the earlier discussion relating to the factors controlling the accumulation of ocean sediments. The following predictions might be made from such a model (see Fig. 24.28C). (i) Glacial sediments will accumulate at the southern edge of the subpolar gyre in high latitudes. (ii) Coral reefs will be found in the subtropical regions; they will extend farther northwards on the western side of the basin, and be very poorly developed, if at all, on the eastern side. (iii) Diatomaceous sediments will accumulate along the convergence between the subpolar and subtropical gyres, particularly on the western side as a result of the high productivity associated with the cold water. (iv) Radiolarian sediments will accumulate under the highly productive subtropical regions of upwelling on the eastern side of the basin. (v) The input of terrigenous sediment from river run-off will be highest at mid-latitudes, whereas wind transport will be more important at low latitudes. Most of the terrigenous sediment, particularly that introduced by river run-off, will be trapped on the continental shelves and will only reach the deep ocean via submarine canyons. (vi) Along the western edge of the deep basin reworking of bottom sediments will occur as a result of the strong western boundary currents. Ridges of sediment will accumulate at the junction between the western boundary currents and the static central masses. (vii) At centres of sea floor spreading along the crestal portions of the mid-ocean ridge, iron and manganese-rich sediments may be found. (viii) Sediments containing relatively high proportions of volcanic debris will occur in the vicinity of volcanic activity.

The predictions of the model agree reasonably well with the observed distribution of sediment in the deep ocean. For example, coral reefs do extend markedly farther northwards and southwards on the western sides of the ocean basins than they do on the eastern sides. Glacially-derived sediments are found in the polar regions and an abundance of siliceous fossils occurs beneath the subpolar convergences. Siliceous fossils are also found in regions of upwelling, including the equatorial belt in the Pacific. Carbonate sediments are found in shallow marginal regions. Although these carbonates have received relatively little attention, with the exception of those of tropical latitudes (see discussion by Ginsburg and James, 1974), they do contribute appreciable quantities of sedimentary material to the system. Outer shelf sediments in higher latitudes (i.e. non-tropical ones) may contain as much as

30% carbonate. This carbonate is composed dominantly of molluscan and algal debris, and foraminiferal contributions increase seaward as net deposition rates and water turbulence decrease. At times of falling sea levels the contribution made by shallow water carbonates to the system diminishes as the shelf area decreases.

Despite this measure of agreement, the model is inadequate in a number of respects. For example, there are two major classes of continental margin provinces resulting from plate tectonics which have been described as "collision" and "trailing edge" margins (Inman and Nordstrom, 1971). These two end members of a geomorphic spectrum lead to very different sedimentary conditions. With trailing edge margins (e.g. those of the North Atlantic) a distinct sedimentary system is formed consisting of broad shelves, stable margins, sediment-infilled slopes and rises, and associated abyssal plains. On collision edge coasts (e.g. that of Chile-Peru) shelves are narrow, slopes are steep and tectonically active, the continental edge is marked by a trench which is the expression of active plate subduction and coastal mountains of high relief border the active margin. In view of the observations by Gibbs (1967), it would be expected that trailing edge margins which form large well-developed drainage basins and coastal plains will receive dissolved load and suspended clays from the lowlands and clastics from the highlands, resulting in a small bedload contribution (10%, or less, of the total transported solid). In contrast, collision edge margins, which front coastal plains which are narrow (or even non-existent), have drainage areas which are steep and limited in extent. These margins receive a sedimentary input only from highlands and the bedload makes up 50%, or more, of the total solid carried. Sediment discharges to the two types of margins are further complicated by the presence of large estuaries on the contemporary trailing edge forms, and by an absence of estuaries on the collision edge varieties. Thus, the trapping of sediments occurs in the coastal belt for the former margin type, and at the base of the slope (trench) for the latter. At more mature stages of coastal development the trailing edge form is represented by prograding coastal plain and barrier deposits (see Swift, 1974; Curray in Stanley, 1969), whereas the collision edge forms continue to be involved in the transfer of sediment to the adjacent deep. The lowering of sea level in glacial times produced little change in the collision edge transport mode, except for a probable increase in the volume of material delivered to the slope base. However, with the trailing edge form transfer of large quantities of sediment to the rise occurs when the sea level falls. Thus, the effects of sea level changes on the two forms are often widely divergent. For the collision margins of the Pacific the principal climatic effects have been the compression of the latitudinal sedimentary belts towards the equator, and an increase in the terrigenous contribution during cold periods; this did not produce any change in the basic patterns of sediment delivery. On the trailing edge margin of the Atlantic, glacial variations had a considerable effect on the delivery of terrigenous sediment to the deep sea. At times when sea level was high, sedimentation was limited and mainly occurred by nepheloid flows and bottom current reworking. In contrast, during periods when sea level was low there was an enhanced deep ocean contribution by far-ranging turbidity currents and slumps (Emery and Uchupi, 1972), which led to an extension of the rises and to further infilling of the abyssal plains.

Another deficiency of the model is revealed when the depth of deep ocean carbonate accumulation is considered. The Pacific Ocean has a much shallower lysocline than does the Atlantic and tends to collect siliceous sediments. The Atlantic has a deep lysocline and the siliceous fossils which accumulate in it are poorly preserved. The model cannot be expected to predict these differences because it involves only a single pole to pole ocean. In reality, of course, the World Ocean consists of several interconnected basins (see Fig. 24.29), of which only the Atlantic Ocean effectively extends into the Arctic.

If the model is modified to the three box version shown in Fig. 24.29 several important consequences result. Firstly, because the Atlantic is the only ocean extending into the Arctic it is also the only ocean which has strong western boundary currents in the northern hemisphere. Secondly, because of the displacement of the land masses into the northern hemisphere it is possible to develop circumpolar flow around Antarctica, and in fact the Circum-Antarctic Current transports the largest volume of water of any current in the ocean (see Chapter 1). Cold, deep ocean water originates in the Norwegian Sea, the Weddell Sea and the Ross Sea, and the circum-Antarctic waters act as a mixing and recooling region. About half the deep water originates in the Norwegian Sea (Broecker, 1974), and there is a net flow of deep water from the Atlantic into the Pacific and Indian Oceans which results in a strong horizontal segregation of nutrients. The accumulation of silica in the Pacific Ocean results from the vertical concentration gradient of silica in ocean waters, and is a direct result of a steady pumping of silica-rich deep water from the Atlantic to the Pacific and of surface water depleted in silica from the Pacific to the Atlantic (Berger, 1970). The reverse effect is observed with carbonate because carbonate dissolution increases with pressure, with decreasing temperature, and with increasing total carbon dioxide concentration. Cold ocean water is rich in oxygen when at the surface, but becomes gradually depleted of it as a result of biological processes. As a consequence of the decrease in oxygen content there is a gradual increase in the total carbon dioxide concentration. Thus, the deep waters reaching the Pacific are much more corrosive to carbonate than they were soon after their sinking in the Atlantic. This causes the lysocline to be lower in the Atlantic (Berger, 1970).

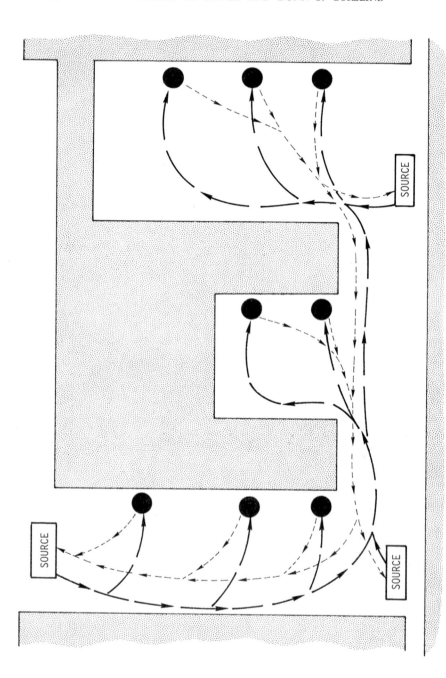

FIG. 24.29. A three box model of ocean circulation. Heavy arrows = path of cold deep bottom water from sources to regions of upwelling. Light arrows = return surface circulation. (Modified from Broecker, 1974.)

### 24.4.2. KINEMATIC MODELS

#### 24.4.2.1. *Dynamics of sedimentation*

If stratigraphical patterns are to be understood, it is necessary when constructing models to take account of the dynamics of the sedimentary system which are not included in the static approach. Sloss (1962) has discussed the various factors involved in stratigraphical modelling and has illustrated them with some representative sequences. He considered that the geometry of a sedimentary body ($S$) is defined by an equation whose elements are the quantity of sedimentary input ($Q$), the dispersal rate from the area of delivery ($D$), the texture and composition of the sedimentary input ($M$) and the rate of subsidence or uplift ($R$). He suggested that one way to view the system is to examine the geometry ($S = f(Q, D, M, R)$), holding all but one factor constant (or equal to zero), and then to examine the sub-elements within each major factor. In all the cases which he defined he was able to construct a stratigraphical model.

Sedimentary systems can also be defined in terms of their response to changes and of the speed at which they return to quasi-equilibrium. An example of this approach has been given by Booth and Gorsline (1974) who considered the necessary factors to be the energy input, the geometry of the shelf, the sediment input (composition and texture) and sea level changes. They have suggested that each of these can be divided into subsets; for example, it is possible to separate the energy term into such factors as waves, tides, winds and currents. In this approach the factors defining the system can be equated for a balanced, or near-equilibrium, condition, and the system can then be disturbed by changing one of the factors. This causes the system to oscillate until it returns to a new equilibrium condition. If the change is repeated then a cyclic oscillation is produced and the product is some form of cyclic sedimentation. If the system is at, or near, equilibrium it is presumed that for a basin continuous uniform sedimentation would result, for a shelf or slope a condition of non-deposition and net transport would be set up.

#### 24.4.2.2. *The effects of sea floor spreading*

Consideration of the dynamics of the sedimentary system, as outlined above, is of value in developing stratigraphical models for fairly limited regions for which the frame of reference can be considered to be essentially fixed for the time period under consideration. However, for larger regions and longer time periods the effects of continental drift must be taken into account. Because of continental drift, the ocean basins have changed their sizes, their shapes, their latitudinal distributions and their depths, throughout geological time. The concept of sea floor spreading permits some of these changes to be readily interpreted in terms of fairly simple models. Thus, for example, sedimentation on the mid-ocean ridge in the South Atlantic can be explained in terms of a simple model involving iron and manganese enrichment at the ridge crest which extends above lysocline, and carbonate dissolution below the lysocline. Thus, examination of a sedimentary section away from the crest of the mid-ocean ridge reveals a sequence of facies changing from clay enriched with iron and manganese to carbonate sediment to deep ocean red clay (Bostrom, 1970; see also Fig. 24.30). This model can be extended further for the subtropical region of the eastern Pacific. Here, the Pacific Plate forming at the crest of the East Pacific Rise migrates northwesterly under the equatorial zone of high productivity. The sediments accumulating on the Pacific plate record its passage under the high productivity zone, and in fact provide an independent check on sea floor spreading. The concept of "plate stratigraphy" has recently been discussed in a lucid exposition by Berger and Winterer (1974). Such ideas can be extended further and used to develop kinematic models for the sedimentary history of the Pacific plate (Berger, 1973; Heezen *et al.*, 1973; Winterer, 1973; van Andel and Moore, 1974).

### 24.4.2.3. *Ocean sediments through time*

The local effects of sea floor spreading through geological time, which have been illustrated in the preceding section, are relatively uncomplicated. On the grander scale, however, there is considerably less certainty. Not only are samples sparse, but a serious limitation is posed by the limits of resolution of biostratigraphy. Resolution of sedimentary events in the deep oceans is probably of the order of $10^4$–$10^5$ years. In offshore continental margin basins (e.g. California Borderland) the resolution may increase in detail to $10^2$–$10^3$ years. In a few anoxic deep bains with high sedimentation rates, such as the Cariaco Basin off Venezuela and the Santa Barbara Basin off Southern California (Hülsemann and Emery, 1961), the resolution may approach one year because bioturbation is eliminated and annual layers are preserved. Thus, the possible error in stratigraphical correlation and interpretation of sedimentary records from the oceans (a function of degree of resolution) may depend on sedimentation rates. This possible error increases with increasing rate. Studies of any given cycle must therefore be based on an appropriate region, and cannot be generally applied to the World Ocean at frequencies of, at best, less than $10^4$ years. In this regard, it should be noted that the large variations in the shallower environments had tended to obscure worldwide climatic effects until the major fluctuations had been shown to exist by examination of deep ocean cores (see Emiliani, 1971; Shackleton and Opdyke, 1973). Following this work, a statistical study of the margin records revealed much more detail and showed that there was a similar

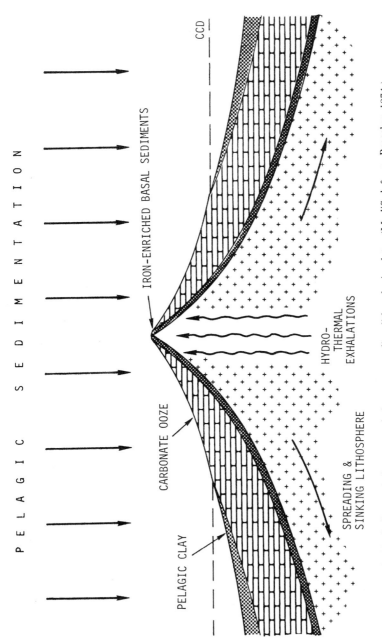

FIG. 24.30. Sediment facies accumulating on spreading lithospheric plates (Modified from Broecker, 1974.)

cyclicity (Gorsline and Prensky, 1975). The resolution will differ in decreasing order of frequency response for biostratigraphical data, palaeomagnetic data and geochemical data (e.g. oxygen isotopic ratios). The shortest responses will be seen in the rare sedimentary structure records of the anoxic basins where annual "varves" are preserved (Hülsemann and Emery, 1961).

Analysis of cores gathered during the Deep Sea Drilling Project indicates gross changes in the composition and rates of accumulation of ocean sediments over the past $120 \times 10^6$ years (see, e.g. Davies and Supko, 1973). On an ocean-wide scale these changes can be interpreted as being the consequence of the gradual evolution of the ocean basins since the middle of the Jurassic, and models for the physical and chemical evolution of Atlantic sediments have indeed been developed (Berggren and Hollister, 1974; Dickinson, 1974). Such models depend on various factors including the changing patterns of ocean circulation as the ocean basin widens and changes shape, the changing importance of the input of material from mid-ocean ridge volcanic activity relative to that of terrigenous material from the ocean margins, the increasing depth of the ocean with time etc.

Worldwide transgressions and regressions, such as those described by Curray (1969) and Rona (1973), may be explained in terms of major changes in ocean volume which resulted from changes in the rate of sea floor spreading. Periods of increased spreading activity led to flooding of the continents by decreasing the volume of the ocean basins. At such times storage of terrigenous sediment occurred in the marginal and epicontinental seas. In periods when the spreading rate decreased and sea level fell the continental contribution to the sediments increased and was delivered almost directly to the deep ocean.

It is implicit in these discussions that the oceans are essentially a closed system. Local changes in nutrient concentration, rates of transport and upwelling, or other such perturbations within the system are short lived with scales of a few tens of thousands of years (Southam, 1974). This leads to the conclusion that major gross changes must be the consequence of changes in the input to the system, or of changes in the relative sizes of parts of the system, for example, a change in the proportion of shelf seas to deep ocean. Thus, the well known occurrence of black shales in the ocean basins at the end of the early Cretaceous reflects not only the poorly developed ocean circulation and the reducing conditions in the bottom waters, but also a very low input of carbonate to the deep sea. This low input may have been caused either by a low rate of erosion on land (low external input), or by an extensive development of shelf seas with a consequent accumulation of shallow water carbonate sediments at the expense of deep ocean ones. It was emphasized at the beginning of this chapter that the only record of the palaeo-oceanic environment is that provided by ancient marine sediments.

Unfortunately, our knowledge of the marine sedimentary system must still be considered to be rudimentary when it is used for the interpretation of changes in sediment accumulation which were the consequences of major changes in the system. Perhaps a geochemical approach of the type outlined in the rest of this volume may one day help to elucidate these matters.

## ACKOWLEDGEMENTS

We would like to emphasize that in preparing this review we have drawn heavily on the ideas and observations of our colleagues. We have attempted to give appropriate recognition in every case and apologize for any omissions or misinterpretations which may have crept in. Special thanks are due to T. L. Vallier for an extensive critical review of the manuscript. W. H. Berger, R. G. Douglas, D. E. Drake, B. D. Edwards and H. A. Karl also read drafts of the manuscript and made many helpful suggestions for improvements.

## REFERENCES

Allen, G. P., ed. (1974). "Proceedings, International Symposium on Interrelationships of Estuarine and Continental Shelf Sedimentation". Institute de Geologie du Bassin d' Aquitane, Bordeaux.

Allen, P. (1964). *J. Sediment. Petrology* **34**, 289.

Arrhenius, G. (1952). *Rep. Swed. Deep Sea Exped.* **5**, 189.

Arrhenius, G. (1963). *In* "The Sea" (M.N. Hill, ed.), Vol. 3, pp. 655–727. Interscience Publishers, New York.

Atwater, T. (1970). *Bull. Geol. Soc. Amer.* **81**, 3513.

Ballard, J. A., Bowles, F. A. and Ruddiman, W. F. (1971). *Trans. Amer. Geophys. Un.* **52**, 243.

Bathurst, R. G. C. (1971). "Carbonate Sediments and their Diagenesis". Elsevier, Amsterdam.

Berger, W. H. (1967). *Science, N.Y.* **156**, 303.

Berger, W. H. (1968). *Deep-Sea Res.* **15**, 31.

Berger, W. H. (1970). *Bull. Geol. Soc. Amer.* **81**, 1385.

Berger, W. H. (1973). *Bull. Geol. Soc. Amer.* **84**, 1941.

Berger, W. H. (1974). *In* "The Geology of Continental Margins" (C. A. Burk and C. L. Drake, eds.) pp. 213–241. Springer-Verlag, New York.

Berger, W. H. and Winterer, E. L. (1974). *In* "Pelagic Sediments on Land and under the Sea" (K. J. Hsü and H. C. Jenkyns, eds.), Special Publication, International Association of Sedimentology, Vol. 1. Blackwell Scientific Publications, Oxford.

Berggren, W. A. and Hollister, C. D. (1974). *Soc. Econ. Palaeontol. Mineral., Spec. Publ. Tulsa* **20**, 126.

Biscaye, P. E. (1965). *Bull. Geol. Soc. Amer.* **76**, 803.

Black, M., Hill, M. N., Laughton, A. S. and Matthews, D. H. (1964). *Quart. J. Geol. Soc. London* **120**, 477.

Blatt, H., Middleton, G. V. and Murray, R. C. (1972). "Origin of Sedimentary Rocks". Prentice-Hall, Englewood Cliffs, New Jersey.

Booth, J. S. (1973). *J. Sediment. Petrology* **43**, 238.

Booth, J. S. and Gorsline, D. S. (1974). *In* "Proceedings, International Symposium on Interrelationships of Estuarine and Continental Shelf Sedimentation" (G. P. Allen, ed.), pp. 145–148. Institute de Geologie du Bassin d' Aquitane, Bordeaux.

Bouma, A. H. (1962). "Sedimentology of some Flysch Deposits—a Graphic Approach to Facies Interpretation". Elsevier, Amsterdam.

Bouma, A. H. and Hollister, C. D. (1973). *In* "Turbidites and Deepwater Sedimentation" (G. V. Middleton and A. H. Bouma, eds.), pp. 79–118. Soc. Econ. Paleontol. Mineral., Pacific Section, Los Angeles.

Boström, K. (1970). *Nature, Lond.* **227**, 1041.

Boström, K. and Peterson, M. N. A. (1966). *Econ. Geol.* **61**, 1258.

Boström, K. and Peterson, M. N. A. (1969). *Mar. Geol.* **7**, 427.

Bramlette, M. N. and Bradley, W. H. (1940). *Prof. Pap. U.S. Geol. Surv.* **196-A**.

Broecker, W. S. (1974). "Chemical Oceanography". Harcourt Brace Jovanovich, New York.

Brown, D. A., Campbell, K. S. W. and Crook, K. A. W. (1968). "The Geological Evolution of Australia and New Zealand". Pergamon Press, London.

Bryan, G. M. (1970). *J. Geophys. Res.* **75**, 4530.

Butze, H. (1955). "Valaströme und Aschenregen". University of Leipzig.

California Department of Water Resources (1967). "Study of Beach Nourishment along the Southern California Coastline". Water Research Progress Report, Sacramento, California.

Calvert, S. E. (1966). *Bull. Geol. Soc. Amer.* **77**, 569.

Calvert, S. E. (1974). *In* "Pelagic Sediments on Land and under the Sea" (K. J. Hsü and H. Jenkyns, eds.), Special Publication, International Association of Sedimentology, Vol. 1, p. 273. Blackwell Scientific Publications, Oxford.

Cann, J. R. and Funnell, B. M. (1966). *Nature, Lond.* **213**, 661.

Chase, T. E. (1975). "Topography of the Oceans" (Map). I.M.R. Technical Report Series. TR57. University of California, San Diego.

Cherkis, N. Z., Fleming, H. S. and Fedin, R. H. (1971). *Trans. Amer. Geophys. Un.* **52**, 243.

Cox, A. (1973). "Plate Tectonics and Geomagnetic Reversals". Freeman, San Francisco.

Crook, K. A. W. (1974). *Soc. Econ. Palaeontol. Mineral., Spec. Publ. Tulsa* **19**, 304.

Curray, J. R. (1969). *In* "New Concepts of Continental Margin Sedimentation" (D. J. Stanley, ed.). American Geological Institute Washington, D.C.

Curray, J. R. and Moore, D. G. (1974). *In* "Geology of Continental Margins" (C. A. Burk and C. L. Drake, eds.), pp. 617–628. Springer-Verlag, New York.

Davies, T. A. (1967). Recent Sedimentation in the Northeast Atlantic. Unpubl. Ph.D. Thesis, Cambridge University.

Davies, T. A. and Edgar, N. T. (1974). *Geophys. Surv.* **1**, 391.

Davies, T. A. and Jones, E. J. W. (1971). *Deep-Sea Res.* **18**, 619.

Davies, T. A. and Laughton, A. S. (1972). *Initial Rep. Deep Sea Drilling Project* **12**, 905.

Davies, T. A. and Supko, P. R. (1973). *J. Sediment. Petrology* **43**, 381.

Davies, T. A., Luyendyk, B. P. *et al.*, (1974). "Initial Reports of the Deep Sea Drilling Project," Vol. 26. U.S. Government Printing Office, Washington, D.C.

Davies, T. A., Weser, O. E., Luyendyk, B. P. and Kidd, R. B. (1975). *Nature, Lond.* **253**, 15.

Davis, W. M. and Ehrlich, R. (1970). *Bull. Geol. Soc. Amer.* **81**, 3537.

Defant, A. (1933). *Details Atlant. Exped. Meteor 1925-26* **2**.

Dickinson, W. R., ed. (1974). *Soc. Econ. Palaeontol. Mineral., Spec. Publ. Tulsa* **22**.

Dietz, R. S. (1961). *Nature, London.* **190**, 854.

Doeglas, D. J. (1946). *J. Sediment. Petrology* **16**, 19.

Douglas, R. G., Roth, P. H. and Moore, T. C. (1973). *Initial Rep. Deep Sea Drilling Project* **16**, 905–909.

Drake, D. E. (1972). Distribution and Transport of Suspended Matter, Santa Barbara Channel, California. Ph.D. Thesis, University of Southern California, Los Angeles.

Drake, D. E. and Gorsline, D. S. (1973). *Bull. Geol. Soc. Amer.* **84**, 3949.

Egloff, J. and Johnson, G. L. (In press). *Can. J. Earth Sci.*

Eittreim, S., Ewing, M. and Thorndike, E. M. (1969). *Deep-Sea Res.* **16**, 613.

Emery, K. O. (1952). *Bull. Geol. Soc. Amer.* **63**, 1105.

Emery, K. O. (1960). "The Sea off Southern California". John Wiley and Sons, New York.

Emery, K. O. (1963a). *In* "The Sea" (M. N. Hill, ed.) Vol. 3, pp. 776–793. Interscience Publishers, New York.

Emery, K. O. (1963b). *Geofisica Internac.* **3**, 11.

Emery, K. O. (1968). *Bull. Amer. Ass. Petrol. Geol.* **52**, 445.

Emery, K. O. and Uchupi, E. (1972). *Mem. Amer. Ass. Petrol. Geol.* **17**.

Emery, K. O., Tracy, J. I. and Ladd, H. S. (1954). *Prof. Pap. U.S. Geol. Surv.* **260A**.

Emery, K. O. Uchupi, E., Gorsline, D. S. and Terry, R. D. (1956). *J. Sediment. Petrology* **27**, 95.

Emiliani, C. (1971). *In* "The Late Cenozoic Glacial Ages" (K. K. Turekian, ed.), pp. 103–197. Yale University, New Haven, Conn.

Ericson, D. B., Ewing, M., Wollin, G. and Heezen, B. C. (1961). *Bull. Geol. Soc. Amer.* **72**, 193.

Ewing, M. and Connary, S. D. (1970). *Mem. Geol. Soc. Amer.* **126**, 41.

Ewing, M., Eittreim, S., Truchan, M. and Ewing, J. I. (1969). *Deep-Sea Res.* **16**, 231.

Fisher, R. L., Bunce, E. *et al.* (1974). "Initial Reports of the Deep Sea Drilling Project," Vol. 24. U.S. Government Printing Office, Washington, D.C.

Fleischer, P. (1970). Mineralogy of Hemipelagic Basin Sediments, California Continental Borderland. Unpubl. Ph.D. Thesis, University of Southern California, Los Angeles.

Folk, R. L. and Ward, W. C. (1957). *J. Sediment. Petrology* **27**, 3.

Fox, P. J., Heezen, B. C. and Harian, A. M. (1968). *Nature, Lond.* **220**, 470.

Frakes, L. A. and Kemp, E. M. (1972). *Nature, Lond.* **240**, 97.

Friedman, G. M. (1958). *J. Geol.* **66**, 394.

Friedman, G. M. (1962). *J. Geol.* **70**, 737.

Garrels, R. M. and Mackenzie, F. T. (1971). "Evolution of Sedimentary Rocks". W. W. Norton, New York.

Gibbs, R. J. (1967). *Bull. Geol. Soc. Amer.* **78**, 1203.

Ginsburg, R. N. and James, N. (1974). *In* "The Geology of Continental Margins" (C. A. Burk and C. L. Drake, eds.), pp. 137–155. Springer-Verlag, New York.

Glass, B. (1967). *Nature, Lond.* **214**, 372.

Goll, R. G. and Bjørklund, K. R. (1971). *Micropalaeontology* **17**, 434.

Gorsline, D. S. and Barnes, P. W. (1972). *Proc. 24th Int. Geol. Congr., Montreal.* **Section 6**, 270.

Gorsline, D. S. and Prensky, S. E. (1975). *Roy. Soc. N.Z. Bull.* **13**,

Gorsline, D. S. and Swift, D. J. P. (1974). *In* "Proceedings, International Symposium on Interrelationships of Estuarine and Continental Shelf Sedimentation" (G. P. Allen, ed.), pp. 235–239. Institute de Geologie du Bassin d'Aquitane, Bordeaux.

Gorsline, D. S. Drake, D. E. and Barnes, P. W. (1968). *Bull. Geol. Soc. Amer,* **79**, 659.

Griffin, J. J. and Goldberg, E. D. (1963). *In* "The Sea" (M. N. Hill, ed.) Vol. 3, pp. 728–741. Interscience Publishers, New York.

Griffin, J. J., Windom, H. W. and Goldberg, E. D. (1968). *Deep-Sea Res.* **15**, 433.

Griffiths, J. C. (1967). "Scientific Method in the Analysis of Sediments". McGraw-Hill, New York.

Gross, M. G. (1970). *Water Resour. Res.* **6**, 927.

Gross, M. G. (1972). "Oceanography—a View of the Earth". Prentice-Hall, Englewood Cliffs, New Jersey.

Haner, B. E. (1971). *Bull. Geol. Soc. Amer.* **82**, 2413.

Hay, W. W., ed. (1974). *Soc. Econ. Palaeontol. Mineral., Spec. Publ. Tulsa* **20**.

Hayes, D. E. and Ringis, J. (1973). *Nature, Lond.* **243**, 454.

Hays, J. D., ed. (1970). *Mem. Geol. Soc. Amer.* **126**.

Heath, G. R. (1974). *Soc. Econ. Palaeontal Mineral., Spec. Publ. Tulsa* **20**, 77.

Heath, G. R. and Moberly, R. (1971). *Initial Rep. Deep Sea Drilling Project* **7**, 991–1007.

Heath, G. R., Moore, T. C., Dauphin, J. P. and Opdyke, N. D. (In press). *Bull. Geol. Soc. Amer.*

Hedgpeth, J. W. (1941). *Amer. Neptune* **5**, 5.

Heezen, B. C. (1959). *Geophys. J. Roy. Astron. Soc.* **2**, 142.

Heezen, B. C. (1963). *In* "The Sea" (M. N. Hill, ed.) Vol. 3, pp. 742–775. Interscience Publishers, New York.

Heezen, B. C. and Hollister, C. D. (1964). *Mar. Geol.* **1**, 141.

Heezen, B. C. and Hollister, C. D. (1971). "The Face of the Deep". Oxford University Press, New York.

Heezen, B. C. and Johnson, G. L. (1963). *Deut. Hydrogr. Z.* **16**, 1.

Heezen, B. C. and Laughton, A. S. (1963). *In* "The Sea" (M. N. Hill, ed.) Vol. 3, pp. 312–364. Interscience Publishers, New York.

Heezen, B. C. Hollister, C. D. and Ruddiman, W. F. (1966). *Science, N.Y.* **151**, 502.

Heezen, B. C., Johnson, G. L. and Hollister, C. D. (1969). *Can. J. Earth Sci.* **6**, 1441.

Heezen, B. C., MacGregor, I. D., Foreman, H. P., Forristal, G., Hekel, H., Hesse, R., Hoskins, R. H., Jones, E. J. W., Kaneps, A., Krasheninnikov, V. A., Okada, H. and Ruef, M. H. (1973). *Nature, Lond.* **241**, 25.

Hess, H. H. (1962). *In* "Petrologic Studies—a Volume in Honor of A. F. Buddington" (A. E. J. Engel, H. L. James and B. F. Leonard, eds.), pp. 599–620. Geological Society of America, Boulder, Colorado.

Hill, M. N., ed. (1963). "The Sea" Vol. 3. Interscience Publishers, New York.

Holeman, J. N. (1968). *Water Resour. Res.* **4**, 737.

Hollister, C. D. and Heezen, B. C. (1967). *Trans. Amer. Geophys. Un.* **48**, 141.

Hollister, C. D. and Heezen, B. C. (1972). *In* "Studies in Physical Oceanography—a Tribute to George Wust on his 80th Birthday" (A. L. Gordon, ed.) Vol. II, pp. 37–66. Gordon and Breach, New York.

Hsü, K. J. and Jenkyns, H., eds. (1974). "Pelagic Sediments on Land and under the Sea". Special Publication International Association of Sedimentology, Vol. 1. Blackwell Scientific Publications, Oxford.

Hülsemann, J. and Emery, K. O. (1961). *J. Geol.* **69**, 279.

Imbrie, J. and Purdy, E. G. (1962). *In* "Classification of Carbonate Rocks" (W. E. Ham, ed.) pp. 253–272. American Association of Petroleum. Geologists, Tulsa, Oklahoma.

Inman, D. L. (1952). *J. Sediment. Petrology* **22**, 125.

Inman, D. L. and Brush, B. M. (1973). *Science, N.Y.* **181**, 20.

Inman, D. L. and Nordstrom, C. E. (1971). *J. Geol.* **74**, 1.

Johnson, D. W. (1919). "Shoreline Processes and Shoreline Development". John Wiley and Sons, New York.

Johnson, D. A. and Johnson, T. C. (1970). *Deep-Sea Res.* **17**, 157.

Johnson, G. L. and Schneider, E. D. (1969). *Earth Planet. Sci. Lett.* **6**, 416–422.

Johnson, G. L., Closuit, A. W. and Pew, J. A. (1969). *Arctic* **22**, 56.

Johnson, G. L., Vogt, P. R. and Schneider, D. E. (1971). *Dent. Hydrogr. Z.* **24**, 49

Jones, E. J. W., Ewing, M., Ewing, J. I. and Eittreim, S. L. (1970). *J. Geophys. Res.* **75**, 1655.

Kaplan, I. R. and Rittenberg, S. C. (1963). *In* "The Sea" (M. N. Hill, ed.), Vol. 3, pp 583–619. Interscience, New York.

Kennett, J. P., Burns, R. E., Andrews, J. E., Churkin, M., Davies, T. A., Dumitrica, P., Edwards, A. R., Galehouse, J. S., Packham, G. H. and van der Lingen, G. J. (1972). *Nature, Lond.* **239**, 51.

Kennett, J. P., Houtz, R. E., Andrews, P. B., Edwards, A. R., Gostin, V. A., Hajós, Hampton, M., Jenkins, D. G., Margolis, S. V., Ovenshine, A. T. and Perch-Nielsen, K. (1974). *Science, N.Y.*, **186**, 144.

Kennett, J. P., Houtz, R. E., Andrews, P. B., Edwards, A. R., Gostin, V. A., Hajós, M., Hampton, M., Jenkins, D. G., Margolis, S. V., Ovenshine, A. T. and Perch-Nielsen, K. (1975). *Initial Rep. Deep Sea Drilling Project* **29**, 1155.

Klein, G. de V. (1963). *Bull. Geol. Soc. Amer.* **74**, 555.

Knox, G. A. (1970). *In* "Scientific exploration of the South Pacific" (W. S. Wooster, ed.), pp. 155–182. National Academy of Sciences, Washington, D.C.

Koblentz-Mishke, O. J., Volkovinsky, V. V. and Kabanova, J. (1970). *In* "Scientific Exploration of the South Pacific" (W. S. Wooster, ed.), pp. 183–192. National Academy of Sciences, Washington, D.C.

Kuenen, Ph. H. (1937). *Leidse Geol. Meded.* **8**, 327.

Kuenen, Ph. H. (1950). "Marine Geology". John Wiley and Sons, New York.

Kuenen, Ph. H. (1951). *Soc. Econ. Palaeontol. Mineral., Spec. Publ., Tulsa* **2**, 14.

Kuenen, Ph. H. (1964). *In* "Turbidites" (A. H. Bouma and A. Brouwer, eds.), *Develop. Sedimentol.* **3**, pp. 1–22. Elsevier, Amsterdam.

Kuenen, Ph. H. (1965). *In* "Chemical Oceanography" (J. P. Riley and G. Skirrow, eds) Vol. 2, pp. 1–22. Academic Press, London and New York.

Kuenen, Ph. H. (1966). *J. Geol.* **74**, 523.

Kuenen, Ph. H. (1970). *N.Z. J. Geol. Geophys.* **13**, 852.

Kuenen, Ph. H. and Migliorini, C. I. (1950). *J. Geol.* **58**, 91.

Lancelot, Y. (1973). *Initial Rep. Deep Sea Drilling Project* **17**, 377.

Langbein, W. B. and Schumm, S. A. (1958). *Trans. Amer. Geophys. Un.* **39**, 1076.

Laughton, A. S., Matthews, D. H. and Fisher, R. L. (1971). *In* "The Sea" (A. E. Maxwell, ed.) Vol. 4, Part II, pp. 543–586. Interscience Publishers, New York.

Laughton, A. S., Berggren, W. A. *et al.* (1972). "Initial Reports of the Deep Sea Drilling Project," Vol. 12. U.S. Government Printing Office, Washington, D.C.

Le Pichon, X. (1968). *J. Geophys. Res.* **73**, 3661.

Lisitzin, A. P. (1972). *Soc. Econ. Palaeontol. Mineral., Spec. Publ., Tulsa* **17**.

Luyendyk, B. P. and Davies, T. A. (1974). *Initial Rep. Deep Sea Drilling Project* **26**, 909.

Margolis, S. V. and Kennett, J. P. (1970). *Science, N.Y.* **170**, 1085.

Markl, R. G., Bryan, G. M. and Ewing, J. I. (1970). *J. Geophys. Res.* **75**, 4539.

Matthews, D. H. (1961). Rocks from the Eastern North Atlantic. Unpubl. Ph.D. Thesis, Cambridge University.

Maury, M. F. (1856). "The Physical Geography of the Sea". Harper Brothers, New York.

Maxwell, A. E., von Herzen, R. P. *et al.* (1970). "Initial Reports of the Deep Sea Drilling Project," Vol. 3. U.S. Government Printing Office, Washington, D.C.

McCave, I. N. (1972). *In* "Shelf Sediment Transport" (D. J. P. Swift, D. B. Duane and O. H. Pilkey, eds), pp. 225–248. Dowden, Hutchinson and Ross, Stroudsbury, Pennsylvania.

McKenzie, D. P. and Sclater, J. G. (1971). *Geophys. J. Roy. Astron. Soc.* **25**, 437.

Meade, R. H. (1974). *In* "Proceedings, International Symposium on the Inter-relationships between Estuarine and Continental Shelf Sedimentation" (G. P. Allen, ed.) pp. 207–214. Institute de Geologie du Bassin d'Aquitane, Bordeaux.

Menard, H. W. (1955). *Bull. Amer. Ass. Petrol. Geol.* **39**, 236.

Menard, H. W. (1961). *J. Geol.* **69**, 154.

Menard, H. W. (1964). "Marine Geology of the Pacific". McGraw-Hill, New York.

Merriam, D. and Bandy, O. (1965). *J. Sediment. Petrology.* **35**, 911.

Middleton, G. V. (1966a). *Can. J. Earth Sci.* **3**, 523.

Middleton, G. V. (1966b). *Can. J. Earth Sci.* **3**, 627.

Middleton, G. V. (1967). *Can. J. Earth Sci.* **4**, 475.

Moore, D. G. (1970). *Spec. Pap. Geol. Soc. Amer.* **107**, 35.

Moore, T. C., van Andel, Tj. H., Sancetta, C. and Pisias, N. (In press). *In* "Marine Plankton and Sediments" (W. R. Riedel and T. Saito, eds.), American Museum of Natural History, New Tork.

Moore, T. C., van Andel, Tj. H. Sancetta, C. and Pisias, N. (In press). *In* "Marine Plankton and Sediments" (W. R. Riedel and T. Saito, eds.)

Murray, J. and Renard, A. F. (1891). *Rep. Challenger Exped.*

Neumann, G. and Pierson, W. J. (1966). "Principals of Physical Oceanography". Prentice-Hall, Englewood Cliffs, New Jersey, 545 pp.

Newell, N. D. and Rigby, J. K. (1957). *Soc. Econ. Palaeontol. Mineral., Spec. Publ. Tulsa* **5**, 15.

Newell, N. D., Purdy, E. G. and Imbrie, J. (1960). *J. Geol.* **68**, 481.

Ninkovich, D. and Heezen, B. C. (1965). *In* "Submarine Geology and Geophysics—Proceedings of the 17th Symposium of the Colston Research Society, University of Bristol" (W. F. Whittard and R. Bradshaw, eds.), pp. 413–454. Butterworth, London.

Normark, W. R. (1970). *Bull. Amer. Ass. Petrol. Geol.* **54**, 2170–2195.

Pasho, D. W. (1973). Character and Origin of Marine Phosphorites. Unpubl. M.S. Thesis, University of Southern California, Los Angeles.

Paterson, I. S. (In press). *Can. J. Earth Sci.*

Pearson, E. A., ed. (1960). "Waste Disposal in the Marine Environment, Proceedings of the First International Conference on Ocean Waste Disposal, Berkeley, California." Pergamon Press, London.

Peterson, M. N. A. (1966). *Science, N.Y.* **154**, 1542.

Peterson, M. N. A., Edgar, N. T. *et al.* (1970). "Initial Reports of the Deep Sea Drilling Project," Vol. 2. U.S. Government Printing Office, Washington, D.C.

Pettersson, H. (1957). *Rep. Swed. Deep Sea Exped.* **1**, 1.

Picard, M. D. (1971). *J. Sediment. Petrology* **41**, 179.

Poldervaart, A. (1955). *Spec. Pap. Geol. Soc. Amer.* **62**, 119.

Pratt, R. M. (1961). *Deep-Sea Res.* **8**, 152.

Reed, W. E., LeFever, R. and Moir, G. J. (1975). *Bull. Geol. Soc. Amer.* **86**, 1328.

Reineck, H. E. and Singh, I. B. (1973). "Depositional Sedimentary Environments". Springer-Verlag, Berlin.

Reineck, H. E., Dorjes, J., Gadow, S. and Hertweck, G. (1968). *Senekenbergiana Lethaea* **49**, 261.

Revelle, R. R. (1944). *Carnegie Inst. Wash. Publ.* **556**.

Rex, R. W. and Goldberg, E. D. (1958). *Tellus* **10**, 153.

Richards, A. F. (1958). *Deep-Sea Res.* **5**, 29.

Riedel, W. R. 1959). *Soc. Econ. Palaeontol. Mineral., Spec. Publ., Tulsa,* **7**, 80–91.

Riedel, W. R. and Funnell, B. M. (1964). *Quart. J. Geol. Soc. London.* **120**, 305.

Rodolfo, K. S. (1964). Suspended Sediment in Southern California Waters. Unpubl. M.S. Thesis, University of Southern California, Los Angeles.

Rodolfo, K. S. (1969). *Bull. Geol. Soc. Amer.* **80**, 1203.

Rona, P. A. (1973). *Bull. Geol. Soc. Amer.* **84**, 2851.

Ross, D. A. (1970). *J. Sediment. Petrology.* **40**, 906.

Schlanger, S. O., Douglas, R. G., Lancelot, Y., Moore, T. C. and Roth, P. H. (1973). *Initial Rep. Deep Sea Drilling Project* **17**, 407–428.

Schneider, E. D., Fox. C. D., Hollister, C. D., Needham, D. H. and Heezen, B. C. (1967). *Earth Planet. Sci. Lett.* **2**, 351.

Scholl, D. W. and Marlow, M. S. (1974). *Soc. Econ. Palaeontol. Mineral., Spec. Publ., Tulsa* **19**, 193.

Sclater, J. C., von der Borch, C. C., Veevers, J. J., Hekinian, R., Thompson, R. W., Pimm, A. C., McGowran, B., Gartner, S., Johnson, D. A. (1974). *Initial Rep. Deep Sea Drilling Project* **22**, 815–831.

Sclater, J. C. and Fisher, R. L. (1974). *Bull. Geol. Soc. Amer.* **85**, 683.

Seibold, E., Exon, N., Hartmann, M., Kögler, F. C., Krumm, H., Lutze, G. F., Newton, R. S., and Werner, F. (1971). *In* "Sedimentology of parts of Central Europe", pp. 209–235. Guidebook for the 8th International Sedimentological Congress, Heidelberg. Verlag Waldemar Kramer, Frankfurt/Main.

Shackleton, N. and Opdyke, N. D. (1973). *Quaternary Res.* **3**, 39.

Shepard, F. P. (1932). *Bull: Geol. Soc. Amer.* **43**, 1017.

Shepard, F. P. and Emery, K. O. (1941). *Spec. Pap. Geol. Soc. Amer.* **31**.

Sibuet, J. C., Pautot, G. and Le Pichon, X. (1971). *In* "Histoire Structurale du Golfe de Gascogne" (J. Debyser, X. Le Pichon and L. Montadert, eds.) Vol. VI (16), pp. 1–16. Editions Technip, Paris.

Simpson, E. S. W., Schlich, R. *et al.* (1974). "Initial Reports of the Deep Sea Drilling Project", Vol. 25. U.S. Government Printing Office, Washington, D.C.

Sloss, L. L. (1962). *J. Sediment. Petrology* **32**, 415.

Southam, J. R. (1974). *Abstr. Progm. Geol. Soc. Amer.* **6**, 962.

Southard, J. B. and Cacchione, D. A. (1972). *In* "Shelf Sediment Transport". (D. J. P. Swift, D. P. Duane and O. H. Pilkey, eds.), pp. 83–98. Dowden, Hutchinson and Ross, Stroudsberg, Pennsylvania.

Stanley, D. J., ed. (1969). "The New Concepts of Continental Margin Sedimentation". American Geological Institute, Washington, D.C.

Stanley, D. J., ed. (1975). "Continental Margin Sedimentation". Dowden, Hutchinson and Ross, Stroudsberg, Pennsylvania.

Stride, A. H. (1963). *Quart. J. Geol. Soc. London* **119**, 175.

Sverdrup,·H. U., Johnson, M. W. and Fleming, R. C. (1941). "The Oceans". Prentice-Hall, Englewood Cliffs, New Jersey.
Swift, D. J. P. (1974). In "The Geology of Continental Margins" (C. A. Burk and C. L. Drake, eds.), pp. 117–135. Springer-Verlag, Berlin.
Swift, D. J. P., Duane, D. B. and Pilkey, O. H., eds. (1972). "Shelf Sediment Transport". Dowden, Hutchinton and Ross, Stroudsburg, Pennsylvania.
Swift, D. J. P., Stanley, D. J. and Curray, J. R. (1971). J. Geol. 79, 322.
Tanner, W. F. (1958). J. Sediment. Petrology 28, 372.
Twenhofel, W. H. (1932). "Treatise on Sedimentation". Williams and Wilkins, New York.
van Andel, Tj. H. (1968). Science, N.Y. 160, 1419.
van Andel, Tj. H. and Moore, T. C. (1974). Geology 1, 87.
van Andel, Tj. H., Heath, G. R. and Moore, T. C. (In press). Mem. Geol. Soc. Amer. 143.
van Straaten, L. M. J. (1950). Konl. Ned. Aardrijkskae. Genoot. 67, 94.
Veevers, J. J., Heirtzler, J. R. et al. (1974). "Initial Reports of the Deep Sea Drilling Project," Vol. 22. U.S. Government Printing Office, Washington, D.C.
Vincent, E. (In press). Allen Hancock Fdn. Monogr. Ser.
Vine, F. J. (1966). Science, N.Y. 154, 1405.
Visher, G. S. (1969). J. Sediment. Petrology 39, 1074.
von der Borch, C. C., Sclater, J. et al. (1974). "Initial Reports of the Deep Sea Drilling Project," Vol. 22. U.S. Government Printing Office, Washington, D.C.
von Rad, U. and Rösch, H. (1974). In "Pelagic Sediments on Land and under the Sea" (K. J. Hsü and H. Jenkyns, eds.), Special Publication International Association Sedimentology, Vol. 1, 1, p. 327. Blackwell Scientific Publications, Oxford.
Walker, R. G. (1970). Spec. Pap. Geol. Ass. Can. 7, 219.
Walker, R. G. (1973). In "Evolving Concepts in Sedimentology" (R. N. Ginsburg, ed.), pp. 1–37. John Hopkins, Baltimore.
Wells, J. W. (1957a). Mem. Geol. Soc. Amer. 67, 609.
Wells, J. W. (1957b). Mem. Geol. Soc. Amer. 67, 1084.
Weser, O. E. (1970). Initial Rep. Deep Sea Drilling Project, 5, 569.
Winterer, E. L. (1973). Bull. Amer. Ass. Petrol. Geol. 57, 265.
Winterer, E. L., Ewing, J. I. et al. (1973). "Initial Reports of the Deep Sea Drilling Project", Vol. 17. U.S. Government Printing Office, Washington, D.C.
Whitaker, J. H. (1974). Soc. Econ. Palaeontol. Mineral., Spec. Publ., Tulsa 19, 106.
Whitmarsh, R. B., Weser, O. E., Ross, D. A. et al. (1974). "Initial Reports of the Deep Sea Drilling Project," Vol. 23. U.S. Government Printing Office, Washington, D.C.
Worthington, L. V. (1970). Deep-Sea Res. 17, 77.
Worthington, L. V. and Volkmann, G. H. (1965). Deep-Sea Res. 12, 667.
Wright, L. D. and Coleman, J. M. (1973). Bull. Amer. Ass. Petrol. Geol. 57, 320.

Chapter 25

# Weathering of the Earth's Crust

## G. D. NICHOLLS

*Department of Geology, University of Manchester, Manchester, England*

## 25.1. INTRODUCTION

Most of the solid detritus carried into the ocean basins from adjacent land masses has passed through that part of the geological cycle known as the weathering stage. Weathering, according to Holmes (1965, p. 37) is "the total effect of all the various sub-aerial processes that co-operate in bringing about the decay and disintegration of rocks, provided that no large-scale transport of the loosened products is involved". Loughnan (1969, p. 1) gives a similar interpretation of the term, observing that rocks at, or near, the earth's surface may be transformed by various physical and chemical processes until the resultant products bear little resemblance to the parent rocks. He explained the term weathering as follows: "Since the breakdown of the rocks results from their direct contact with the prevailing atmospheric conditions or weather the term *weathering* is used to cover these processes". In both of these interpretations weathering is associated with the interface between the earth's surface and the atmosphere, and it is to be noted that the term cannot properly be applied to the interaction between ocean water and solid rocks outcropping on the ocean floor. The processes and reaction products of the alteration of deep-sea basalts in that environment differ from those of sub-aerial weathering and will not be discussed in this chapter.

Weathering processes may be broadly subdivided into two main groups, though neither group operates to the complete exclusion of the other. These are: (1) physical (or mechanical) weathering, which is essentially a physical disaggregation of the pre-existing rock, and (2) chemical weathering, which involves some chemical change in the source rock material. Some writers (e.g. Garrels and Mackenzie, 1971) have mentioned a third group, called biological weathering, though the activities of organisms in breaking down parent rock can usually be referred either to a physical disaggregation of rock material or to the chemical effects of secreted organic compounds. Although many recent writers have emphasized the predominant effect of chemical weathering (Garrels and Mackenzie, 1971, p. 136; Stumm and Morgan, 1970, p. 395), the importance of physical mechanisms in increasing the reactive surface area of the parent material should not be overlooked. Conversely, processes of physical weathering are facilitated by weakening of the parent rocks during the early stages of chemical weathering.

The relative importances of the two groups of weathering processes are, in part, dependent upon the local environment. Quite apart from major climatic controls, which are discussed later in this chapter, considerable variations can be introduced through the interplay of weathering and transportation processes. Theoretically, the ultimate end of all chemical weathering reactions between pre-existing rocks and percolation waters should be complete dissolution of the rocks. However, long before this is attained the solid weathering products are removed by erosion and transported to other depositional environments. Erosional removal of material from the weathering site may occur before there has been any significant chemical weathering, or may be delayed until only the most persistent products of weathering remain, or may transport material already partly altered but still capable of further alteration. The possibility of erosional interruption at any stage of the weathering process influences the relative importance of physical and chemical weathering at any particular locality as much as rock type or even major climatic conditions.

## 25.2. PHYSICAL WEATHERING

Most processes of physical weathering require the existence of some fissures or fractures within the rock along which water can penetrate, and almost all rocks display these on a micro- if not macroscopic scale. Simple unloading of rocks, consequent upon removal of overburden through erosional processes, is a factor in opening up joints and partings originally established either during the primary cooling of igneous rocks or during compaction and/or recrystallization of sedimentary and metamorphic rocks. Tectonic activity may also result in the fracturing of rocks, although the effects which it produces are normally much more localized than those caused by unloading. Whatever initiates weathering some avenues of entry of water into the rock are necessary for physical weathering to proceed on any significant scale.

An important process included under the heading of physical weathering is frost shattering of rocks. When water fills cracks and crevices in rocks and freezes it can exert a pressure of up to 150 kg cm$^{-2}$. Since the tensile strengths of most rocks are below 120 kg cm$^{-2}$, stresses generated by the water freezing can exceed the tensile strength of the host rock and this will cause the extension of cracks and fissures deeper into the rock surface. Repeated cycles of freezing and thawing will lead eventually to rupturing of the rocks. At the low ambient atmospheric temperatures of an environment subject to cyclical freeze-thaw conditions, most chemical reactions will proceed very slowly and the liberated fragments of parent rocks are therefore likely to be little affected by chemical weathering. The effectiveness of freeze-thaw disruption of rocks is well shown in temperate mountainous areas where extensive screes represent the accumulations of rock fragments liberated in this way. The process is no less effective in regions having a more subdued topography; however, its effects tend to be masked where the parent rock is covered by a blanket of weathered rock fragments.

Somewhat similar in effect to disruption by freeze-thaw alternation is the widening and extension of cracks in rocks by the rootlets of plants growing within them. Although in individual cases this is an extremely slow process the overall effect of rock disruption by root growth is far from negligible, particularly since the overall length of the root system of individual plants can be measured in kilometres. The area of the earth's surface naturally covered by vegetation greatly exceeds that where repeated freeze-thaw cycles are sufficiently frequent for frost shattering to be effective.

Laboratory experiments have demonstrated that the stresses set up by cyclical heating and cooling of the surfaces of rocks, to an extent similar to diurnal temperature changes, are normally too small to cause fracture of fresh massive rocks. However, in the presence of water, which leads to the slow hydrolysis of the rock material, alternations of heating and cooling can eventually lead to rupture. In nature, "fatigue" developed after many millions of repeated heating-cooling cycles, may be a contributory factor in failure of the rock before chemical alteration is clearly perceptible. Within arid regions it would seem that the diurnal heating-cooling cycle is almost the only mechanism by which rock disaggregation might occur, although even here the combined cooling and hydrolysis caused by infrequent thunder showers may have an effect out of all proportion to its duration.

D

To some extent certain erosional processes may be regarded as contributing to physical weathering. The abrasion of rock surfaces by the passage over them of rock-loaded moving ice results in considerable physical disaggregation of rocks, the products of which are eventually dumped at the melting ice-front. These products are largely unaffected by chemical weathering when first dumped, but many of them are in such a finely divided condition that they are particularly susceptible to subsequent attack. However, continental and valley glaciers at present cover less than 12% of the land area, and the amount of glacially-derived material delivered to the oceans is less than one-tenth of the total amount of particulate matter introduced from the continental areas.

An analogous biological mechanism is afforded by the activities of earth-worms and other burrowing organisms. Though principally acting upon material already broken down to a substantial degree by other processes, burrowing organisms further comminute inorganic material, thus increasing the specific surface area susceptible to weathering. Organic material excreted by the organisms may have a concurrent chemical effect and here, as through-out the whole of the weathering stage of the geological cycle, it is difficult to disentangle the physical and chemical aspects.


## 25.3. CHEMICAL WEATHERING

### 25.3.1. GENERAL

Chemical weathering describes a series of changes in the chemistry of rock material at, or near, the exposed surface of the earth resulting from the reaction between the lithosphere on the one hand and the hydrosphere and atmosphere on the other. Mineral components of rocks that crystallized under magmatic, metamorphic or sedimentary diagenetic conditions are likely to be unstable under the physicochemical conditions currently prevailing at or near the lithosphere-atmosphere interface. They are thus vulnerable to various exothermic reactions. The common rock forming minerals are the solid reactants in these reactions and, fortunately, they are relatively few in number and type despite the enormous variety of recognized rock types. The surface area they present to the aqueous phase and the rate at which they react are important factors in determining their relative vulnerability to chemical weathering. The significant features of the aqueous phase are its composition and the length of time it is in contact with the mineral species (i.e. the rate of percolation through the weathering profile). The aqueous reactants in the reactions are not pure water. Rain falling on the surface of the earth has already acquired dissolved constituents during its passage through the atmosphere. Dissolution of atmospheric gases, e.g. carbon dioxide and oxygen, endows rain water with slightly acidic and oxidizing properties and solution of atmospheric particulate solids introduces detectable amounts of elements such as Na, K, Mg, Ca etc. Garrels and Mackenzie (1971, p. 107) give the following "best" estimate of the average composition of rain water assuming a pH of 5·7 and equilibrium with the atmosphere: $Na^+$ 1·98 ppm; $K^+$ 0·30 ppm; $Mg^{2+}$ 0·27 ppm; $Ca^{2+}$ 0·09 ppm; $Cl^-$ 3·79 ppm; $SO_4^{2-}$ 0·58 ppm; $HCO_3^-$ 0·12 ppm. Where rain falls directly upon a rock outcrop the aqueous phase initially involved in chemical weathering reactions will have this composition (or one very close to it).

The dissolved carbon dioxide content of rain water is particularly important in weathering. Rain water in equilibrium with the present day atmosphere would have a pH of 5·7 and should contain about 1 ppm of total dissolved carbon dioxide. Recorded pH values for rainfall are frequently below 5·7 and this is generally attributed to the solution of sulphurous gases released from the combustion of fossil fuels into the atmosphere. Frequently the pH of rainwater rises during a shower from abnormally low values at the start to more normal values as the atmosphere is cleansed of its "contaminating" gases and particles. Conversely, land-derived fine-grained calcium carbonate dust carried into the atmosphere by winds may be dissolved in falling rain water with the result that the pH increases.

In the majority of weathering profiles rain water passes through soil before taking part in the weathering reactions and is often changed markedly in composition in doing so. Soil water compositions change during percolation of water down through the various soil horizons and it would be meaningless to attempt to present an average soil water composition analogous to that derived for rain water. In the upper parts of the soil, where a major chemical reaction is the oxidation of humus and other organic matter releasing carbon dioxide, the pH of the soil water can be as low as 4·0 to 4·5, and the concentration of dissolved $CO_2$ is often substantially higher than that in rain water. An even more acid micro-environment is developed immediately adjacent to the roots of living plants, and pH values as low as 2·0 have been measured in their proximity. In contrast, in parts of the humid tropics the oxidation of decaying organic matter takes place before burial and at such a fast rate that the carbon dioxide produced is lost to the atmosphere. In soil waters from such areas the pH values and concentrations of dissolved carbon dioxide approach those found in rain water. Oxidation of organic matter consumes dissolved oxygen and thus reduces the redox potential of the waters which may be further reduced by bacterial action. Waters of low redox potential can carry more iron in solution than can those of normal oxygen content and for this reason they have correspondingly greater leaching powers for this element.

The nature of the reactions taking place between the aqueous phase and the individual minerals forming rocks depends on the properties of the minerals themselves. Some rock-forming minerals, especially those originally formed by evaporation of solutions (the evaporite minerals) dissolve congruently but most others dissolve incongruently.

### 25.3.2. CONGRUENT SOLUTION DURING WEATHERING

Congruent solution during weathering is well exemplified by halite, NaCl, a compound with nearly ideal ionic bonding. Consideration of the composition of ground waters pumped from important sandstone aquifers indicates that relatively soluble constituents such as halite must be present in very small amounts in many rocks, although in many instances the exact form in which they occur is unknown. Indeed it is not, at present, feasible to exclude the possibility that the trace amount of NaCl in such rocks is present in solution as concentrated brine, forming films on the surfaces of other rock forming minerals. Halite is readily soluble and its solubility at 25°C is about $350 \, \text{g kg}^{-1}$ of water. The calcium sulphate minerals, gypsum and anhydrite, also dissolve congruently although the solubility of gypsum is much less than that of halite ($\sim 2 \, \text{g kg}^{-1}$ of water) and the rate of solution much slower; for anhydrite the rate of solution is even slower. When these minerals dissolve the $Na^+$, $Ca^{2+}$, $Cl^-$ and $SO_4^{2-}$ ions simply pass into solution and further modify the ground water composition.

Theoretically, silica in the form of quartz, opal, chalcedony, tridymite or cristobalite dissolves congruently. In water of pH 7 and at a temperature of 25°C the solubility of quartz is theoretically about $6 \cdot 5 \, \text{mg l}^{-1}$, whereas freshly precipitated silica gel has a solubility of $115 \, \text{mg l}^{-1}$. The non-quartz varieties of silica found in nature have solubilities between these two extremes, and therefore in nature they should slowly pass into solution and be re-precipitated as quartz. At pH values above $9 \cdot 0$ the solubilities of silica minerals are pH dependent and at values $> 9 \cdot 8$ increase sharply with increasing pH. This is so because the dominant dissolved form of silica in water of pH above $9 \cdot 8$ is $H_3SiO_4^-$, whereas below this pH it is $H_4SiO_4$, ($pK_1 = 9 \cdot 8$). However, this enhanced solubility of silica minerals at high pH is of little importance in weathering since the pH of the aqueous phase in weathering profiles rarely approaches $9 \cdot 8$.

In the context of weathering the important feature of the solubility of quartz is the extreme slowness of the reaction between quartz and water. When water percolates through the weathering profile it is usually in contact with quartz for such a short time relative to the reaction time that dissolution of the quartz is quite negligible. Data published by Miller (1961) on the composition of water draining from a quartzite in New Mexico suggest that $< 0.1 \, \text{mg} \, l^{-1}$ of the dissolved silica in this water can be attributed to dissolution of quartz even though the total dissolved silica concentration is well below the equilibrium value for the quartz-water system. Dissolution of silica minerals is almost imperceptible under waterlogged conditions since the volume of the interstitial water is small and the solubility of quartz is low. Not surprisingly, some authors have regarded quartz as completely stable with respect to chemical weathering processes. Quartzites and quartz-cemented quartz sandstones are weathered principally by physical mechanisms.

Precipitation of quartz from silica-bearing aqueous solutions is also a very slow process. Silica liberated into solution by the other weathering reactions discussed later in this chapter can build up to concentrations well in excess of the solubility for quartz; thus, water draining from granite may have concentrations of silica four or five times greater than the theoretical maximum (White et al., 1963; Wolff, 1967), and the silica content of the ground water in felspathic sandstones is usually well in excess of that which would be in equilibrium with quartz. Where other weathering reactions build up the dissolved silica content to such levels in the aqueous solutions, solution of quartz is no longer even theoretically possible and the mineral is then, of course, stable against chemical weathering processes.

Other oxide and hydroxide minerals which would be expected to dissolve congruently have extremely low solubilities and are, in consequence, virtually unaffected by the aqueous phases. For both the aluminium bearing minerals, gibbsite and boehmite, and the titanium oxides (e.g. rutile, anatase) the solubility is strikingly dependent upon pH, the minimum solubilities being within the pH range 5–6. The hydrous iron (III) oxides goethite and lepidocrocite and the iron (III) oxides, haematite and maghemite, are also of low solubility in oxidizing waters, though their solubility increases significantly with decreasing pH (see Figure 25.1). The solubility of these iron compounds is also strikingly increased as the redox potential of the water falls (also shown by Figure 25.1), and reduction of the redox potential of the water may be more significant than reduction of pH in increasing the solubility of iron in natural waters. The mobility of iron in waters of low oxidation potential is of particular importance when chemical weathering takes place beneath a cover of decaying vegetation, e.g. peat.

The carbonate minerals, the principal components of limestones, also dissolve congruently, although with these minerals the situation is complicated by the inter-relationship between the dissolved carbon dioxide concentration of the aqueous phase (which may vary as described earlier) and the amount of carbonate mineral that can be dissolved by this phase. Dissolved carbon dioxide reacts with water molecules to form $HCO_3^-$ ions and $H^+$ ions. The $H^+$ ions react with $CO_3^{2-}$ ions released by the solution of

G. D. NICHOLLS

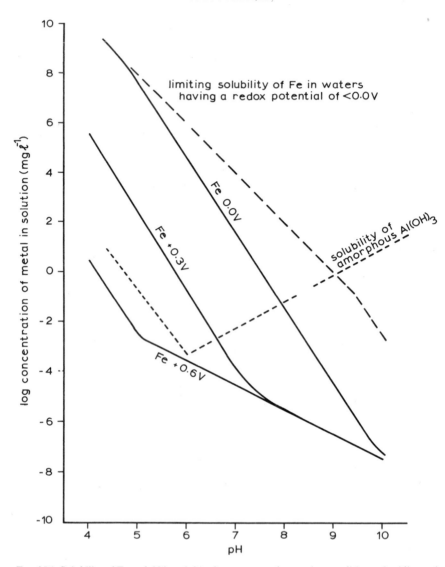

Fig. 25.1. Solubility of Fe and Al in sulphur-free waters under varying conditions of acidity and redox potential.

solid carbonate to yield $HCO_3^-$. This promotes the release of further $CO_3^{2-}$ ions by carbonate solution to replace those converted to $HCO_3^-$ ions. Taking calcite ($CaCO_3$) as a representative carbonate mineral a general reaction for the solution of carbonates can be written in this form.

$$CaCO_3 + CO_2 + H_2O \rightleftharpoons Ca^{2+} + 2HCO_3^-$$

solid     gas     liquid     in solution     in solution

For this reaction the thermodynamic equilibrium constant, ($K$), is given by the expression

$$K = \frac{a_{Ca^{2+}} \cdot a^2_{HCO_3^-}}{a_{CO_2(g)}}$$

where $a$ is the activity of the appropriate ion in the aqueous solution and $a_{CO_2(g)}$ is the activity of the $CO_2$ in the gas phase in the weathering profile. Garrels and Mackenzie (1971) cite a value of $10^{-5.8}$ for $K$ at room temperature (25°C?). Since each ion of $Ca^{2+}$ entering the solution results in the conversion of one molecule of $CO_2$ to an ion of $HCO_3^-$ and the formation of one ion of $HCO_3^-$ from the dissolved carbonate this expression can be rewritten (assuming that in very dilute solutions calcium activity can be replaced by its molar concentration ($x$)).

$$\frac{(x)(2x)^2}{P_{CO_2}} = 10^{-5.8}$$

or

$$x^3 = 10^{-6.4} \cdot P_{CO_2}$$

where $P_{CO_2}$ is the partial pressure of $CO_2$ in the solution. For rain water passing through the present-day atmosphere in which $P_{CO_2} = 10^{-3.5}$ atm $x = 10^{-3.3}$ and the concentration of $Ca^{2+}$ in such water saturated with $CaCO_3$ would be 20 ppm. If, however, the partial pressure of $CO_2$ in the aqueous phase in the weathering profile had been raised, as described earlier by oxidation of organic matter, to $10^{-2.5}$ atm (a 10-fold increase) the equilibrium concentration of $Ca^{2+}$ would be raised to 44 ppm and in an atmosphere of nearly pure $CO_2$ it would be about 300 ppm.

The solubilities of calcite, aragonite, dolomite and magnesite are nearly the same under weathering conditions, but dolomite and magnesite dissolve much more slowly than do the calcium carbonates. The relative resistance of dolomite rich rocks to the weathering processes is the result of the extremely slow rate of solution of the mineral dolomite rather than a consequence of lack of solubility, and in this case, as with quartzite, the rock stability is reaction-rate dependent. Siderite and rhodochrosite, two other rock forming carbonates, are much less soluble than are the carbonates of calcium and magnesium. Since the cations liberated from the former carbonates by weathering (iron and manganese) are, under appropriate conditions, subject to oxidation and precipitation, the solution of siderite and rhodochrosite should, strictly, be regarded as incongruent.

25.3.3. INCONGRUENT SOLUTION DURING WEATHERING

Incongruent solution is exemplified by minerals containing oxidizable cations of which various iron-bearing minerals may be taken as examples. e.g. magnetite ($Fe_3O_4$), siderite ($FeCO_3$), pyrite ($FeS_2$), in which some or all of the iron present is in the ferrous condition. During the weathering process the iron (II) liberated from the parent mineral lattices is oxidized to the 3 + oxidation state by the dissolved oxygen of the aqueous phase in the weathering profile. In this form it is highly insoluble and precipitates out as one of the hydrous iron (III) oxide species, goethite ($\alpha FeO \cdot OH$) and lepidocrocite ($\gamma FeO \cdot OH$) or as haematite ($\alpha Fe_2O_3$) or maghemite ($\gamma Fe_2O_3$). There is still much uncertainty about the factors which determine the exact nature of the iron (III) precipitates. All have extremely low solubilities and the differences between them are so small that there should be little tendency for one to be converted to another within the time span of most weathering cycles. There is perhaps some tendency for haematite to form under warm, damp conditions and for goethite to form elsewhere. However, Schellman (1964) has described the co-existence of goethite and maghemite in a weathering profile developed on serpentinite at Kalimantan in Borneo and the co-existence of haematite and goethite has been reported by Loughnan and Bayliss (1961), Bayliss and Loughnan (1963), and Craig and Loughnan (1964). Calculations indicate that these iron (III) weathering products should be thermodynamically stable relative to magnetite and siderite under any conditions in which the partial pressure of oxygen in the associated gas phase exceeds $10^{-60}$ atm and that of carbon dioxide is less than 1 atm. The weathering of iron sulphide involves the oxidation of sulphur as well as iron according to the equation

$$4FeS_2 + 15O_2 + 8H_2O \rightleftharpoons 2Fe_2O_3 + 8SO_4^{2-} + 16H^+$$

this leads to the generation of very acidic conditions in the aqueous phase. As shown earlier the iron (III) hydroxides are much more soluble under conditions of low pH and the iron (III) sulphate mineral, jarosite, may be precipitated either with, or instead of, the more usual oxides and hydroxides. However the jarosite is, in turn, slowly dissolved by the percolating waters and is therefore only an interim product of the weathering of the sulphide minerals.

Almost all silicate minerals undergo incongruent dissolution, the principal exception being quartz if this is to be regarded as a silicate mineral. If quartz is included, the silicate minerals make up more than three-quarters of the material suffering weathering, and even if it is excluded, they are the major reactants in the weathering processes. Garrels and Mackenzie (1971, p. 154) have claimed that the concentrations of many components of natural waters are controlled by equilibria involving one or more silicate minerals.

Loughnan (1969) has recognized that three simultaneous processes are involved in the weathering of silicate minerals, namely (1) breakdown of the parent mineral structure leading to the release of cations and silica, (2) removal of some of the released constituents in solution, and (3) reconstitution of the residue with components from the atmosphere to form new minerals which are stable or metastable in the environment. Jenny (1950) has pointed out that unsatisfied charges must exist at the surface of any crystal and that when these are in contact with water hydration occurs through attraction of the water dipoles to the surface. Unsatisfied negative charges on the crystal surface attract the positive poles of the water dipoles and unsatisfied positive charges attract the negative pole. The attractive forces at the crystal surface may be strong enough to polarize the water dipoles to such an extent that dissociation occurs, leaving an $OH^-$ ion bonded to a surface cation of a crystal surface having unsatisfied positive charges, or a $H^+$ ion bonded to a surface oxygen when the unsatisfied charge is negative.

A further stage involves the breaking of structural bonds in the pre-existing mineral. In its simplest form this reaction may be expressed diagrammatically thus:

breaks
↓
—cation—O—

an hydroxyl ion attaches to the cation and a hydrogen ion to the anion, to give

– cation – OH        H – O –

and a water molecule in the aqueous phase dissociates to replace the hydroxyl and hydrogen ions used in this reaction. The bonds most readily broken are those of relatively low strength. Rock-forming silicate minerals possess element-oxygen structural bonds and Nicholls (1963) has presented a table showing the approximate relative strengths of various element – oxygen bonds (see Table 25.1). From this table it is apparent that the Ca – O bond is broken more readily in most minerals than is the Fe – O bond and so on to the Si – O bonds which are the most stable and least likely to be broken. This table really applies only to elements in 12-fold or less co-ordination and not to elements in 14-fold co-ordinated positions such as those which occur between the layers of various sheet silicates (the interlayer positions), Wear and White (1951) have examined the relative stability of various ions in 14-fold co-ordinated positions by calculating the maximum value for the oxygen radius necessary to produce a stable configuration of 14 oxygens around the ion under consideration and comparing it with the actual radius of the oxygen ion in the lattice (1·40 Å). For potassium the calculated value was almost identical with the actual one, indicating that potassium would be stable in such a co-ordinated position. Other elements

TABLE 25.1. Relative strengths of various element-oxygen bonds for positions of 12-fold co-ordination or less in rock-forming minerals.*

| Bond | Approximate relative strength |
|------|-------------------------------|
| Si–O | 2·4 |
| Ti–O | 1·8 |
| Al–O | 1·65 |
| $Fe^{3+}$–O | 1·4 |
| Mg–O | 0·9 |
| $Fe^{2+}$–O | 0·85 |
| Mn–O | 0·8 |
| Ca–O | 0·7 |
| Na–O | 0·35 |
| K–O | 0·25 |

* From Nicholls (1963).

in order of decreasing stability in this position are Rb, Na, Cs and Li. When sufficient bonds are broken the cation is liberated from the mineral lattice into solution, where it may possibly take part in further chemical reactions, although it is more likely to be transported from the weathering profile in the percolating waters.

With orthosilicates, such as olivine, release of the cations ($Mg^{2+}$, $Fe^{2+}$ etc.) by rupture of the relatively weak $Mg^{2+}$ – O or $Fe^{2+}$ – O bonds releases silica tetrahedra, which can be taken into solution as $H_4SiO_4$ molecules. Nevertheless, as Garrels and Mackenzie (1971, p. 154) have pointed out, the cations and tetrahedra will go into solution at different rates and the surface of each silicate grain in contact with the reacting aqueous phase will become armoured by a coating having a composition slightly different from the original one. Establishment of this coating hinders further release of ions from the unaltered interior of the mineral, and these must thereafter escape by diffusion through the disordered coating. Polymerization of the silica tetrahedra in the coating to form sheets may initiate the formation of new or secondary mineral lattices, e.g. serpentine $Mg_3Si_2O_5(OH)_4$, the formation of which involves the retention of some of the liberated Mg in the coating. With further alteration the serpentine may be replaced by saponite (Loughnan, 1969, p. 77), which can be regarded as a sheet silicate of the general formula $Mg_6Si_8O_{20}(OH)_4 \cdot nH_2O$ in which some of the Si has been replaced by Al, charges being compensated by introduction of $Na^+$ and $Ca^{2+}$ into the interlayer positions. The various chemical reactions may be written:

$$2Mg_2SiO_4 + H_2O + 2H^+ \rightleftharpoons Mg_3Si_2O_5(OH)_4 + \quad Mg^{2+}$$
$$\text{Serpentine} \qquad\qquad \text{expelled into solution}$$

$$4Mg_3Si_2O_5(OH)_4 + 12H^+ \rightleftharpoons Mg_6Si_8O_{20}(OH)_4 + 6Mg^{2+} + 12H_2O$$
$$\text{Saponite}$$

The utilization of $H^+$ ions in these reactions results in an increase in the pH of the aqueous phase. Most of the iron expelled from the original mineral lattices of natural iron-bearing olivines is oxidized to the $3+$ oxidation state under most conditions of weathering and is subsequently precipitated as haematite or maghemite.

With the chain and band silicates (pyroxenes and amphiboles) the cations linking the chains or bands (e.g. Ca, Mg, Fe(II) etc.) are the first to be liberated by hydrolysis and the released chains polymerize into sheets that eventually form smectite and/or chlorite clay minerals. Iron (II), under most conditions of weathering, is oxidized to the $3+$ oxidation state in which form it is precipitated; however, if the conditions in the weathering profile are reducing its behaviour may resemble that of magnesium. Titanium liberated from the lattice forms the oxide mineral anatase. Some $Ca^{2+}$ and $Mg^{2+}$ ions are incorporated into the newly forming smectite and/or clay minerals, though some may be carried away in solution. When the participating aqueous phase is rich in dissolved carbon dioxide, some of the expelled $Ca^{2+}$ ions may be precipitated as calcite, $CaCO_3$ but this is only a transitory phase and is eventually redissolved as weathering proceeds. If the smectite, and/or chlorite, clay minerals are not removed from the weathering site, but are subjected to further attack by the percolating waters, destruction of the lattices brought about by further leaching of $Ca^{2+}$ and $Mg^{2+}$ is accompanied by break-up of the sheets. The freed silica tetrahedra and temporarily liberated aluminium recombine to form kaolinite. Consequently, strong weathering of the chain and band silicates eventually produces kaolinite, iron (III) oxides and hydrous oxides and anatase, with calcium, magnesium and some silica being removed in solution.

The sheet silicates are relatively stable lattices with respect to weathering hydrolysis. Interlayer ions or molecules are removed first by leaching, and only under conditions of extreme weathering are the layers in the structures disrupted. Muscovite (white mica) is almost as stable against chemical weathering as is quartz. Observational data leave no doubt that the tri-octahedral micas (biotites) are more susceptible to chemical alteration than are the di-octahedral ones (muscovites), and that they weather to the tri-octahedral smectite-like mineral, vermiculite. Walker (1949) has suggested that, concomitantly with the release of potassium from between the double "mica-sandwich" sheets (layers) under hydrolysis, iron (II) in octahedral co-ordination within the sheets is oxidized to the 3 + oxidation state without being removed from the lattice, thus reducing the net negative charge on the layers and preventing any re-entry of potassium into the inter-layer positions. Other explanations of the marked difference in behaviour of the two types of mica include differences in the orientation of the O – H bond of the hydroxyl groups in the lattices (Bassett, 1960) and preferential dissolution of octahedrally co-ordinated $Fe^{2+}$ and $Mg^{2+}$ from the tri-octahedral micas relative to the octahedrally co-ordinated $Al^{3+}$ from the di-octahedral micas (Loughnan, 1969).

Leaching of the interlayer ions (or molecules) may be restricted to certain levels in the lattices; this results in the production of a single grain which possesses the characters of a smectite in one part and of a mica in another. Such grains are referred to as mixed-layered minerals or degraded micas. They may be removed from a weathering site while still in this "interim" condition.

Eventually, as in the sheet silicates produced as intermediate alteration products of ortho, chain, band and framework silicates, the sheets themselves may be disrupted by weathering hydrolysis to liberate some silica into solution with combination of the remainder with any available aluminium to yield kaolinite. Bauxite formation clearly demonstrates that even the kaolinite lattice may eventually become unstable, release silica and be transformed to gibbsite.

In three-dimensional framework silicate lattices, e.g. the felspars, there are limits to the number of a particular type of bond that can be broken without rupture of others, and account must be taken of all the bonds in assessing the overall stability of a mineral lattice. The calcium-rich plagioclases are less stable to weathering than are the sodium-rich plagioclases despite the fact that the Ca – O bond is stronger than the Na – O bond. This is because in the calcium plagioclases there are more Al—O bonds, some of which occupy positions analogous to those tenanted by the much stronger Si – O bonds in sodium-rich plagioclases.

Eventually, the surface layers of the minerals are rendered unstable by progressive bond rupture and fragments of the – Si – O – framework, with or without Al replacing some of the Si, are liberated into the surrounding water. De Vore (1957) has argued that these fragments may be quite substantial. He has pointed out that if decomposition of the felspar structure releases chains from the (100) or (010) surfaces of the mineral, and if these chains retain the original Si – Al ordering of the tetrahedra, then they could polymerize directly into sheets having the composition $(AlSi_3O_{10})$ identical with those found in the layered silicate products of felspar weathering. If, however, breakdown of the felspar framework into individual tetrahedra occurs, aluminium originally in tetrahedral co-ordination would be expected to assume its more stable octahedral co-ordination in the aqueous environment. In this case it would be difficult to understand the formation, as weathering products of felspars, of sheet silicates with at least part of their aluminium in tetrahedral co-ordination, e.g. illite. Instead, kaolinite might be expected as the weathering product. Clearly the two products, illite and kaolinite, could be sequential with the kaolinite produced under more extreme chemical weathering conditions.

In recent years as more thermodynamic data about rock-forming minerals have become available there has been an increasing tendency to discuss weathering reactions in terms of thermodynamic equilibrium relationships. Although this approach possess a superficially attractive veneer of exactitude for a subject which was previously essentially qualitative in nature, it has severe limitations when applied to weathering reactions. Until recently basic thermodynamic data were only available for pure "end-member" compounds, whereas the majority of natural rock-forming minerals are intermediates of the solid solution series. For layer silicates (sheet silicates) the situation has been substantially improved by the publication of a method of estimating the Gibbs energies of formation of layer silicates (Tardy and Garrels, 1974). This now permits estimates to be made for layer silicates of any intermediate composition. However, uncertainty about critical thermodynamic values still exists for the breakdown of the parent rock-forming silicates. A second limitation lies in the application of an approach designed to consider whole systems to processes operating on a micro-scale in micro-environments. There is no place for mixed layer silicates in any thermodynamic treatment of reactions of the type occurring in weathering. Equilibrium as understood in thermodynamics is virtually unattainable in weathering processes as erosion is likely to intervene and disrupt the whole weathering regime long before thermodynamic equilibrium is attained. Most of the products of the weathering process at the earth's surface at the present time are thermodynamically metastable intermediates which owe their transitory existence to their extremely low reaction rates.

Despite these limitations the thermodynamic approach is valuable in showing which types of minerals would be unstable (or stable) in association with an aqueous phase of a specified composition or with the ground water existing at a particular locality, and thus indicating the sorts of chemical changes that might be expected to affect a specific parent mineral. Consideration of the system $K_2O - Al_2O_3 - SiO_2 - H_2O$ will illustrate how this approach can be used. For the reaction:

$$2 \cdot 9 KAlSi_3O_8 + 11 \cdot 2H_2O + 2H^+ \rightleftharpoons K_{0 \cdot 9}Al_{2 \cdot 9}Si_{3 \cdot 1}O_{10}(OH)_2$$

crystalline potassium felspar        illite

$$+ 5 \cdot 6H_4SiO_4 + 2K^+$$

the Gibbs free energy of the reaction ($\Delta G$) obtained from published data on free energies of formation is 68·0 kJ at 25°C. Then from log $k = -\Delta G(2\cdot303\ RT$, the log of the equilibrium constant ($k$) is 68·0/$-5\cdot70 = -11\cdot9$. Inserting this value into the equation for the equilibrium constant, and assuming that the solution is sufficiently dilute for activity to be equated with concentration and that the activities of the solids and water are unity, we can write for equilibrium at 25°C:

$$\log k = -11\cdot9 = \log\left[\frac{[K^+]^2[H_4SiO_4]^{5\cdot6}}{[H^+]^2}\right]$$

rearranging

$$-11\cdot9 = 2\log\left[\frac{[K^+]}{[H^+]}\right] + 5\cdot6\log[H_4SiO_4]$$

This is plotted as boundary 1 in Fig. 25.2.

For the equilibrium between illite and kaolinite a similar equation can be derived:

$$8\cdot09 = 1\cdot8\log\frac{[K^+]}{[H^+]} + 0\cdot4\log[H_4SiO_4],$$

This is plotted as boundary 2 in Fig. 25.2.

Other boundaries, obtained similarly, can be plotted to produce an activity-activity diagram showing the stability fields of various minerals in this system at 25°C and a pressure of 1 atmosphere. Other activity-activity diagrams relevant to weathering have been published by Stumm and Morgan (1970) and Garrels and Mackenzie (1971).

Activity-activity diagrams, such as that shown in Fig. 25.2, can be used to predict the phase which is thermodynamically stable with an aqueous solution of stated composition. If the ground water in the weathering profile of a felspathic sandstone contains 6 mg l$^{-1}$ of dissolved $SiO_2$ and 1·23 mg l$^{-1}$ of dissolved potassium and has a pH of 7·5 (i.e. log $[K^+]/[H^+] = 3\cdot0$; log $[H_4SiO_4] = -4\cdot0$) the stable alteration product of the felspar should be kaolinite and chemical changes upon weathering should be directed towards the production of this mineral. The presence of illite or gibbsite would indicate that equilibrium had not been attained.

Such diagrams are also of value in indicating the changes which are likely to occur when material is transported by water away from the weathering site. However, discussion of such applications is beyond the scope of the present chapter.

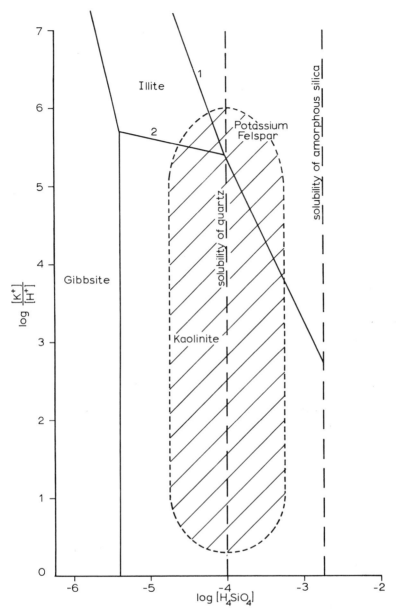

FIG. 25.2. Activity-activity diagram for a part of the system $K_2O-Al_2O_3-SiO_2-H_2O$ at 25°C and 1 atm. The stability fields of kaolinite and illite, two possible products of felspar weathering, are shown as functions of the activities of $K^+$, $H^+$ and dissolved silica. The diagonally shaded area encloses a compositional field for ground waters in felspathic sandstones. (After Garrels and Mackenzie, 1971; it is assumed that activities can be equated with concentrations, see p. 96.)

The broad overall picture presented by chemical weathering is one of progressive leaching of the parent material and of concentration of aluminium and, to a lesser degree, iron and titanium in the solid residues. Most of the many and varied solid products of weathering encountered in weathering profiles, (which tend to reflect the nature of the underlying parent rock) are metastable intermediates in the sequence of rock alteration. As the degree of chemical weathering increases kaolinite and iron(III) and titanium oxides and hydroxides tend to develop from a wide range of original parent rocks, and bauxitic and lateritic residues are the eventual products if weathering is continued to a very advanced stage.

## 25.4. ENVIRONMENTAL CONTROLS ON WEATHERING

In the light of the preceding discussion it is apparent that climate will have a significant effect upon weathering. At high latitudes, the low temperature and the existence of water in the form of ice for much of the year causes physical weathering, especially ice abrasion, to predominate over chemical weathering. The fine-grained products of degradation of the land surface produced under these conditions consist largely of finely ground minerals of all the varieties found in the rocks of the area (often called "glacial rock flour").

In temperate regions, physical and chemical weathering proceed simultaneously, although erosion and transportation frequently intervene before the process of chemical decay is far advanced. In such regions the effect of topography is marked. In areas of high relief and steep slopes much of the rainfall runs off at the surface and relatively small amounts percolate into the rock to take part in chemical weathering. High rates of run-off promote erosional attack upon the rocks, and thus the same factor hinders chemical weathering and leads to early removal of physically degraded material down the transport paths to the sediment depositional areas. In very flat areas the rate of run-off is much reduced; however, the sub-surface percolation tends to be sluggish and the percolation rates are low enough to permit the concentration levels of the leached constituents to build up and this causes a deceleration of the weathering reactions with the passage of time. Swamp conditions in which the water table is at or near the surface may develop in some areas, and the resultant accumulation and decay of vegetation may lead to the establishment of localized reducing conditions. Mobility of iron under such conditions may bring about considerable changes in the compositional trends followed in the weathering cycle. Rolling topography offers the most favourable conditions for substantial chemical weathering of the parent material before the weathering process is terminated by erosion and transportation. In such environments a higher proportion of the rainfall percolates into the surface than in areas of high relief, but escape of ground water through springs in the valleys tends to develop a steady sub-surface drainage system. Movement of the weathered products downhill by soil creep is slow compared with transportation in running water. The products of physical degradation of the parent rocks therefore remain for longer periods in environments which are favourable to chemical weathering. Under these conditions the incidence of rainfall can have an important influence on chemical weathering. Long dry periods following seasonal heavy rain, can result in ground water being drawn upwards through the weathering profile for much of the year to be lost at the surface by evaporation. The consequent concentration of dissolved salts in the ground water may inhibit chemical weathering. In contrast, steady rainfall evenly distributed throughout the year maintains a steady downward percolation of a very dilute aqueous phase and leads to the promotion of chemical weathering reactions, and even to the slow but continued removal of elements which are extremely immobile in the weathering profile. In view of these facts it is not surprising that the suspended solid loads of streams in the temperate regions are so varied in mineralogy and character.

In arid regions, water that does penetrate into the rocks after infrequent thunderstorms tends to be drawn back to the surface by capillary action and in this way high concentration levels of dissolved salts are built up in the ground water by evaporation. This tends further to inhibit chemical weathering in an environment which is already, for other reasons, unfavourable to such reactions, and physical weathering is the more prominent form of rock degradation under these conditions. Any products of chemical weathering found in such areas tend to be of the metastable intermediate type, e.g. illite, chlorite, smectites. Kaolinite is not common unless it is a component of the parent rock.

In tropical regions, the relatively high temperatures and rainfall, which is often evenly distributed through much of the year, promote strong chemical weathering. Physical disintegration is more frequently due to biological agencies (roots etc.) than to any other cause, and profuse vegetation may lead to extensive accumulations of decaying organic matter at the surface. Percolating water may develop low values of pH and redox potential and this increases the probability of leaching, especially of iron. Protection of the weathering profile against erosion is afforded by the vegetative cover, and, under favourable conditions, chemical weathering can continue for long periods. Consequently, it tends to reach advanced stages and the products include kaolinite and the hydroxides and oxides of aluminium, iron and titanium, rather than the metastable intermediates encountered in the temperate regions.

## 25.5. CONCLUSIONS

The nature of both the parent material and the environment greatly influence the nature of the weathering products produced at different places on the earth's surface. However, it is clear that some broad generalizations can be made about the types of weathering products contributed from different parts of the earth's surface to the world ocean. At high latitudes glacial rock flour would be expected to be important, and any chemically altered material delivered to the ocean is likely to be a metastable intermediate product, e.g. chlorite. In those parts of the ocean receiving material transported by rivers from catchments in temperate areas, the delivered detrital material is likely to be of great variety ranging from finely divided parent rock minerals, to kaolinite, and even gibbsite. When the material supplied to the oceans originates largely from arid regions it is likely to show relatively minor chemical alteration, in sharp contrast to that delivered from low latitude tropical areas. Detrital material delivered from the latter areas consists mainly of kaolinite, quartz, iron and titanium oxides and hydroxides and even gibbsite. Some interest attaches to the fact that the concentration ratios of $H^+ : K^+ : H_4SiO_4$ of most ocean waters lie within the field of stability of illite and for this reason, kaolinite (the silicate mineral towards which a whole plexus of chemical weathering reactions tends to lead), should in the sea suffer transformation to illite, a very common sheet silicate of sediments and sedimentary rocks. Kaolinite delivered to the oceans could have been derived as a product of advanced weathering from a very wide range of parent rock-forming aluminosilicate minerals. The abundant illite of the sedimentary group of materials need not have all come from the alteration of potassium bearing aluminosilicates (felspars and micas) and, almost certainly, some of it did not do so.

## REFERENCES

Bassett, W. A. (1960). *Bull. Geol. Soc. Amer.* **71**, 449.
Bayliss, P. and Loughnan, F. C. (1963). *Amer. Mineral.* **48**, 410.
Craig, D. C. and Loughnan, F. C. (1964). *Aust. J. Soil Res.* **2**, 218.
De Vore, G. W. (1967). *Proc. Nat. Conf. Clays and Clay Minerals,* **6**, 26.
Garrels, R. M. and Mackenzie, F. T. (1971). "Evolution of Sedimentary Rocks". W. W. Norton, New York.
Holmes, A. (1965). "Principles of Physical Geology". Nelson, London.
Jenny, H. (1950). *In* "Origin of Soils–Applied Sedimentation" (Trask, P. D. ed.), p. 41. John Wiley & Sons, New York.
Loughnan, F. C. (1969). "Chemical Weathering of the Silicate Minerals". Elsevier, New York, London and Amsterdam.
Loughnan, F. C. and Bayliss, P. (1961). *Amer. Mineral.* **46**, 209.

Miller, J. P. (1961). *U.S. Geol. Surv., Water-Supply Pap.* **1535–F.**
Nicholls, G. D. (1963). *Sci. Progr. (London)* **51,** 12.
Schellman, W. Von (1964). *Geol. Jahrb., Hannover,* **81,** 645.
Stumm, W. and Morgan, J. J. (1970). "Aquatic Chemistry". John Wiley & Sons, New York.
Tardy, Y. and Garrels, R. M. (1974). *Geochim. Cosmochim. Acta,* **38,** 1101.
Walker, G. F. (1949). *Mineral. Mag.* **28,** 693.
Wear, J. I. and White, J. L. (1951). *Soil Sci.* **71,** 1.
White, D. E., Hem, J. D. and Waring, G. A. (1963). *U.S. Geol. Surv. Prof. Pap.* **440–F.**
Wolff, R. G. (1967). *Amer. J. Sci.* **265,** 106.

Chapter 26

# Lithogenous Material in Marine Sediments

## HERBERT L. WINDOM
*Skidaway Institute of Oceanography,*
*P.O. Box 13687, Savannah, Georgia 31406, U.S.A.*

## 26.1. INTRODUCTION

The detrital fraction of sediments comprises materials delivered to the oceans as solid phases which undergo little alteration during their transport and final deposition. According to the classification suggested by Goldberg (1954), the detrital fraction of marine sediments can be divided into the lithogeneous and the cosmogeneous components, the former being quanti-

103

tatively far more important than the latter. In this classification, the litho-geneous components are defined as those arising from land erosion, submarine vulcanism and underwater weathering, when the solid phase undergoes no major chemical change during its residence in sea water. Thus, these compo-nents contain not only minerals transported from the continents, but also those injected directly into the ocean as a result of submarine vulcanism, or indirectly via the atmosphere from terrestial vulcanism.

Ambiguities exist in the classification of the lithogenous component of marine sediments since some materials can have either a detrital or a non-detrital origin. For example, montmorillonite results from chemical weather-ing on the continents and also from submarine weathering of marine volcanic materials (Ross and Hendricks, 1945; Bonatti, 1963). Chlorite, a primary mineral of metamorphic rocks on the continents and also a product of continental weathering, may form under certain circumstances in the marine environment. For example, Swindale and Fan (1967) have suggested that gibbsite may alter to chlorite in the sediments off the coast of Kauai, Hawaii.

Following customary practice, montmorillonite and chlorite will, regardless of their origins, be considered to be lithogeneous minerals. This is the only practical approach since continentally derived montmorillonite and chlorite are analytically indistinguishable from their counterparts which originate by submarine weathering. In addition, no distinction is made between clay minerals having these two origins, except on the basis of their distribution patterns.

Although the relative proportions of most components of marine sediments (e.g. biogenous and hydrogenous varieties) are dependent on the physical, chemical and biological conditions prevailing in the area of the ocean in which they are deposited, special factors influence the proportion of the lithogenous fraction; these include the source of the material and the mode of transport to the site of deposition. For this reason, its distribution will be determined by that of the so-called regions of lithogenesis which are defined on the basis of their characteristic weathering regimes (Strakhov, 1960). These regions are shown in Fig. 26.1, and it will be demonstrated that their distribution together with the prevailing regional transport mechanism controls the distribution of the lithogeneous component in marine sediments.

## 26.2. CLAY MINERALS

The $<2 \, \mu m$ fraction of pelagic clays makes up to 50–70% of the total sedi-ment and may contain examples of all the various components occurring

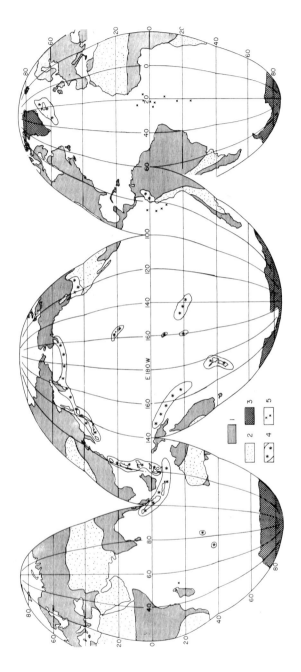

FIG. 26.1. Major regions of lithogenesis according to Strakhov (1960): 1. regions of humid lithogenesis; 2. regions of arid lithogenesis; 3. regions of polar (ice) lithogenesis; 4. volcanoes and volcanic regions; 5. single volcanoes.

in marine sediments. The relative concentration of each component depends on the characteristics of the site of deposition. In all pelagic sediments, however, the dominant materials making up the lithogenous component of the clay-sized fraction are the common clay minerals montmorillonite, illite, kaolinite and chlorite, together with lesser amounts of quartz and feldspars. The relative proportions of these major clay minerals in a sediment will vary according to the prevailing transport processes and the proximity to particular zones of lithogenesis. In addition to these four minerals, their mixed-layer derivatives have been identified in marine sediments, as also have palygorskite and sepiolite, which may be of either authigenic or detrital origin.

Several important studies of the distribution of clay minerals in the world ocean have been made. Biscaye (1965) determined the $<2\,\mu m$ clay mineralogy of $\sim 500$ surface sediment samples from the Atlantic and parts of the Antarctic and Indian Oceans and their adjacent inland seas. Griffin et al. (1968) utilized Biscaye's data together with an additional $\sim 500$ analyses of surface sediment samples, primarily from the Pacific and Indian Oceans, and described the distribution of clay minerals in the World Ocean. A similar survey was subsequently made by Rateev et al. (1969) who used Biscaye's data together with analyses of the clay minerals in the $<1\,\mu m$ sized fraction of 300 to 400 additional samples from the Pacific, Antarctic and Indian Oceans. Since the data of Griffin et al. (1968) were obtained for the $<2\,\mu m$ fraction they are probably more comparable with those of Biscaye (1965) than with those of Rateev et al. (1969). Despite this, the distribution patterns described by the latter workers are quite similar to those found in the earlier study.

The distribution of the clay minerals given in the present chapter is primarily based on data from Griffin et al. (1968) (which includes that of Biscaye (1965)) and has been augmented with recent data such as that of Goldberg and Griffin (1970) and Griffin and Goldberg (1969) which refer to the detailed clay mineralogy of Indian Ocean and Caribbean Sea sediments respectively. The observations of Rateev et al. (1969) were also used to obtain a better understanding of the clay mineral distributions around Antarctica.

The most recent data of Venkatarathnam and Biscaye (1973) for approximately a hundred cores from the eastern Indian Ocean have not been included in the clay mineral distributions shown in Figs. 26.2–26.4, although their results generally agree with those of Goldberg and Griffin (1970). However, the density of sampling used by the latter authors was too low to permit a detailed understanding of many of the sediment features which have been identified by Venkatarathnam and Biscaye (1973) who have delineated particular sediment provinces influenced by various source areas and transport pathways.

The following sections discuss the distribution characteristics of the different clay minerals ocean by ocean.

26.2.1. KAOLINITE

Kaolinite minerals are hydrous aluminium silicates. They are dioctahedral
1:1 clay minerals (i.e. the structure is a 2-layer sheet composed of one tetrahedral and one octahedral layer), the different species being characterized
by the manner of stacking of the 7 Å layer units. Of all the clay minerals
occurring in marine sediments, kaolinite is the one most clearly derived from
the continents where it is formed by the chemical weathering of primary
silicate minerals. Since it forms from solutions containing alkali and alkaline
earth ions only under conditions of relatively low pH, it is unlikely that this
mineral is formed in the oceans (Arrhenius, 1963).

The distribution of kaolinite in marine sediments reflects the intensity of
chemical weathering in the source areas and the prevailing transport pathways. It is therefore produced in the greatest amounts in the areas shown in
Fig. 26.1 as regions of humid lithogenesis. In these regions kaolinite production increases towards the equator. This pattern is reflected in marine sediments, in which kaolinite is enriched in equatorial regions. Because of its
abundance in tropical areas Griffin et al. (1968) describe it as the low latitude
clay mineral.

The average kaolinite concentrations in the sediments of the major ocean
basins are given in Table 26.1, and their distribution is illustrated in Fig. 26.2.

26.2.1.1. Atlantic Ocean

Although the latitudinal zoning of kaolinite in the Atlantic Ocean was
recognized by Yeroshchev-Shak (1961) and Goldberg and Griffin (1964),
it was not until the work of Biscaye (1965) that the strong association of

TABLE 26.1

*Kaolinite concentrations in sediments of the major ocean basins. (Data
taken from Griffin et al., 1968, Goldberg and Griffin, 1970, and Venkatarathnam and Biscaye, 1973)*

| Area | No. of Samples | Average % |
|---|---|---|
| North Atlantic | 193 | 20 |
| Gulf of Mexico | 38 | 12 |
| Caribbean Sea | 54 | 24 |
| South Atlantic | 196 | 17 |
| North Pacific | 170 | 8 |
| South Pacific | 151 | 8 |
| Indian | 245 | 16 |
| Bay of Bengal | 51 | 12 |
| Arabian Sea | 29 | 9 |

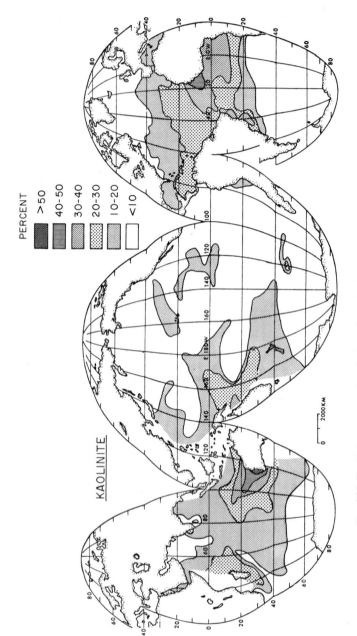

FIG. 26.2. Kaolinite concentrations in the <2 μm size fraction of sediments in the World Ocean. Data from Griffin et al. (1968) and Goldberg and Griffin (1970).

kaolinite (and gibbsite) with tropical land masses was clearly established. In the sediments of the Equatorial Atlantic, kaolinite generally comprises 20 to 30% of the total clay minerals present. In the Eastern Equatorial Atlantic near the African coast the concentration of kaolinite in the clay minerals can be as high as 50%, probably because of the input of sediments from the Niger River which drains the kaolinite-rich soils of Equatorial Africa (Horn et al., 1967; Jungerius and Levelt, 1964) and because of the wind transport of kaolinite-rich desert soils from North Africa. Further, on the basis of the high kaolinite concentrations of dust collected on Barbados, Delaney et al. (1967) have suggested that the relatively well defined decreasing kaolinite concentration gradient away from the African coast could be a result of this wind transport. This is borne out by the results of Chester et al. (1972) who found relatively high concentrations of this mineral in dusts collected over the equatorial Atlantic.

In the Western Equatorial Atlantic, detritus from South American rivers is the principle sediment source. In general, the detritus carried by these rivers has a low kaolinite content relative to that of other minerals such as illite (Gibbs, 1967; Depetris and Griffin, 1968); the São Francisco is an exception because it contributes significant amounts of kaolinite to the sediments near the South American Coast (Horowitz and Makitie, 1963).

The concentrations of kaolinite in Carribean Sea sediments are similar to those of the Central Atlantic Ocean, but are much higher than are those of the adjacent Gulf of Mexico. According to Griffin and Goldberg (1969), kaolinite is more abundant in the eastern part of the Caribbean than in the western areas adjacent to Central America and western Cuba. This could be due to the transport to this area of Amazon river detritus (Jacobs and Ewing, 1969) which is richer in kaolinite than is detritus from North American rivers. Thus, the Gulf of Mexico sediments are apparently dominated by the montmorillonite-rich supply of detritus from the Mississippi (Pinsak and Murray, 1960).

### 26.2.1.2. Pacific Ocean

Pacific Ocean sediments have particularly low kaolinite concentrations. This is especially evident in the Eastern and South Central Pacific where the effect of river discharge, and therefore the supply of continental detritus, is minimal. It is probable that any contribution of wind transported dust from the African deserts would also be small because of the scavenging of air masses by precipitation as they traverse the mountainous regions of Central and South America.

The kaolinite content of the sediments of the Eastern Pacific Ocean is low compared with that of the Western Pacific. The reason for this can be clearly seen from an examination of the distribution of the regions of lithogenesis (Fig. 26.1) around the Pacific Ocean basin. Along the Eastern Pacific coast relatively arid source areas with low run-off supply very little kaolinite to the surrounding sediments. In contrast, regions of intense chemical weathering and high run-off along the Western Pacific coast act as important sources of kaolinite.

The Australian continent appears to be an important source of kaolinite which is transported by prevailing winds to the South Pacific where it is subsequently incorporated in the sediments. In support of this, Griffin et al. (1968) found the western areas of the Australian deserts to be relatively enriched in kaolinite. Further, Windom (1969) showed that dust collected from snow-fields of New Zealand was rich in kaolinite and apparently had been transported from the Australian deserts. Heath (1969) has suggested that kaolinite in North Pacific sediments may also be aeolian because it is associated with wind-transported quartz.

### 26.2.1.3. *Indian Ocean*

Indian Ocean sediments have kaolinite concentrations intermediate between those of the Atlantic and Pacific Oceans. The most detailed study is that by Goldberg and Griffin (1970) whose data, together with those of Griffin et al. (1968), were used to delineate the distribution pattern shown in Fig. 26.2. These authors found that, in comparison with the sediments of the rest of the Indian Ocean, those of the Bay of Bengal and the Arabian Sea were relatively depleted in kaolinite. In the Bay of Bengal this apparently results from the discharge of kaolinite-poor detritus by the Ganges-Brahmaputra river system which drains the montmorillonite-rich basaltic terrain of the Deccan Traps (Roy and Barde, 1962). A similar input of kaolinite-poor detritus in the river run-off is also apparent in the sediments of the eastern Arabian Sea. The sediments in the rest of the Arabian Sea have a clay mineral assemblage dominated by wind-transported kaolinite-poor material from the deserts of Northern India and West Pakistan.

The most striking feature of the kaolinite distribution in the Indian Ocean is the gradient of increasing concentration towards Western Australia. The work of Griffin et al. (1968) has shown that the deserts of Western Australia are very rich in kaolinite and apparently supply a significant amount of material to the south-eastern Indian Ocean sediments by atmospheric transport in the prevailing wind system; this conclusion has subsequently been confirmed by Venkatarathnam and Biscaye (1973).

### 26.2.2. CHLORITE

Chlorite is a common clay-sized mineral consisting of a three layer "sandwich" of two tetrahedral layers and an octahedral layer with a brucite layer between each 3-layer unit, the whole forming a unit 14 Å in thickness. Most chlorites are trioctahedral, but some have both dioctahedral and trioctahedral layers.

The distribution of chlorite in deep-sea sediments is the inverse of that of kaolinite; the mineral is however, more uniformly distributed throughout the World Ocean than is the latter (Table 26.2). It occurs as a primary mineral in both igneous and metamorphic rocks, and is relatively uniformly distributed on the continents. In areas in which chemical weathering is minimal, mechanically

TABLE 26.2

*Chlorite concentrations in sediments of the major ocean basins. (Data taken from Griffin et al., 1968; Goldberg and Griffin, 1970 and Venkata-rathnam and Biscaye, 1973)*

| Area | No. of Samples | Average % |
|------|----------------|-----------|
| North Atlantic | 183 | 10 |
| Gulf of Mexico | 38 | 18 |
| Caribbean Sea | 55 | 11 |
| South Atlantic | 208 | 11 |
| North Pacific | 170 | 18 |
| South Pacific | 140 | 13 |
| Indian | 245 | 10 |
| Bay of Bengal | 51 | 14 |
| Arabian Sea | 29 | 18 |

weathered chlorite will be delivered to the oceans in an unaltered state. At lower latitudes, where chemical weathering predominates, chlorite may be modified or even destroyed during the weathering process. Latitudinal variations in the severity of chlorite weathering have been demonstrated by Griffin *et al.* (1968) who observed that the 001 faces of chlorites from high latitudes give a sharper diffraction peak than do those from lower latitudes. Because chlorite is strongly concentrated in sediments of the Polar regions of the World Ocean it has been termed the *high latitude* clay mineral by Griffin *et al.* (1968).

Chlorite occurs in marine sediments, not only in the $<2\,\mu m$ size fraction, but also in coarser sizes, reflecting its size distribution in the source areas. For example, at high latitudes chlorite is delivered to marine sediments through glacial action with little, or no, alteration of the initial size distribution.

Griffin *et al.* (1968) have suggested that the chlorites in deep-sea sediments are of the magnesium-iron variety in which 2·5–4·5 heavy atoms occupy the six octahedral sites in the brucite layer. This implies an FeO content of up to $\sim 20\%$. Although Copeland *et al.* (1971) have found Mid-Atlantic Ridge sediments containing chlorites having even higher iron concentrations, the high iron content and the lanthanide enrichment of these chlorites indicates that their source was the greenstone of the Mid-Atlantic Ridge. It is likely that local sources of chlorite may also exist in other ocean basins. It has been suggested that, in addition to these sources and the major continental source, authigenic formation of chlorite in marine sediments can occur under restricted conditions (Swindale and Fan, 1967). The distribution of chlorite in the deep-sea sediments of the World Ocean is shown in Fig. 26.3, and the average percentage concentrations in the various ocean basins are given in Table 26.2.

### 26.2.2.1. Atlantic Ocean

In contrast to kaolinite, the chlorite of Atlantic sediments shows a gradient of increasing concentration towards the polar regions. This was first recognized by Biscaye (1965) who also noted that the kaolinite to chlorite ratio decreased by an order of magnitude from equatorial to high latitude sediments. The high concentration of chlorite in the sediments of the polar regions can be related to the probable source areas. For example, Berry and Johns (1966) found that the chlorite in the bottom sediments of the North Atlantic-Arctic Ocean was similar in abundance and crystallinity to that found in the surrounding continental regions. Griffin et al. (1968) suggested that the chlorites of the chlorite-rich zone of sediments which extends out from the St. Lawrence River originated in the soils of the river basin (Forman and Brydon, 1965).

In South Atlantic deep-sea sediments, chlorite increases systematically towards Antarctica, (Rateev et al., 1969), where its source is apparently the Antarctic continent (Goodell et al., 1962) and its most important mode of transport is ice-rafting.

### 26.2.2.2. Pacific Ocean

The distribution of chlorite in deep-sea sediments of the Pacific Ocean is similar to that in the Atlantic Ocean, i.e. there is an increasing concentration towards higher latitudes. In the Pacific, the highest concentrations of chlorite occur in the sediments adjacent to the Alaskan and Canadian Coasts; these chlorites appear to be highly crystalline (Gorbunova, 1963). Griffin et al. (1968) have found that the major rivers draining into this area contain relatively high concentrations of chlorite in their detritus, this suggests that the chlorite originates from Alaska, Western Canada and the northwestern United States. The Asian continent is also important in supplying sediment to the northwestern Pacific. Chlorite supplied from the Asian and North American mainlands is subsequently redistributed by the Oyashio and California currents. The pattern of increasing chlorite concentration in sediments near Antarctica is similar to that found in the Atlantic Ocean and is apparently the result of ice rafting.

Relatively high concentrations of chlorite are found in the sediments of the South Pacific adjacent to New Zealand and Australia, and may be the result

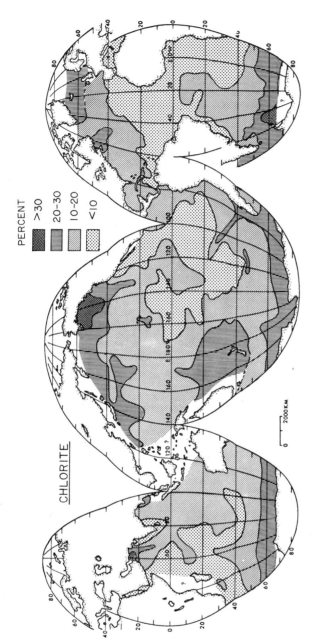

FIG. 26.3. Chlorite concentrations in the <2 μm size fraction of sediments in the World Ocean. Data from Griffin et al. (1968) and Goldberg and Griffin (1970).

of atmospheric transport of dust from the Australian continent; in this context, Conolly (1965) found relatively high chlorite contents in the soils of southeastern Australia, and Windom (1969) observed relatively high concentrations of this mineral in New Zealand dusts which apparently originated in Australia. These results appear to support the view that movement through the atmosphere is the principal mode of supply of chlorite to southwestern Pacific sediments.

### 26.2.2.3. *Indian Ocean*

Indian Ocean sediments are generally impoverished in chlorite, with the exception of those of the northern Arabian Sea which are supplied with chlorite by the major rivers (such as the Indus) or by wind transport (Stewart *et al.*, 1965; Gorbunova, 1966; Goldberg and Griffin, 1970).

### 26.2.3. MONTMORILLONITE

The montmorillonite (or smectite) group of clay minerals has a basic structural unit which consists of two tetrahedral layers linked by a single octahedral layer. Water molecules lie between these basic three-layer units, and give montmorillonite its expandable nature. The term nontronite has been used by some authors to describe the iron-rich montmorillonites of marine sediments (Bonatti, 1965; Arrhenius, 1963).

This mineral is abundant in deep-sea sediments because of the wide-spread occurrence in the marine environment of volcanic material which can alter to montmorillonite. Montmorillonite is also produced by the continental weathering of basic igneous rocks and by the leaching of potassium from illite or muscovite. However, since high contents of montmorillonite are found in areas associated with vulcanism, Griffin *et al.* (1968) have characterized it as an indicator of a *volcanic regime*. The distribution of montmorillonite in the sediments of the World Ocean is shown in Fig. 26.4, and its average concentrations in the various ocean basins of the world are given in Table 26.3.

### 26.2.3.1. *Atlantic Ocean*

The unusually high input of continental debris rich in illite and chlorite probably results in the sediments of the North Atlantic Ocean being less rich in montmorillonite than are those of any other ocean basin.

The montmorillonite-rich detritus discharged from the Mississippi River has an important influence on the composition of the sediments of the Gulf of Mexico and the Caribbean (Pinsak and Murray, 1960). There is little evidence to indicate that the montmorillonite in North Atlantic sediments was formed by the alteration of volcanic materials (Goldberg and Griffin, 1964; Biscaye, 1965); however, the montmorillonite in the sediment adjacent to

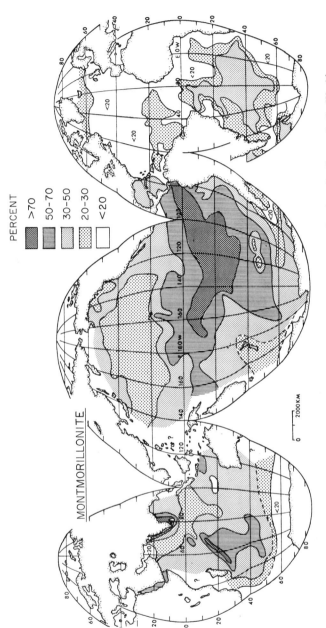

FIG. 26.4. Montmorillonite concentrations in the <2 μm size fraction of sediments in the World Ocean. Data from Griffin *et al.* (1968) and Goldberg and Griffin (1970).

TABLE 26.3

*Montmorillonite concentrations in sediments of the major ocean basins.*
*(Data taken from Griffin et al., 1968, Goldberg and Griffin, 1970 and*
*Venkatarathnam and Biscaye, 1973)*

| Area | No. of Samples | Average % |
|------|----------------|-----------|
| North Atlantic | 192 | 16 |
| Gulf of Mexico | 38 | 45 |
| Caribbean Sea | 49 | 27 |
| South Atlantic | 214 | 26 |
| North Pacific | 170 | 35 |
| South Pacific | 145 | 53 |
| Indian | 245 | 47 |
| Bay of Bengal | 51 | 45 |
| Arabian Sea | 29 | 28 |

the Lesser Antilles Arc is exceptional, as Goldberg (1964) has shown that it is related to volcanic debris.

The influence of the Amazon River on the montmorillonite concentrations of the sediments in the southern North Atlantic is clearly evident (see Fig. 26.4), and Gibbs (1967) has found that the detritus of this river contains 26 to 33% of montmorillonite. After injection into the Atlantic some of this material may be transported into the Caribbean (Jacobs and Ewing, 1969).

South Atlantic sediments have higher concentrations of montmorillonite than do those of the North Atlantic. Griffin *et al.* (1968) have suggested that this results either from an aeolian input of montmorillonite (or volcanic material) or from the alteration of volcanic material from the Mid-Atlantic Ridge. The source of the atmospherically transported montmorillonite could be the active volcanic areas of the South Pacific and/or South America. This material will be diluted along the continental margins by sediments discharged by rivers (see Fig. 26.4) Furthermore, Biscaye (1965) did not find any relationship between the contents of montmorillonite and volcanic material in these sediments. The detection of montmorillonite in dusts collected in the South Atlantic by Chester *et al.* (1972) indicates that at least a proportion of this mineral is supplied by atmospheric transport. However, Chester *et al.* (1974) have shown that the inorganic particulate matter in the upper water layers of the South Atlantic contains only about half as much montmorillonite as does the underlying sediment; this suggests an *in situ* mechanism for the formation of part of the montmorillonite in the sediments.

### 26.2.3.2. Pacific Ocean

The highest montmorillonite concentrations in deep-sea sediments are found in the Pacific. In general, the concentration is relatively high over the whole of the ocean floor and is highest in the South Central Pacific sediments. Pacific Ocean sediments can be regarded as containing uniformly large concentrations of montmorillonite which are diluted in certain areas by continental run-off or wind-transported continentally derived materials. Griffin et al. (1968) have provided clear evidence for the dilution of montmorillonite with continentally derived illite in the North Pacific, and have also shown that the suspended sediments of the rivers emptying into the Pacific along the coast of North America (with the exception of those draining the Cascade Range) are low in montmorillonite.

In the South Pacific high concentrations of montmorillonite are found in the sediments in the proximity of areas of intense volcanic activity. Since these areas are remote from continental sources there is little dilution of the montmorillonite with other clay minerals, and this is reflected in the low sedimentation rates of non-carbonate material in this area (Goldberg and Koide, 1962).

The distribution pattern of montmorillonite in western South Pacific sediments (Fig. 26.4) suggests dilution by atmospheric inputs of other clay minerals. Dilution by continentally derived materials is also apparent along the coasts of southern South America and Antarctica.

### 26.2.3.3. Indian Ocean

High montmorillonite concentrations in the sediments around India have been attributed to river discharges (Goldberg and Griffin, 1970). Concentrations are also high in the sediments of the south-western Indian Ocean in the vicinity of the Mid-Indian Ocean Rise which suggests that these sediments result from the degradation of volcanic material. Sediments which probably have a similar origin are found near Java in the eastern Indian Ocean in another volcanic province (Menard and Smith, 1966).

Around the margins of the Indian Ocean, with the exception of the coast of India, evidence of dilution of montmorillonite by continentally derived material containing other clay minerals is clear. Dilution of montmorillonite by material transported from the Antarctic continent apparently also occurs.

### 26.2.4. ILLITE

The term illite refers to the family of fine grained micas described by Bradley and Grim (1961). The most common illites are predominently dioctahedral and have approximately half as much aluminium substituting for silicon in the tetrahedral sheets as does muscovite (Weaver, 1964). In illites, the basic three layer structural units have potassium ions lying between them in intersheet positions.

The characteristic feature of the distribution of illite in the World Ocean is that its abundance is greater in the sediments of northern latitudes than it is in those of southern ones. The fact that the majority of the world's land mass is concentrated in the northern hemisphere suggests that the illite has a continental origin (Weaver, 1958), and confirmatory evidence for the detrital origin of the illite in recent marine sediments is provided by potassium-argon dating which shows it to be several hundreds of millions of years old (Hurley et al., 1963).

It is clear that river discharge strongly influences the pattern of distribution of illite in pelagic sediments (see Fig. 26.5). However, in the North and South Pacific wind transport plays the principal role in controlling its distribution in the deep-sea sediments. The average concentrations of illite in the major ocean basins are listed in Table 26.4.

TABLE 26.4

*Illite concentrations in sediments of the major ocean basins. (Data taken from Griffin et al., 1968, Goldberg and Griffin, 1970 and Venkatarathnam and Biscaye, 1973)*

| Area | No. of Samples | Average % |
|---|---|---|
| North Atlantic | 181 | 56 |
| Gulf of Mexico | 38 | 25 |
| Caribbean | 56 | 36 |
| South Atlantic | 208 | 47 |
| North Pacific | 170 | 40 |
| South Pacific | 146 | 26 |
| Indian Ocean | 129 | 30 |
| Bay of Bengal | 25 | 29 |
| Arabian Sea | 29 | 46 |

26.2.4.1. *Atlantic Ocean*

Illite is the dominant clay mineral in the $<2\,\mu m$ fractions of the Atlantic Ocean sediments and probably also in the $2$–$20\,\mu m$ fractions (Biscaye, 1965). The highest concentrations of it occur in the North Atlantic, in which its concentration in the $<2\,\mu m$ fraction may reach 70% of the total clay minerals. Griffin et al. (1968) have suggested that the high concentrations found in the sediments of the relatively confined West European Basin adjacent to Spain are the result of atmospheric transport. Biscaye (1965) postulated that the Mid-Atlantic Ridge, by acting as a transport barrier, is responsible for the mineralogical variations found in North Atlantic Ocean sediments. However, Murray (1970) found no differences in the mineralogy

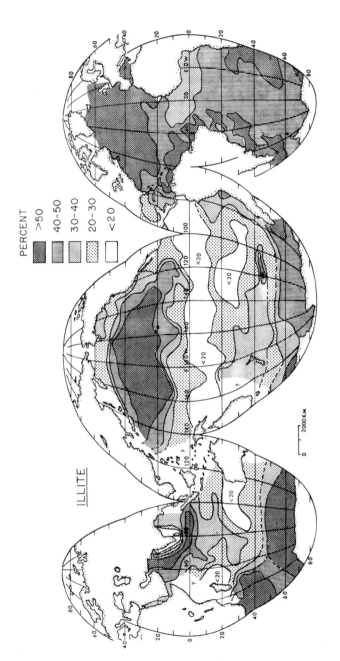

Fig. 26.5. Illite concentration in the <2 μm *size* fraction of sediments in the World Ocean. Data from Griffin *et al.* (1968) and Goldberg and Griffin (1970).

of cores taken in a profile across the Mid-Atlantic Ridge, and concluded that the Ridge exerted no control over sediment distribution.

Pinsak and Murray (1960) found relatively low concentrations of illite in the sediments of the Gulf of Mexico and attributed this to the dominant influence of the montmorillonite-rich detritus of the Mississippi. A gradient of increasing illite in the sediments is found from the Gulf of Mexico into the Caribbean Sea. This is apparently the result of run-off from the Magdelena River (Griffin et al. 1968), with further possible contributions due to transport into the Carribean of material discharged by the Amazon and Orinoco Rivers (Jacobs and Ewing, 1969).

In the South Atlantic, the influences of the Orinoco, the Amazon, the São Francisco and the Paraná rivers are observed as tongues of high illite concentration extending out into the South Atlantic sediments (see Fig. 26.5), as high concentrations of illite are characteristic of the suspended loads of rivers draining South America (Gibbs, 1967; Depetris and Griffin, 1968).

The high concentrations of illite in the sediments surrounding the southern tip of Africa are thought by Griffin et al. (1968) to be aeolian in origin. Van der Merwe and Heystek (1955) found high illite concentrations in the desert soils of Southwest Africa which could serve as the source of these sediments.

The relatively low concentration of illite adjacent to the equatorial coast of west Africa is apparently the result of the diluting influence of both the kaolinite-rich sediments discharged by the Niger and Congo Rivers, and the wind-transported dust from the neighbouring continent.

### 26.2.4.2. Pacific Ocean

The most striking feature in the illite distribution in Pacific Ocean sediments (Fig. 26.5) is the band of high concentrations stretching across the North Pacific. This band, which was observed by Griffin and Goldberg (1963) and by Rateev et al. (1966), corresponds with the band of quartz enrichment reported by Rex and Goldberg (1963), and on the basis of this, Griffin et al. (1968) suggested that it resulted from atmospheric transport. This hypothesis is supported both by the finding of high illite concentrations in dust collected from North American snowfields (Windom, 1969) and by the similarity of the oxygen isotope ratios of quartz isolated from these sediments to that of quartz collected from the probable source areas (Clayton et al., 1972). In the coastal areas of the North Pacific, continental run-off of river detritus, poor in illite, causes the illite concentrations to be lower than that found in other areas of the North Pacific.

Sediments of the South Pacific have illite concentrations considerably lower than those of the North Pacific. However, a band of sediments with a relatively high illite content occurs at latitudes underlying the southern Westerlies.

On the basis of a comparison of sediments from the South Pacific with atmospheric dusts collected in New Zealand Windom (1969) suggested that these sediments are probably derived, in part, from Australia via atmospheric transport. This conclusion is substantiated by recent data given by Clayton *et al.* (1972) for the oxygen isotopic composition of quartz in Pacific sediments. The clay mineralogy of soils from the Australian deserts (Griffin *et al.*, 1968) indicates that the illite concentration is sufficiently high to explain the concentrations found in South Pacific sediments.

High concentrations of illite consistently occur in the sediments off the Pacific section of the Antarctic Ocean (Rateev *et al.*, 1969). This material is apparently derived from the Antarctic continent and transported to the site of deposition by ice rafting.

### 26.2.4.3. *Indian Ocean*

The most prominent feature of the illite distribution in Indian Ocean sediments is the well defined concentration gradient around the Indian Peninsula. This trend of decreasing illite toward the Indian coast was observed by Goldberg and Griffin (1970), and is caused by the diluting effect of montmorillonite-rich river detritus from the Deccan Traps of India. The high concentrations of illite in the Arabian Sea sediments are apparently derived from sources in the north (including the deserts of North India and West Pakistan) from which it is probably transported via the atmosphere. Venkatarathnam and Biscaye (1973) found that the sediments of the Bay of Bengal and the eastern Indian Ocean have relatively high illite concentrations as a result of the influence of the Ganges run-off. The rest of the Indian Ocean basin appears to be relatively poor in illite. However, the Antarctic Ocean sediments south of the Indian Ocean are relatively rich in this mineral (Rateev *et al.*, 1969).

### 26.2.5. OTHER CLAY MINERALS

### 26.2.5.1. *Mixed layer clays*

The most common mixed layer clay is composed of randomly intermixed layers of illite and montmorillonite (Weaver, 1956). However, other mixed layer clays, such as chlorite-montmorillonite and illite-chlorite-montmorillonite, also occur.

The only relatively complete study of mixed layer clays in marine sediments is that by Biscaye (1965). Using data from this investigation Rateev *et al.* (1969) outlined the distribution of mixed layer clays in the Atlantic Ocean (Fig. 26.6). On the basis of this Biscaye suggested that these minerals have a continental origin rather than an *in situ* diagenetic one, as suggested by Berry (1963). Biscaye (1965) observed that mixed layering was detectable in more

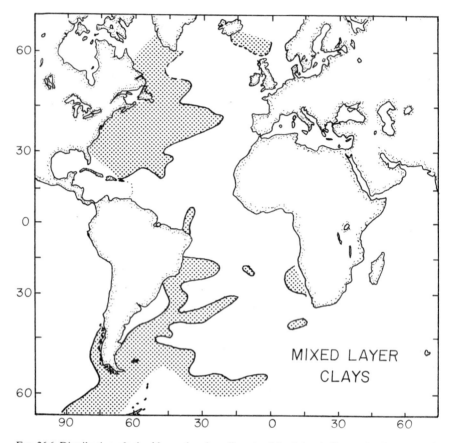

Fig. 26.6. Distribution of mixed layer clays in sediments of the Atlantic Ocean after Rateev *et al.*
(1969) using the data of Biscaye (1965). The cross-hatched areas indicate sediments containing
mixed-layers clays

samples in the 2–20 μm fraction than in the <2 μm fraction. He also found
abundant 10·5 Å mixed layer material in 5 out of 6 samples from the Gulf of
Aden.

### 26.2.5.2. *Palygorskite (attapulgite) and sepiolite*

Palygorskite (attapulgite)* and sepiolite are clay minerals having a chain
structure. Palygorskite has five octahedral positions filled with Mg and Al
ions in an approximately 1:1 ratio. Sepiolite has either eight or nine octahedral sites filled largely with Mg ions. There is relatively little substitution
at the tetrahedral sites of either of these minerals.

* It is uncertain whether attapulgite is a separate mineral from palygorskite.

Palygorskite and sepiolite, which have only rarely been identified in marine sediments, appear to be mainly authigenic in origin, and should not be considered to be lithogeneous materials. However, since they occur on the continents they may be delivered to marine sediments along with other lithogenous materials.

Palygorskite (attapulgite) has been observed in the sediments of the Gulf of Aden and the Red Sea (Heezen *et al.*, 1965). It has also been found in deep Atlantic sediments by Bonatti and Joensuu (1968) who considered that this mineral, together with sepiolite, was probably formed by magnesium-rich hydrothermal solutions acting on montmorillonitic clays. However, Sabatier (1969) has suggested that the palygorskite found by Bonatti and Joensuu might have been derived from deposits rich in this mineral occurring on adjacent land masses (Müller, 1961). Bowles *et al.* (1971) have reported the occurrence of relatively high concentrations of palygorskite and sepiolite in sediments of the Atlantic Ocean, and have postulated that the massive occurrence indicates an authigenic origin for both minerals. According to Bonatti and Joensuu (1969), if the Mg-silicates of the palygorskite-sepiolite group are formed authigenically, then they should occur frequently in those deep-sea sediments which are associated with active ridges and fracture zones.

Windom (1969) has found evidence for the occurrence of land-derived sepiolite. Samples of dust collected in New Zealand which originated from Australia were found to contain a mineral with a 12 Å *d*-spacing which was tentatively described as sepiolite. At least three samples of sediments from the South Pacific were also found to contain this mineral.

## 26.3. QUARTZ

The occurrence of quartz in deep sea sediments has been recognized for many years and was first described in detail by Murray and Renard (1891). Its occurrence as a primary mineral in igneous rocks, especially those on the continents, and its resistance to chemical and mechanical weathering account for its presence in marine sediments. Quartz is common in sediments of all the basins of the World Ocean, and is especially abundant in those of the continental shelves. In these nearshore areas some quartz sand deposits are of economic importance (Emery, 1965).

Various lines of evidence have shown that the major source of the quartz in deep-sea sediments is the continents. However, quartz of submarine volcanic origin has also been identified in these sediments by Peterson and Goldberg (1962). The principal transport processes responsible for the distribution patterns of quartz in marine sediments include wind transport, river discharge and ice-rafting. Quartz has, therefore, been of interest to many workers since its distribution can reveal much about the mechanics of transport and deposition of the lithogeneous components in marine sediments.

The distribution of quartz in the World Ocean is shown in Fig. 26.7 and its average concentrations in the major ocean basins are given in Table 26.5.

TABLE 26.5

*Quartz concentrations in sediments of the major ocean basins. (Data taken from Goldberg and Griffin, 1964, Beltagy and Chester (in Riley and Chester, 1971, p. 333); Ellis, 1972; Griffin and Goldberg, 1969; and Windom (unpublished data)*

| Area | No. of Samples | Average % |
|------|----------------|-----------|
| North Atlantic | 103 | 13 |
| South Atlantic | 117 | 11 |
| North Pacific | 120 | 10 |
| South Pacific | 32 | 4 |
| Indian | 117 | 8 |

With the possible exception of the South Pacific, for which there is little data, there is a striking correlation between the average concentration of quartz in the sediments of the major ocean basins and that of illite in the $<2\,\mu m$ size fraction (Fig. 26.8). This is clearly a reflection of the detrital nature of both minerals and of their similar modes of transport to marine sediments.

### 26.3.1. ATLANTIC OCEAN

Data on the concentration of quartz in the deep-sea sediments of the Atlantic Ocean have been published by several workers: Goldberg and Griffin (1964), for both the North and South Atlantic; Beltagy and Chester (given in Riley and Chester, 1971, p. 333), for the North Atlantic; Ellis (1972) for the South Atlantic; Griffin and Goldberg (1969), for the Caribbean Sea.

Both North Atlantic and South Atlantic sediments contain relatively high concentrations of quartz. These are attributable to the relatively high ratio of the area of the drainage basins on the adjacent continents to the area of the ocean as compared with the corresponding ratios for the other oceans. The high concentrations of quartz in North Atlantic sediments of higher latitudes are apparently related to its abundance in the granitic terrain of the adjacent continents. River discharge and ice-rafting are probably the major mechanisms which deliver it to these sediments.

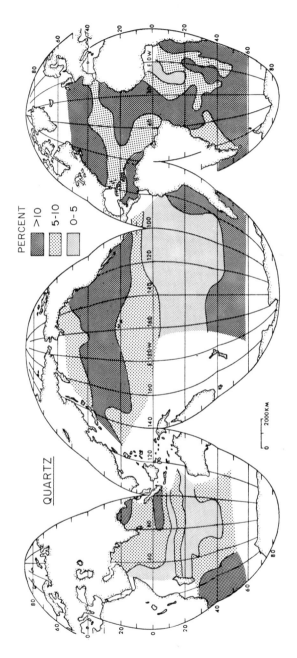

FIG. 26.7. Quartz concentrations in sediments of the World Ocean on a carbonate free basis. Data taken from Goldberg and Griffin (1964), Beltagy and Chester (in Riley and Chester, 1971, p. 333), Ellis (1972), Griffin and Goldberg (1969) and Windom (unpublished data).

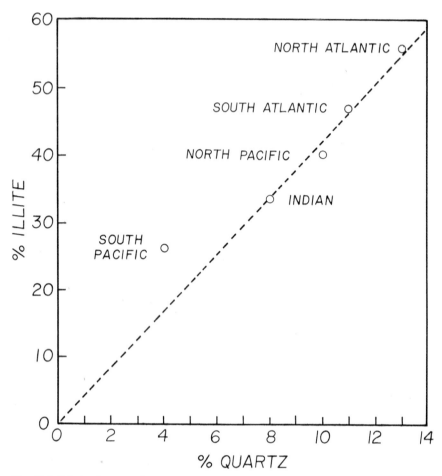

FIG. 26.8. Relation between the average concentrations of quartz and illite (in <2 μm fraction) in sediments of the major ocean basins.

Indications of a large input of quartz off the coast of West Africa by wind transport are apparent. Studies by Beltagy and Chester (see Riley and Chester, 1971, p. 333) have shown a well defined concentration gradient of quartz in the sediments increasing toward the coast of West Africa. The influence of aeolian transport on the sediments in this area has been substantiated by direct observations of dust falls in the vicinity of the African coast (Game, 1964). Quartz-bearing dust transported from the African continent across the Tropical North Atlantic has been collected on Barbados by Delaney *et al.* (1967) and by Prospero (1968). The particle sizes of the quartz in sediments of the Tropical North Atlantic were found by Windom (1969) to range from 1 to 30 μm. Quartz in dust samples collected in the Central American snowfields was found to have generally the same size distribution. The maximum in the size distribution of quartz in Tropical Atlantic sediments shifts to smaller sizes on going from East to West (Beltagy and Chester, 1970; in Riley and Chester, 1971, p. 333). The relatively high concentrations of quartz in Caribbean Sea sediments are apparently due to the discharge of the Magdalena River which drains a terrain composed of metamorphic and sedimentary rocks (Griffin and Goldberg, 1969).

The quartz distribution in the South Atlantic has no clear pattern diagnostic of any single major transport mechanism. According to Ellis (1972) the major sources of quartz in this area are the South American and African continents, river run-off being the dominant transport mechanism. Additional sources also occur on the Antarctic continent from which ice-rafting is probably the principal transport mechanism. A component of atmospherically transported quartz from the South African deserts is probably also present in South Atlantic sediments. Run-off of detritus from the Niger and Congo rivers apparently dilutes the relatively high quartz concentrations found in most of the sediments of the South Atlantic Ocean.

### 26.3.2. PACIFIC OCEAN

The most extensive study of the distribution of quartz in Pacific Ocean sediments was made by Rex and Goldberg (1958); this revealed a strong latitudinal variation of quartz in the deep sea sediments which correlated with the distribution of arid regions in the surrounding land areas. Additional data on the size distribution of quartz in Pacific sediments has been obtained by Windom (1969).

In general, North Pacific deep-sea sediments have concentrations of quartz similar to those of the Atlantic, whereas those of the South Pacific are poor in this mineral, reflecting the ratios of the land drainage areas to those of the corresponding ocean basins. There is a band of high quartz concentration extending across the North Pacific which corresponds to the band of high illite concentration in the <2 μm fraction of the sediments. The highest concentrations of quartz were found by Rex and Goldberg (1958) to be in the Central North Pacific which suggests a wind-borne origin for this mineral. Across the entire North Pacific the size distribution of the quartz in these sediments has a maximum at ~5 μm (Windom, 1969). Dusts from snowfields on the North American continent contain quartz with a similar size distribution.

Windom (1969) has suggested that the quartz in South Pacific sediments has an aeolian origin; a conclusion which is based on a comparison of the size distribution of quartz in the sediments with that in atmospheric dust originating from the Australian deserts. Other transport mechanisms are probably ice-rafting from Antarctica, and wind transport from the Atacama Desert of South America. The atmospheric transport of quartz to the Pacific sediments has recently been substantiated by Clayton *et al.* (1972) using a technique based on oxygen isotope composition. The atmospheric transport of quartz from South American sources to equatorial Pacific sediments is supported by results obtained by Prospero and Bonatti (1969).

### 26.3.3. INDIAN OCEAN

A detailed study of the quartz distribution in Indian Ocean sediments was made by Goldberg and Griffin (1970). Their data indicate that the quartz in Arabian Sea sediments results largely from wind transport, whereas that in the Bay of Bengal is transported to it by the Ganges-Brahmaputra river system. The relatively low concentrations of quartz in sediments adjacent to the Australian continent are unexpected because the distribution of kaolinite clearly indicates that movement through the atmosphere is the dominant transport mechanism in this area. The high concentrations of quartz in the sediments near the tip of South Africa are apparently due to both ice-rafting from Antarctica and transport from the South African continent, either by wind or rivers.

## 26.4. FELDSPARS

Feldspars are ubiquitous in the sediments of the World Ocean, with plagioclase of intermediate to calcic composition occurring in higher concentrations than do the alkali feldspars. The feldspars in marine sediments originate from both marine vulcanism and continental weathering.

Those from the former source are primarily intermediate to calcic plagioclases (of oligoclase to labradorite varieties); in contrast, the latter are characterized by the more sodium-rich members of the plagioclase series together with potassium feldspars (primarily orthoclase and microcline). The distribution of feldspars in the sediments of the World Ocean therefore reflects not only the continental weathering and transport regimes, but also marine volcanic activity. However, because of the similarity of the feldspar species that are supplied by both continental weathering and marine vulcanism, the distribution of this mineral often yields little information about its source or mode of transport.

### 26.4.1. ATLANTIC OCEAN

Biscaye (1965) identified plagioclase and microcline in all the sediment samples ($\sim 500$) from the Atlantic and Indian Ocean which he examined. However, no attempt was made to establish either their relative abundance or details of their composition. The feldspars in these sediments were predominantly between 2 and 20 μm in size. Goldberg *et al.* (1963) have studied the distribution of feldspars relative to that of quartz in a profile of sediment samples taken across the North Atlantic. Their results indicate a close similarity in the size distribution of the two minerals. In addition, the ratio of quartz to alkali feldspar appears to be relatively constant across the North Atlantic between North Africa and the northern coast of South America. However, the ratio of quartz to plagioclase decreases westward from the North African coast. This was interpreted as indicating an aeolian origin for the feldspars like that of the quartz, with the supply of these minerals to the sediments decreasing westward; this contrasts with plagioclase, the supply of which remains relative constant. Subsequent studies by Delaney *et al.* (1967) of the mineralogy of dust samples collected on Barbados, support the conclusion that the alkali feldspars have an aeolian origin.

Systematic latitudinal variations in the plagioclase:quartz ratio in Atlantic sediments were observed by Goldberg and Griffin (1964). In the equatorial Atlantic the ratio is ~1, increasing by a factor of 4 in sediments of higher latitudes both to the north and to the south. The decrease in the ratio in low latitude sediments is thought to result from an increased input of quartz from tropical areas. The increase in the ratio in sediments of high latitudes may be due to both an enhanced supply from marine volcanic sources and lower input from the continents.

Griffin and Goldberg (1969) found the feldspar:quartz ratios in Caribbean sediments to average ~1, with lower values in sediments near the coast of South America. Adjacent to the coast of Central America and in the vicinity of the Lesser Antilles higher feldspar:quartz ratios are common, reflecting the dominance of volcanic debris in these sediments.

### 26.4.2. PACIFIC OCEAN

A positive correlation between quartz and sodic to intermediate plagioclase in the 2 to 20 μm size fraction was observed for North Pacific Ocean sediments by Heath (1969). He suggested that the plagioclase has been transported by wind because its ratio to quartz was essentially the same as that found by Windom (1969) for dust collected from snowfields in the northern hemisphere. A relationship between plagioclase and pyroxene was observed for Equatorial Pacific sediments by Heath, (1969). The latter worker was able to identify three distinct mineral associations containing feldspar in North and Central Pacific sediments. These were:

> (1) an acidic, or continental association, composed of sodic to intermediate plagioclase and potassium-bearing feldspars;

(2) an island arc association containing pyroxene and plagioclase of intermediate composition;

(3) an oceanic association characteristized by intermediate to calcic plagioclase and by smaller amounts of pyroxene.

Peterson and Goldberg (1962) have made an extensive study of feldspars in South Pacific sediments, and have concluded that most of them were derived from within the ocean basin. The plagioclases in the sediments range in composition from oligoclase to intermediate and basic varieties such as andesine and labradorite. Potassium-rich feldspars, such as sanidine and anorthoclase, were also observed. Alkali-rich feldspars are more abundant in the sediments in the vicinity of the East Pacific Rise, whereas the more calcic plagioclases increase in proportion in the sediments away from the rise.

### 26.4.3. INDIAN OCEAN

Sufficient data on feldspar concentrations in Indian Ocean sediments exist to establish its distribution in this basin (J. J. Griffin, unpublished data). Figure 26.9 indicates that the feldspar, which is predominantly plagioclase, is associated with volcanic areas. For example, the gradient of plagioclase off the Indonesian coast suggests that it originated in a volcanically active area. A band of high plagioclase content is also found off Madagascar in the same position as the band of montmorillonite-rich sediments (Fig. 26.4); this coincides with the Madagascar Rise. Another area of high plagioclase concentration appears to be associated with a continental source along the South African coast.

## 26.5. OTHER MINERALS

In addition to the major lithogeneous minerals in marine sediments, many less common minerals also occur, including amphiboles, pyroxenes, olivines and various heavy minerals. Some of these minerals may originate from either continental or marine volcanic sources. Other minerals, such as gibbsite and some heavy minerals, are essentially of continental origin (primary minerals of metamorphic and igneous rocks or the products of chemical weathering). Because of their relatively low concentrations in marine sediments, these minerals have received relatively little attention. However, the more important of them are reviewed below.

### 26.5.1. GIBBSITE

Biscaye (1965) found a very strong association between kaolinite and gibbsite

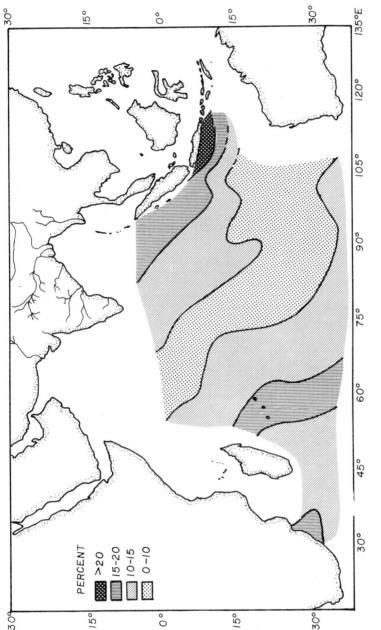

FIG. 26.9. Feldspar concentration in sediments of the Indian Ocean on a carbonate-free basis (J. J. Griffin, unpublished data).

in the deep-sea sediments of the Atlantic Ocean, and both minerals were found to be most abundant in sediments of the tropical Atlantic Ocean. The occurrence of gibbsite together with kaolinite in weathering profiles of tropical land areas is strong evidence for the continental origin of this mineral.

### 26.5.2. AMPHIBOLES

Amphiboles are a common constituent of the marine sediments of all the major oceans basins (Biscaye, 1965; Goldberg and Griffin, 1964; Stewart *et al.*, 1965; Heath, 1969), and can originate from both terrestrial and oceanic sources. Minerals of the amphibole group are apparently relatively stable in the marine environment.

In a detailed study of the distribution of amphiboles in Atlantic Ocean sediments Biscaye (1965) found a distribution pattern similar to that of chlorite; this suggests that it has a continental origin. Windom (1969) found amphiboles to be common constituents of North Pacific sediments and also of dusts present in snowfields in adjacent areas. This suggests that in North Pacific sediments this mineral may have a continental origin, and that it is primarily wind transported.

### 26.5.3. PYROXENES AND OLIVINES

Pyroxenes and olivines are two minerals commonly present in basaltic volcanic rocks, particularly those from oceanic areas. Because of their instability to chemical weathering, olivines are only common in sediments near active volcanic areas of the oceans. In contrast, pyroxenes appear to be widespread in marine sediments as a result of the common occurrence of pyroclastic debris from volcanic areas. Goldberg and Griffin (1964) found a latitudinal trend in the abundance of pyroxene in Atlantic Ocean sediments. This mineral increases in concentration in sediments of higher latitudes, perhaps as a result of the higher production of volcanic material in these areas.

### 26.5.4. MINOR MINERALS

Numerous other minerals have been observed in minor amounts in marine sediments. These include goethite, rutile, zircon, tourmaline, kyanite, sillimanite, anatase, pyrophyllite, garnet, epidote, staurolite, andalusite, titanite, cassiterite, monazite, spinel and many others. Because of their low abundance in continental rocks, and/or their relative instability in the marine environment, these minerals are not common in deep-sea sediments. However, many of them are found in minor amounts in the sediments near their source (e.g. on continental shelves and in the vicinity of oceanic ridges or volcanic areas). Numerous studies have been made of the distributions of many of these minerals in near-shore areas and continental shelf sediments, but little is known of their occurrence in deep-sea sediments.

REFERENCES

Arrhenius, G. O. S. (1963). *In* "The Sea" (Hill, M. N., ed.), Vol. 3, pp. 655–727. Interscience Publishers, New York.

Berry, R. W. (1963). Ph.D. Thesis, Washington University, St. Louis.

Berry, R. W. and Johns, W. D. (1966). *Bull. Geol. Soc. Amer.* **77**, 183.

Biscaye, P. E. (1965). *Bull. Geol. Soc. Amer.* **76**, 803.

Bonatti, E. (1963). *N.Y. Acad. Sci. Trans.* **28**, 938.

Bonatti, E. (1965). *Bull. Volcanol.* **28**, 3.

Bonatti, E. and Joensuu, O. (1968). *Amer. Mineral.* **53**, 975.

Bonatti, E. and Joensuu, O. (1969). *Amer. Mineral.* **54**, 568.

Bowles, F. A., Angino, E. A., Hosterman, J. W. and Galle, O. K. (1971). *Earth Planet. Sci. Lett.* **11**, 324.

Bradley, W. F. and Grim, R. E. (1961). *In* "X-Ray Identification and Crystal Structures of Clay Minerals" (Brown, G., ed.), p. 208. Mineralogical Society, London.

Chester, R., Elderfield, H., Griffin, J. J., Johnson, L. R. and Padgham, R. C. (1972). *Mar. Geol.* **13**, 91.

Chester, R., Stoner, J. H. and Johnson, L. R. (1974). *Nature, Lond.* **249**, 335.

Clayton, R. N., Rex, R. W., Syers, J. K. and Jackson, M. L. (1972). *J. Geophys. Res.* **77**, 3907.

Conolly, J. R. (1965). *J. Proc. Roy. Soc. N.S.W.* **98**, 111.

Copeland, R. A., Frey, F. A. and Wones, D. R. (1971). *Earth Planet. Sci. Lett.* **10**, 186.

Delaney, A. C., Delaney, Audrey C., Parkin, D. W., Griffin, J. J. Goldberg, E. D. and Reinmann, B. E. F. (1967). *Geochim. Cosmochim. Acta,* **31**, 885.

Depetris, P. J. and Griffin, J. J. (1968). *Sedimentology,* **11**, 53.

Ellis, B. D. (1972). M.S. dissertation, Oregon State University, Corvalis.

Emery, K. O. (1965). *U.S. Geol. Surv. Prof. Pap.* **325-C**, C137.

Forman, S. A. and Brydon, J. E. (1965). *Geol. Pedol. Eng. Studies, Roy. Soc. Can., Spec. Publ.* **3**, 140.

Game, P. M. (1964). *J. Sediment. Petrology,* **34**, 355.

Gibbs, R. J. (1967). *Bull. Geol. Soc. Amer.* **78**, 1203.

Goldberg, E. D. (1954). *J. Geol.* **62**, 249.

Goldberg, E. D. (1964). *Trans. N.Y. Acad. Sci.* **27**, 7.

Goldberg, E. D. and Griffin, J. J. (1964). *J. Geophys. Res.* **69**, 4293.

Goldberg, E. D. and Griffin, J. J. (1970). *Deep-Sea Res.* **17**, 513.

Goldberg, E. D. and Koide, M. (1962). *Geochim. Cosmochim. Acta,* **26**, 417.

Goldberg, E. D., Koide, M., Griffin, J. J. and Peterson, M. N. A. (1963). *In* "Isotopic and Cosmic Chemistry" (Craig, H., Miller, S. L. and Wasserburg, G. J., eds.), p. 211. North-Holland, Amsterdam.

Goodell, H. G., McKnight, W. M., Osmond, J. K. and Gorsline, D. S. (1962). *Contrib. Dep. Geol., Florida State Univ., Tallahassee,* **2**, 52 pp. (Unpublished manuscript).

Gorbunova, Z. N. (1963). *Litol. Polez. Iskop.* **1**, 28.

Gorbunova, Z. N. (1966). *Oceanology,* **6**, 215.

Griffin, J. J. and Goldberg, E. D. (1963). *In* "The Sea" (Hill, M. N., ed.) Vol. 3, pp. 728–741. Interscience Publishers, New York.

Griffin, J. J. and Goldberg, E. D. (1969). *Amer. Assoc. Petrol. Geol. Mem.* **11**, 258.

Griffin, J. J., Windom, H. L. and Goldberg, E. D. (1968). *Deep-Sea Res.* **15**, 433.

Heath, G. R. (1969). *Bull. Geol. Soc. Amer.* **80**, 1997.

Heezen, B. C., Nesteroff, W. D., Oberlin, A. and Sabatier, G. (1965). *C.R. Acad. Sci. Paris* **260**, 5819.

Horn, M. E., Hall, V. L., Chapman, S. L. and Wiggins, M. M. (1967). *Proc. Soil Sci. Soc. Amer.* **31**, 108.

Horowitz, A. and Makitie, O. A. (1963). *Amer. Chem. Soc. Abstr.* **63**, 145836.

Hurley, P. M., Heezen, B. C., Pinson, W. H. and Fairbairn, H. W. (1963). *Geochim. Cosmochim. Acta,* **27**, 393.

Jacobs, M. B. and Ewing, M. (1969). *Science, N.Y.* **163**, 805.

Jungerius, P. D. and Levelt, J. W. M. (1964). *Soil Sci.* **97**, 89.

Menard, H. W. and Smith, S. M. (1966). *J. Geophys. Res.* **71**, 4305.

Müller, G. (1961). *Neues Jahrb. Mineral., Abh.* **97**, 272.

Murray, J. W. (1970). *Earth Planet. Sci. Lett.* **10**, 39.

Murray, J. and Renard, A. F. (1891) "The Scientific Results of the Exploring Voyage of H.M.S. *Challenger,* Deep-Sea Deposits", H.M. Stationery Office, London.

Peterson, M. N. A. and Goldberg, E. D. (1962). *J. Geophys. Res.* **67**, 3477.

Pinsak, A. P. and Murray, H. H. (1960). *Proc. 7th Nat. Conf. Clays and Clay Minerals,* Pergamon Press, Oxford, pp. 162–177.

Prospero, J. M. (1968). *Bull. Amer. Meteorol. Soc.* **49**, 645.

Prospero, J. M. and Bonatti, E. (1969). *J. Geophys. Res.* **74**, 3362.

Rateev, M. A., Gorbunova, Z. N., Lisitzin, A. P. and Nosov, G. E. (1966). *Litol. Polez. Iskop.* **3**, 3.

Rateev, M. A., Gorbunova, Z. N., Lisitzyn, A. P. and Nosov, G. L. (1969). *Sedimentology,* **13**, 21.

Rex, R. W. and Goldberg, E. D. (1958). *Tellus,* **10**, 153.

Rex, R. W. and Goldberg, E. D. (1963). *In* "The Sea" (Hill, M. N., ed.) Vol. 3, pp. 295–304. Interscience Publishers, New York.

Riley, J. P. and Chester, R. (1971). "Introduction to Marine Chemistry", 455 pp. Academic Press, London and New York.

Ross, C. S. and Hendricks, S. B. (1945). *U.S. Geol. Surv., Prof. Pap.* **B205**, 23.

Roy, B. B. and Barde, N. K. (1962). *Soil Sci.* **93**, 142.

Sabatier, G. (1969). *Amer. Mineral.* **54**, 567.

Stewart, R. A., Pilkey, O. H. and Nelson, B. W. (1965). *Mar. Geol.* **3**, 411.

Strakhov, N. M. (1960). "Principles of the Theory of Lithogenesis" Vol. 1. *Izv. Akad. Nauk, Moscow.*

Swindale, L. D. and Fan, P. (1967). *Science, N.Y.,* **157**, 799.

Van der Merwe, C. R. and Heystek, H. (1955). *Soil Sci.* **81**, 399.

Venkatarathnam, K. and Biscaye, P. E. (1973). *Deep-Sea Res.* **20**, 727.

Weaver, C. E. (1956). *Amer. Mineral.* **41**, 202.

Weaver, C. E. (1958). *Bull. Am. Assoc. Petrol. Geol.* **42**, 254.

Weaver, C. E. (1964). *In* "The Fundamental Aspects of Petroleum Geochemistry" (Nagy, B. and Columbo, U., eds.), pp. 37–75. Amsterdam and London, Elsevier.
Windom, H. L. (1969). *Bull. Geol. Soc. Amer.* **80**, 761.
Yeroshchev-Shak, V. A. (1961). *Dokl. Akad. Nauk SSSR*, **137**, 695.

Chapter 27

# Hydrogenous Material in Marine Sediments; Excluding Manganese Nodules

## H. ELDERFIELD

*Department of Earth Sciences, The University of Leeds, England*

## 27.1. INTRODUCTION

The components of marine sediments may be classified according to the geosphere in which they originate and such a classification is often useful in interpreting the origins of particular minerals. For example, according to Arrhenius (1963) the grouping together of sediment components originating from weathering of the lithosphere is useful in that the distribution and composition of the *lithogenous* component of marine sediments may be used to evaluate both processes acting on the lithosphere and those involved in

137

transportation of minerals into the pelagic environment. Similarly, a study of the *hydrogenous* material in marine sediments aids evaluation of the transport of elements across the water–sediment interface and the mechanisms of mineral genesis. as well as providing a background for interpreting the physico-chemical state of the palaeo-ocean.

The hydrogenous material in marine sediments plays a vital role in maintaining the geochemical balance and must be considered in geochemical models postulated for sea water (see Chapter 5). Such material also has an important influence on sediment geochemistry.

In this account emphasis has been placed on an evaluation of hydrogenous processes, and of the mechanisms by which hydrogenous material is added to marine sediments. No attempt has been made to give a bibliographic survey of the location and composition of sediments rich in hydrogenous material. and the descriptions of the various mineral types comprising this sediment component are limited to examples which demonstrate the processes by which the hydrogenous component is accumulated in marine sediments.

## 27.2. CLASSIFICATION OF HYDROGENOUS MATERIAL IN MARINE SEDIMENTS

The hydrogenous (authigenic) material in marine sediments is formed in sea water by inorganic reactions. Chester and Hughes (1967) have subdivided the hydrogenous component of deep-sea sediments into primary material. formed directly from sea water, and secondary material, resulting from submarine alteration of pre-existing minerals. This classification is difficult to apply rigorously to all marine sediments since several of the so-called primary materials (e.g. phosphates) have not been shown unequivocally to form in the absence of pre-existing minerals. and others (e.g. oxides and hydroxides, sorbed components) may rely on mineral surfaces for nucleation and siting of exchange reactions. Despite this difficulty. this broad distinction is a useful one and the classification used here is analogous to that of Chester and Hughes (1967), but attempts to define these two main mineral groups more precisely.

Hydrogenous materials are classified as either *halmyrolysates*. which are formed as a result of reactions between sediment components (usually silicates) and sea water, or *precipitates*. which are primary inorganic compounds formed by the direct removal of elements from sea water. pre-existing sediments playing no active chemical role.

Halmyrolysis (originally called "submarine weathering") is a term embracing all the chemical and physico-chemical processes which occur as a result of reactions between sediment components and sea water subsequent to *in situ* weathering and prior to diagenesis. This term has been particularly applied to the alteration of minerals in the suspended load of rivers when they encounter sea water and while they are undergoing transportation to the site of deposition. However, it is also used to describe the submarine weathering of marine volcanic assemblages and the processes occurring during the transportation and weathering of marine particulates of any origin. The boundaries between weathering and halmyrolysis, and halmyrolysis and diagenesis, are difficult to define. Since the diagenetic processes that occur in the upper layers of marine sediments are partly controlled by the chemistry of the overlying water column, the marine pre-burial stage of diagenesis may be defined as the last stage of halmyrolysis. Similarly, the processes occurring in fresh-water transportation and weathering are very important in their influence on later reactions in sea water, and the fresh water-sea water boundary may be regarded as the initial site of halmyrolysis.

Precipitation processes during marine sedimentation are of three main types. The simplest is that in which precipitates form as a result of over-saturation of sea water with respect to particular compounds. Such material is often nucleated by pre-existing seed crystals, as in the formation of aragonite oolites, but is classified as a precipitate rather than a halmyrolysate because the nucleating surface has a physical rather than chemical effect on the nature of the hydrogenous phase. When precipitation results from chemical reaction with a solid phase, as in the formation of hydroxy-magnesium interlayers in clay minerals, the material is classified as a halmyrolysate. Other precipitates (e.g. halite, anhydrite and gypsum) are formed when otherwise undersaturated species in sea water become super-saturated as a result of evaporation. Likewise, supersaturation may occur by a change in the oxidation state of an element in sea water resulting in the precipitation of a hydrogenous phase; the commonest example of this is the formation of ferro-manganese nodules. However, ferro-manganese minerals may also be classified as halmyrolysates because of the roles of sorption, volcanism and diagenesis in the supposed mechanisms of their formation.

The classification of hydrogenous material in marine sediments is summarized in Fig. 27.1. In this figure, three main groups of hydrogenous material are recognized. (1) Hydrogeneous products of the reactions of lithogenous material with sea water. These lithogenous halmyrolysates are formed by the reaction of river and other land-derived particulates with elements dissolved in sea water which under certain circumstances may enter the structure of the lithogenous minerals and effect mineralogical transformations. (2) Hydrogenous products of submarine volcanism. These can be either, *volcanic* halmyrolysates (generally formed as a result of lava–sea water inter-

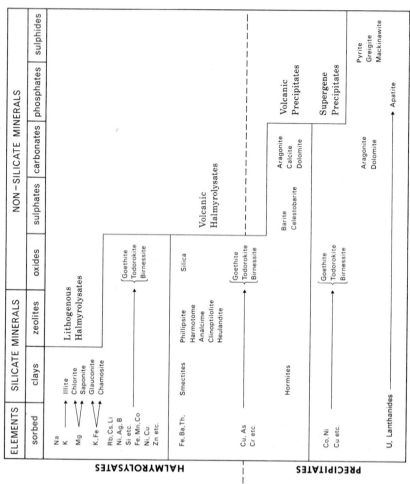

FIG. 27.1. A classification of the hydrogenous material in marine sediments.

action), or *volcanic* precipitates which are formed as a result of the introduction of excess elements into sea water by volcanism. (3) *Supergene precipitates* which consist of hydrogenous material precipitated from sea water or from interstitial solutions of sediments. There are some exceptions to this classification and a hydrogenous mineral may be present in more than one group.

This review is concerned chiefly with the inorganic reactions that take place within the water column and at, or near, the water-sediment interface. For this reason little mention is made of the authigenic phases which have been identified in deeper sections of the sediment column as a result of the Deep Sea Drilling Project. In addition, emphasis is placed on the first two groups of hydrogenous material classified above, especially those involved in halmyrolytic reactions. Certain topics are not covered here or have received only superficial treatment as they are reviewed in other contributions to this volume and the subsequent ones, or elsewhere in the literature (see e.g. Stewart, 1963; Borchert, 1965; Tooms et al., 1969; Bathurst, 1971; Berner, 1971), and the reader is referred to these sources for further information.

## 27.3. HALMYROLYSIS OF LITHOGENOUS MATERIAL IN THE OCEANS

### 27.3.1. SITES AND MECHANISMS OF SORPTION AND ION-EXCHANGE

Cation exchange is the first important chemical reaction involving solid material to occur at the fresh water–sea water boundary. This results in the incorporation of a hydrogenous phase on the surface of the particulates suspended in estuarine waters. Many of the hydrogenous reactions in the oceans are ultimately linked to the surface charge on sediment components and to the sorption of cations on charged surface sites. Apart from the *geochemical transformations* resulting from sorption processes there is some evidence that *mineralogical transformations* occur as a result of the incorporation of sorbed metal ions into mineral structures. The possible genesis of such authigenic mineral phases is considered in Section 27.3.5.

The majority of the fine-grained inorganic suspended load of rivers comprises aluminosilicate debris, often difficult to characterise mineralogically, which is formed by the decomposition of silicate rocks during weathering. Three types of exchange site have been recognised for the clay minerals of this debris. (Bolt et al., 1963). These are: (1) basal surface sites; (2) edge–interlayer sites; (3) interlayer sites. Basal surface and interlayer sites mainly result from isomorphous substitutions within the lattice structure, whereas edge-interlayer sites, sometimes called "frayed edges" (Sawhney, 1972), are caused by broken bonds arising from weathering processes. Broken bonds tend to be on non-cleavage surfaces and hence are parallel to the $c$-axis of clay minerals. The electrical charges on these types of sites are neutralized by sorbed counter-ions of the mobile layer, thus forming the electrical double layer. In this type of double layer selective as well as non-selective sorption can occur.

Potassium fixation (see Section 27.3.5) is an example of selective sorption by sediments. The theories of ion-exchange selectivity proposed by Eisenman (1962a, b) and Ling (1962) convincingly explain the selectivity sequences observed for a wide range of exchange media. However, the low field strength of the fixed groupings of clay minerals allows their selectivity to be treated more simply in terms of Gregor's (1948) theory which is based on ion hydration. Thus, fixation of cations with low hydration energy (e.g. $K^+$, $NH_4^+$, $Rb^+$, $Cs^+$ : see Table 27.1) leads to interlayer dehydration and layer collapse

TABLE 27.1

*Hydration energies of ions at 25° C\**

| | $-\Delta G_h^i$ | | $-\Delta G_h^i$ | | $-\Delta G_h^i$ |
|---|---|---|---|---|---|
| $Cs^+$ | 66 | $Ba^{2+}$ | 310 | $Mg^{2+}$ | 450 |
| $Rb^+$ | 74 | $Sr^{2+}$ | 341 | $Zn^{2+}$ | 479 |
| $NH_4^+$ | 75 | $Pb^{2+}$ | 353 | $Co^{2+}$ | 480 |
| $K^+$ | 79 | $Ca^{2+}$ | 373 | $Ni^{2+}$ | 491 |
| $Na^+$ | 97 | | | $Cu^{2+}$ | 491 |
| | | $Cd^{2+}$ | 425 | | |
| $Ag^+$ | 113 | $Hg^{2+}$ | 431 | $Be^{2+}$ | 577 |
| $Li^+$ | 121 | $Mn^{2+}$ | 433 | | |
| | | $Fe^{2+}$ | 448 | $Fe^{3+}$ | 1025 |

* Data from Vasil'ev (1960), units are kcal mol$^{-1}$.

(Sawhney, 1972); in this way cations can be *fixed* in interlayer positions. In contrast, cations with high hydration energy (e.g. $Ca^{2+}$, $Mg^{2+}$, $Sr^{2+}$) produce hydrated expanded interlayers and these ions remain exchangeable. The result is an affinity sequence for ion exchange—the Hofmeister series

$$Cs^+ > K^+ > Na^+ > Li^+$$
$$Ba^{2+} > Sr^{2+} > Ca^{2+} > Mg^{2+}$$

which is obeyed by most clays except when Coulombic interactions between the fixed and mobile layers are strong compared with those arising from hydration (Eisenman, 1962b). Selective sorption appears to result from the high selectivity of edge–interlayer sites for cations with low hydration energy. Bolt *et al.* (1963) have studied potassium fixation by the clay mineral illite and have found that the selectivity for $K^+$ over $Ca^{2+}$ on edge–interlayer sites was 250-fold greater than that for basal surface sites. Several other investigators (e.g. Jackson, 1963; Rich, 1964) have stressed the importance of edge-interlayer sites and have suggested that selective sorption occurs because of the presence on minerals of "frayed edges" caused by weathering processes. According to this concept, sorption at low solution concentration is initially onto frayed edges (edge–interlayer sites) which have a high selectivity for ions of low hydration energy because of the ease of collapse of such sites (Sawhney, 1972). At higher concentrations, edge–interlayer sites become saturated and sorption occurs on non-selective sites, thus resulting in an overall decrease in selectivity. The fixation of $K^+$, etc., occurs by the collapse of frayed edges which are converted into collapsed interlayer sites from which cation removal is controlled by solid (or film) diffusion (Mortland and Ellis, 1959). The extent of cation fixation is also related to the nature of the mineral structure, particularly to the siting of the charge layer. The mineralogical consequences of ion-fixation are discussed in Section 27.3.5.

Another way by which sorbed cations may become fixed by clay minerals is by cation migration from exchange sites into the silicate skeleton. This has been observed (see for example: Hoffman and Klemen, 1950; Greene-Kelly, 1955; Glaeser and Mering, 1967) on thermal treatment of dioctahedral clays saturated with cations of small ionic radius (less than 0·7 Å), particularly for lithium, magnesium, aluminium, nickel and beryllium. It is not certain whether all the cations migrate into vacant octahedral sites as was originally proposed, and some workers (e.g. Tettenhorst, 1962; Calvet and Prost, 1971) believe that cations migrate into the hexagonal holes of the Si–O networks, but do not fully penetrate deeper into the octahedral layer. Infrared analysis (Calvet and Prost, 1971; Elderfield, in prepn.) has revealed the development of areas of trioctahedral character in these small cation clays which is consistent with at least some migration to the octahedral layer. Migration has been observed following heat treatment at 200–300°C, but has not been detected at lower temperatures. It would appear, therefore, that such incorporation of small cations into clay mineral lattices is likely to occur post-depositionally, under conditions of deep-burial diagenesis. The small ionic radius of boron favours its incorporation into clay mineral lattices by an analogous process of anion migration. Because of the extensive debate concerning boron in clays this topic is treated separately in Section 27.3.3.3.

Sorption and exchange reactions involving clay minerals result from counter-ions balancing surface charges in the double layer in response to crystallographic or weathering parameters. However, sorption and cation fixation by hydrous oxide minerals results from another type of double layer in which the fixed layer is comprised of potential-determining ions sorbed from solution. This occurs because the surface charge originates from surface chemical reactions involving ionisable functional groups with the result that the charge is strongly pH dependent and is balanced by ions in the mobile layer. At low pH values the fixed layer has a positive charge and the mobile layer is composed of anions; conversely, cation sorption occurs at high pH values when the fixed layer charge is negative. The double layer on some sites on clay minerals is of this type but, in general, the double layer of clays results from an interior lattice charge rather than the adsorption of potential-determining ions. For this reason, the zero point of charge of most clay minerals occurs at low pH values except in instances in which crystallographic factors cause the mineral (e.g. kaolinite) to have significant anion-exchange properties; under these circumstances the Z.P.C. occurs at somewhat higher pH values (Table 27.2; see also Chapter 4).

TABLE 27.2

*Zero point of charge of clay minerals and oxides**

| Mineral | $pH_{ZPC}$ |
|---|---|
| Montmorillonite | 2·5 |
| Kaolinite | 4·6 |
| $Fe(OH)_3$ amorphous | 8·5 |
| Goethite | 6·7 |
| $Al(OH)_3$ amorphous | 8·0 |
| Gibbsite | 5·0 |
| $MnO_2$ amorphous | 2·8 |
| $MnO_2$ dehydrated | 4·3 |

*Data from Stumm and Morgan (1970).

Because of their strongly pH-dependent charge characteristics the sorptive properties of hydrous oxides are of particular interest. Sorption of Group Ia and Group IIa cations takes place predominantly in the diffuse double layer, whereas transition metal ions become specifically attached to the oxide surface (Stumm and Morgan, 1970). For example, Morgan and Stumm (1964) observed that $Mn^{2+}$, $Ni^{2+}$ and $Zn^{2+}$ ions are sorbed on $MnO_2$ much more effectively than are $Mg^{2+}$ and $Ca^{2+}$ ions. The sorption of alkali metal and alkaline earth ions on hydrous oxides may generally be interpreted in terms of ion-exchange (e.g. $H^+$, or other cations, being released as the metal ions are sorbed). However, the specific siting of transition elements means that their desorption is not by simple ion-exchange but, as for selectively-sorbed cations on clay minerals, is controlled by diffusion from surface sites through the diffuse double layer into solution.

27.3.2. MAJOR-ELEMENT EXCHANGE REACTIONS

Since the overall selectivity of minerals during sorption reactions tends to decrease at high solute concentrations a model based on non-selective sorption may be used as a starting point for describing the ion-exchange behaviour of the major elements in sea water. The non-selective sorption of cations from solution by most minerals is usually rapid and can be represented by some sort of mass action relationship. In view of this, it can be predicted from a comparison of the average major cation contents of river and sea waters (Table 27.3) that river clays with $Ca^{2+}$ and $K^+$ as exchange-

TABLE 27.3

*Activities and activity ratios of major cations in average river water and sea water**

|  | activity $\times 10^3$ | |
|---|---|---|
|  | River water | Sea water |
| $Na^+$ | 0·26 | 332 |
| $K^+$ | 0·06 | 6·2 |
| $Ca^{2+}$ | 0·30 | 2·39 |
| $Mg^{2+}$ | 0·14 | 13·5 |
| $Na^+/K^+$ | 4·5 | 53·5 |
| $Mg^{2+}/Ca^{2+}$ | 0·45 | 5·7 |
| $(Na^+ + K^+)/(Ca^{2+} + Mg^{2+})$ | 0·73 | 21·3 |

\* Data (for 25°C) from Berner (1971).

able ions will convert to $Mg^{2+}$ and $Na^+$ forms shortly after they are transported into sea water. This has been confirmed by numerous studies (see for example: Potts, 1959; Keller, 1963; Carroll and Starkey, 1960; Russell, 1970) and is very well illustrated by Russell's data (Table 27.4). Berner (1971)

TABLE 27.4

*Exchangeable cations on clays from Rio Ameca and adjacent marine sediments**

|  | River clay | | Marine clay |
|---|---|---|---|
|  | Original composition | After 2 weeks in sea water | |
| $Ca^{2+}$ | 48 | 17 | 6 |
| $Mg^{2+}$ | 10 | 38 | 31 |
| $Na^+$ | 6 | 7 | 6 |
| $K^+$ | 12 | 4 | 3 |
| total | 76 | 66 | 46 |
| $Mg^{2+}/Ca^{2+}$ | 0·21 | 2·2 | 5·1 |
| $Na^+/K^+$ | 0·50 | 1·8 | 2·0 |

\* Data from Russell (1970): concns in meq/100 g.

has used Russell's data to show that the Mg/Ca ratio of a river clay on exposure to sea water for two weeks in the laboratory is quantitatively compatible with what could be deduced from a simple mass action model, assuming ideal solution behaviour for the exchangeable ions occupying interlayer exchange sites of a montmorillonite-type clay. However, the marine sediments adjacent to the river which Russell studied have a higher Mg/Ca ratio than would be deduced from the ideal solution model, and Berner has suggested that this reflects the loss of specific exchange sites which results in changes in the cation exchange properties of the clays. A limitation of this treatment is that the effects of monovalent–divalent ion exchange are ignored. Consistent with the difference between the $(Na^+ + K^+)/(Mg^{2+} + Ca^{2+})$ activity ratios of river water and sea water (Table 27.3) Russell's results show that halmyrolysis produces a net decrease in divalent–cation surface alkalinity which is balanced by an increase in monovalent–cation alkalinity (Table 27.5). From this it can be seen that changes in the cationic composition of river clays, during the initial halmyrolysis reaction, conform with those predicted from a combination of an ideal solution model involving equilibria of similarly charged ions and a net calcium-potassium exchange followed by cation fixation (Table 27.5). Although a decrease in cation exchange capacity occurs during halmyrolysis, the total cation content remains constant because ions are rearranged rather than lost to sea water.

TABLE 27.5

*Halmyrolysis reactions of Rio Ameca clays\**

| Cation (meq/100 g) | $Na^+$ | $K^+$ | $Mg^{2+}$ | $Ca^{2+}$ |
|---|---|---|---|---|
| River EX. | $\left.\begin{matrix}6\\13\end{matrix}\right\}19$ | $\left.\begin{matrix}12\\19\end{matrix}\right\}31$ | $\left.\begin{matrix}10\\132\end{matrix}\right\}142$ | $\left.\begin{matrix}48\\3\end{matrix}\right\}51$ |
| Clay FIX. | | | | |
| Marine EX. | $\left.\begin{matrix}6\\23\end{matrix}\right\}29$ | $\left.\begin{matrix}3\\40\end{matrix}\right\}43$ | $\left.\begin{matrix}31\\120\end{matrix}\right\}151$ | $\left.\begin{matrix}6\\14\end{matrix}\right\}20$ |
| Clay FIX. | | | | |
| Cation change during halmyrolysis | $+10$ | $+12$ | $+9$ | $-31$ |
| Halmyrolysis reactions: | | | | |
| 10 K-clay → 10 Na-clay | $+10$ | $-10$ | | |
| 9 Ca-clay → 9 Mg-clay | | | $+9$ | $-9$ |
| 22 Ca-clay → 22 K-clay | | $+22$ | | $-22$ |
| $(21K_{FIX} + 1K_{EX})$ | | | | |

\* From Neal and Elderfield (in prepn.), based on data from Russell (1970); EX = exchangeable, FIX = fixed (non-exchangeable).

### 27.3.3. TRACE-ELEMENT SORPTION

#### 27.3.3.1. *Adsorption–desorption reactions*

The high selectivity of edge–interlayer sites on clay minerals at low solute concentration, and the specific siting of transition metals on surface sites of hydrous oxides, are particularly relevant with regard both to the sorption of trace elements from fresh water by mineral particulates, and to the ion-exchange behaviour of these particulates with major cations in sea water. During fresh-water weathering and transport, the suspended load of rivers gains trace elements by sorption. Because of the low concentration of dissolved material in fresh waters trace elements are sorbed with a high degree of selectivity, initially on edge-interlayer sites. Since these sites are produced by weathering processes it is likely that the weathering environment will play a major role in determining the sorptive capacity of the suspended load, and calculations based on the relative sorptive capacities of various well-crystalline clay minerals are probably meaningless in estimating trace element uptake. In support of this, Tiller *et al.* (1963) have shown that the clay fractions of a variety of soils of widely different mineralogies and origins adsorbed cobalt by specific exchange processes to a remarkably similar extent. All but one of the soils which Tiller and his co-workers examined fitted a single isotherm, and prior reactions with heavy metal cations or the presence of organic matter and amorphous aluminosilicates had little influence over sorption. This suggests that crystalline structure is not the most important factor controlling the reactivity of minerals with trace elements, and it is probable that the weathering environment with its influence on surface properties is of greater significance.

When the suspended load of rivers crosses the fresh water–sea water boundary, trace element desorption may theoretically take place. If thermodynamic (mass action) criteria are applied, as with *non-selective* sorption of major cations (Section 27.3.2), it can be shown that desorption will occur: (1) if the concentration of the trace element in river water is greater than that in sea water; (2) if the sorbed trace element is displaced by major elements present in sea water; or (3) if the mole fraction of the sorbed trace element on the river particulates is greater than that in equilibrium with the trace element dissolved in sea water. However, the trace metals sorbed on clay minerals primarily occupy edge-interlayer sites which fix metals of low hydration energy (*selective* sorption), and so desorption processes should be modified because such metals are fixed in interlayer sites in river clays and can only slowly diffuse when the clays are transferred to sea water. The metals Rb, Cs, Li and Ag particularly would be expected to be concentrated in marine clays by this mechanism (Table 27.1). This agrees with a laboratory study by Wahlberg and Fishman (1962) who reported that caesium adsorbed

F

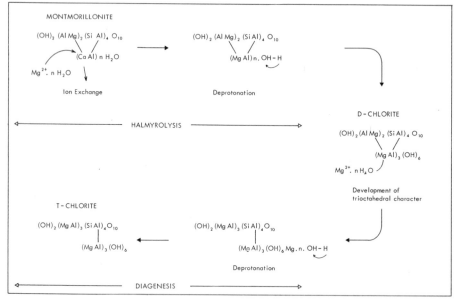

FIG. 27.2. A schematic model for the transformation of montmorillonite to trioctahedral chlorite during halmyrolysis and diagenesis.

on clay minerals was not easily exchanged when the clays were transferred from a medium of high caesium concentration to one of lower, as thermodynamic considerations predict. In addition, Kharkar et al. (1968), using a radiotracer technique, showed (Table 27.6) that 20–70% of the cobalt adsorbed on to clay minerals from distilled water medium was released when the clays were suspended in sea water, whereas only 20–30% of the adsorbed silver was released under these conditions. In contrast, no Co or Ag was released from freshly precipitated $Fe(OH)_3$, and no Co was released from powdered $MnO_2$. Kharkar and his co-workers interpreted this data in terms of trace element desorption as a result of displacement by $Mg^{2+}$, $Na^{2+}$, etc., in sea water. However, the desorptions which they observed must partly be a reflection of the release of trace elements into sea water free of $^{60}Co^{2+}$ and $^{110}Ag^+$ ions. Comparison of the partition of these elements between the mineral phases and solution (Table 27.6) shows, in fact, that silver is more strongly sorbed in sea water than in distilled water when allowance is made for this dilution factor and for desorption due to other than differences in solute concentrations is observed only for cobalt from montmorillonite and illite and silver from manganese dioxide. Rather than reflecting exchange of trace elements for major elements in sea water, the partition data appear to reflect the site selectivity processes discussed in Section 27.3.1. They are the

TABLE 27.6

Adsorption–desorption reactions for Co and Ag on clay minerals and Fe and Mn oxides*

| | Co | | Ag | | $Kd_{Co}$ | | $Kd_{Ag}$ | |
| | % ads. in fresh water | % des. in sea water | % ads. in fresh water | % des. in sea water | Fresh water | Sea water | Fresh water | Sea water |
|---|---|---|---|---|---|---|---|---|
| Montmorillonite | 90 | 70 | 30 | 30 | 9·0 | 0·43 | 0·43 | 2·3 |
| Illite | 90 | 20 | 20 | 20 | 9·0 | 4·0 | 0·25 | 4·0 |
| Kaolinite | 40 | 60 | 10 | 0 | 0·67 | 0·67 | 0·11 | ∞ |
| $MnO_2$ | 25 | 0 | 85 | 95 | 0·33 | ∞ | 5·7 | 0·05 |
| $Fe(OH)_3$ | 95 | 0 | 60 | 0 | 19·0 | ∞ | 1·5 | ∞ |

* Based on data from Kharkar et al. (1968); ads. = adsorbed. des. = desorbed; distribution coefficient. Kd = % sorbed/% in solution.

consequence of the rapid exchange of trace elements of high hydration energy which occupy non-selective sites on the clay minerals. In contrast, there is little or no release of trace elements occupying selective sites on the clays nor of most trace elements sorbed on hydrous oxide minerals.

Because of their strong sorptive character the hydrous oxides in river particulates are particularly important in controlling heavy metal fixation in estuarine sediments. The estuarine environment can act as a trap for heavy metals because of their slow desorption from some sediment types. However, the decrease in pH which often occurs during estuarine sedimentation may cause a release of sorbed metals into interstitial solutions because of the pH-dependent surface properties of the hydrous oxides. In addition, the low redox potential of rapidly accumulating estuarine sediments will cause a reduction of iron and manganese present in oxide phases with a consequent release of associated trace elements into pore waters. The subsequent fate of the heavy metals will be determined by the balance between their diffusion into the overlying waters and their precipitation as sulphides or hydrous oxides in the estuarine sediment.

The laboratory study made by Kharkar et al. (1968) is one of several experimental investigations which have demonstrated that adsorption can occur on sediments in fresh water (see also: Sayre et al., 1963; Gross and Nelson, 1966) and desorption can occur in sea water (e.g. Sonnen, 1965; Johnson et al., 1967). However, studies of trace element reactions based on measurements of the composition of estuarine waters are scarce. Turekian (1971) has recently summarized the results of his studies of the compositions of a group of rivers draining Connecticut and N. Carolina, U.S.A. (Table 27.7).

TABLE 27.7

*Co and Ag in Connecticut rivers, estuaries and sea water\**

| | | $\mu g\,l^{-1}$ | |
| --- | --- | --- | --- |
| | | Co | Ag |
| River water | At site of industrial injection of metals | 7·7 | 1·15 |
| | After sorption by the suspended load | 0·11 | 0·46 |
| Estuarine water | | 0·20 | 0·80 |
| Adjacent sea water | | 0·06 | 0·37 |

\* Based on data from Turekian (1971).

Cobalt and silver which were introduced into some rivers as a result of industrial pollution were rapidly sorbed by the rivers' suspended load. However, a significant increase in dissolved trace elements was observed in the estuarine environment. This was interpreted in terms of desorption from river particulates although a further injection of dissolved Co and Ag from local industrial waste was not totally discounted. Coastal sea water had low Co and Ag levels and trace elements were presumably trapped in the estuarine sediments, to which they were possibly brought initially by the action of organisms. Turekian concluded that most cationic trace metals have very little chance of leaving an estuary in solution and most of the stream input would be ultimately incorporated into the estuarine sediments as hydrogenous phases consisting of sulphides or reduced oxide minerals.

Recently, Elderfield *et al.* (in prepn.) have investigated the stream supply of zinc to sea water from mineralized drainage areas by comparing the zinc concentrations in waters, sediments and particulate matter from associated river, estuarine and marine environments. They showed (Table 27.8) that there was a dramatic increase in soluble zinc in water in the estuary and a decrease in the zinc associated with the suspended load. This evidence pointed to the desorption of zinc from river particulates at the fresh water–sea water boundary. However, the suspended load of the estuary (approx. $50 \, \text{mg} \, l^{-1}$) was considerably greater than that of the river (approx. $5 \, \text{mg} \, l^{-1}$), and it appears that zinc released from tidally-suspended bottom sediments is usually quantitatively more important than that from primary river particulates. It appears therefore, that there is no unequivocal evidence for massive desorption of trace elements from river particulates at the fresh water–sea water boundary and that the estuarine environment is reasonably effective in trapping the hydrogenous phases of the suspended load of rivers. However, remobilization processes can occur in the deposited sediment causing certain metals to be enriched in the surface sediments of the estuary. Desorption may occur from these tidally-suspended sediments and can account for the increase in dissolved trace elements such as zinc in the estuarine water.

### 27.3.3.2. *Hydrogenous trace elements in deep-sea sediments*

The chemical reactivity of particulate matter during transport and sedimentation in the oceans is a question of some debate, and whether inorganic marine particulates have the ability to sorb, or in some way concentrate, trace elements from sea water has an important bearing on the geochemistry of deep-sea sediments. It is now recognized that deep-sea clays are enriched in certain trace elements (e.g. Mn, Co, Ni, Zn) as compared with their near-shore counterparts. Chester and M-Hanna (1970) have shown that this enrichment reflects the fact that the amounts of non-lithogenous sediment

TABLE 27.8

*Partition of Zn between water, particulates and sediment at Conway, N. Wales**

| | Tributary drainage | | | River | Estuary | Sea |
|---|---|---|---|---|---|---|
| | Mineralized | Unmineralized | Mean† | | | |
| Sediment (ppm) | 3700 | 460 | 830 | 530 | 900 | 265 |
| Particulates (ppm) | 4500 | 680 | 1130 | 800 | 130 | 250 |
| Water ($\mu g\,l^{-1}$) | 630 | 22 | 35 | 33 | 57 | 9 |

* Data from Elderfield et al. (in prepn.).
† This value is for the total tributary drainage and so reflects the predominance of unmineralized catchment.

components are greater in deep-sea areas than in near-shore areas in which lithogenous material in the sediment is only in the early stages of adjustment to the marine environment. Several mechanisms have been proposed to explain the formation of this trace element-rich hydrogenous material in deep-sea sediments. Apart from "active ridge" theories which link metal enrichment with submarine volcanism (see Section 27.4.3.2), two principal non-volcanic theories have been put forward: (1) the "differential transport" theory (Turekian, 1967) which assumes that trace element uptake by particulate matter takes place primarily in the river environment and that the chemical activity of marine particulates is of secondary importance; (2) the "trace element veil" theory (Wedepohl, 1960) which explains the observed enrichments purely in terms of precipitation of a metal-rich hydrogenous phase from sea water. Turekian has suggested that the metal enrichment is caused by the differential transport to deep-sea areas of the clay-sized fraction of river particulates, which are rich in metals originally sorbed in the fresh-water environment. In contrast, Wedepohl has invoked a mechanism involving the homogenous precipitation of trace elements from sea water superimposed on marine sediments which are deposited at various rates. An argument against the unreactivity of such river particulates in sea water, which is implicit in the "differential transport" theory is that this material must remain unflocculated so that it can reach deep ocean areas, and as such will retain a surface charge and hence the ability to sorb trace elements from sea water. In addition, the higher content of trace elements in the pelagic fraction of Pacific deep-sea clays, as compared with Atlantic deep-sea clays, argues against differential transport (Chester, 1971). Further, the inferred reactivity of aeolian dust in North Atlantic deep-sea sediments (Chester and Johnson, 1971) supports the "trace element veil" theory. It is not unreasonable to expect that particulate matter will concentrate trace elements from sea water into hydrogenous phases similar to the hydrogenous ferro-manganese oxide phases which become concentrated around nuclei to form manganese nodules. This hypothesis is supported by comparing the manganese/trace element ratios of the hydrogenous fractions of marine sediments with those of manganese nodules (Elderfield, 1972a). The results for cobalt and nickel (Table 27.9) suggest that these elements are present in the sediment in association with a manganese oxide phase which has formed in the same way as manganese nodules, and it is recognized that all stages in the formation of hydrogenous ferro-manganese oxide minerals are present in marine sediments. These stages range from manganese-stained clay particles to fully developed manganese nodules.

### 27.3.3.3. Boron fixation

The major interest in boron uptake by sediments lies in the recognition by

TABLE 27.9

*Manganese/element ratios for Mn nodules and the non-lithogenous fraction of sediments from deep-sea and near-shore areas\**

| | Deep-sea | | Near-shore | |
| | Sediment | Nodule | Sediment | Nodule |
|---|---|---|---|---|
| Co | 79 | 68 | 840 | 860 |
| Ni | 41 | 33 | 50 | 240 |
| Cr | 47 | 19,000 | 6 | 1,750 |

\* Data from Elderfield (1972a).

Goldschmidt and Peters (1932) and Landergren (1945) that marine sediments usually contain more boron than do fresh-water ones, and in the application of this difference to the determination of a palaeosalinity in argillaceous rocks (see for example: Landergren, 1958; Degens et al., 1957, 1958; Frederickson and Reynolds, 1960). Since these early studies, which have been reviewed by Shaw and Bugry (1966) there has been considerable speculation about the mechanism by which boron is incorporated into clay mineral lattices. This has led to serious doubts about the validity of the boron salinity technique, and Harder (1970) has concluded that the large number of variables involved makes it difficult to apply it in an unequivocal manner.

The mechanisms suggested for the incorporation of boron into marine sediments involve the formation of either: (1) a precipitate phase; or (2) a halmyrolysate phase. In mechanism (1) it is proposed that colloidal co-precipitation of boron with silica occurs on suspended solids in estuaries (Levinson and Ludwick, 1966), or that boron replaces silicon in the tetrahedral sheets of authigenic clay minerals (Arrhenius, 1954; Goldberg and Arrhenius, 1958; Landergren and Carvajal, 1969). Mechanism (2) involves the adsorption of boron onto the surface of clays where it is partly fixed on sites having high bonding energies (Lerman, 1966), or it diffuses into the crystal structure either during halmyrolysis (Fleet, 1965; Couch and Grim, 1968) or during later diagenesis (Perry, 1972). The precipitation mechanism requires either the formation of significant amounts of colloidal silica or amorphous silicates in the estuarine environment. If this occurs (see Section 7.3.3.4) such material would comprise much less than 10% of the suspended load (Russell, 1970), or an appreciable neoformation of clay minerals, for which there is little evidence (see Section 27.3.5). Because of this, any precipitate phase would need to contain about 1% boron to account for the observed total sediment values; this seems unlikely in view of the known input of boron to the oceans by river water. For this reason Perry (1972) considers that mechanism (2) is unlikely. Several workers including Levinson and Ludwick (1966), who favour the precipitation mechanism, believe that the initial reaction in the uptake of boron by clays involves adsorption. Couch and Grim (1968) have suggested that sorption occurs by the edge fixation of the $B(OH)_4^-$ ion on "frayed edge" sites (see Section 7.3.1) from which boron atoms migrate into tetrahedral sites through the formation of Frenkel and Schottky crystal defects, and replace tetrahedral silicon and/or aluminium. The time scale of fixation is important, and the work of Couch and Grim suggests that diffusion into the mineral structure is the rate-determining step. Since 10–30 % of the boron in Recent clays can be displaced by suspending the sample in distilled water (Dewis et al., 1972) it is evident that not all the boron is strongly fixed during halmyrolysis. Further, the boron fixed irreversibly on river clays which are transported to a marine environment does not appear to reach the high concentrations found in older sedimentary rocks (Perry, 1972). This evidence favours Perry's suggestion that most of the fixation of boron takes place during deep-burial diagenesis. This is consistent with the general phenomenon for the incorporation of potassium and magnesium into marine sediments (see Sections 27.3.1 and 27.3.5) in which the sorption of these elements by clay minerals may cause partial mineral transformation during halmyrolysis. However, migration of these elements into the mineral lattice, which results in major mineral transformations, is enhanced by the higher temperatures, and more concentrated pore fluids, of the diagenetic rather than the depositional environment.

### 27.3.3.4. Abiological reactions of silicon in sea water

Because of the key role played by silicon in the geochemical cycle its partition between sea water and sediments, and its effect on the geochemical balance has been widely studied. Oceanic budget calculations (see for example: Harriss, 1966; Gregor, 1968; Calvert, 1968) suggest that in general the supply of silicon to the oceans by river water is approximately balanced by biological removal processes. However, the inherent uncertainties of such calculations have led some workers (e.g. Burton and Liss, 1968) to believe that important non-biological processes are operative in removing dissolved silicon from sea water. This latter hypothesis has been strengthened by the belief that silicates exert an ultimate control on the pH and chemical composition of the oceans (see Chapter 5). One suggested abiological removal process is the sorption of dissolved silicon onto suspended matter in oceans and estuaries. Experiments by Mackenzie and Garrels (1965), Mackenzie et al. (1967) and Siever (1968) have shown, not surprisingly, that silicate minerals take up silicon from Si-enriched water, and release it to Si-depleted water. However, Fanning and Schink (1969) have pointed out that none of the minerals used was able to extract silicon from sea water at levels less than

$\sim 5.5$ ppm $SiO_2$ which is much greater than the upper limit of concentrations found in deep waters of the North Atlantic ($1.2$–$3.0$ ppm $SiO_2$), and higher concentrations are only found in this ocean in Antarctic Bottom Water Since silicon can be released by clays into waters containing low concentrations of silicon ($0$–$2.1$ ppm $SiO_2$) Fanning and Schink argue that it is likely that detrital clays will release Si (up to about $0.4\%$ of total Si on the dry-sediment weight) during descent through silicon-poor surface waters (see also Section 27.4.2.5).

The removal of silicon from sea water by river particulates in estuaries during initial mixing processes was first demonstrated by Bien et al. (1958) who concluded that most of the dissolved silicon in waters of the Mississippi River was removed by adsorption. Schink (1967) has reinterpreted this study and has suggested that only $10$–$20\%$ of the silicon was removed; this agrees with Liss and Spencer's (1970) estimate for the estuary of the River Conway, North Wales. Other workers (see for example; Maeda, 1952; Stefansson and Richards, 1963; Schink, 1967; Burton et al., 1970) have been able to detect no appreciable inorganic removal of silicon by this process, and Burton and his co-workers have concluded that the estuarine removal of dissolved silicon is of relatively minor importance in the cycle of this element. The work of Liss and Spencer (1970) suggests that any silicon sorbed by this process is not totally exchangeable with the solution phase, and their experiments indicated that about $75\%$ of the sorbed silicon appeared to be strongly associated with the particulate phase. These experiments were performed using higher silicon concentrations than those found in the Conway estuary, and this probably exaggerates the amount of "fixed silicon" present. Nevertheless, it seems likely that some of the sorbed silicon is deposited together with the suspended load, although the exchangeable silicon is probably removed during phytoplankton blooms when silicon is being rapidly removed from sea water. Such removal decreases even further the net effect of abiological processes on the silicon budget. A recent study by Boyle et al. (in prepn.) casts further doubt on the effectiveness of inorganic removal of silicon in estuaries. These authors have formulated a model comparing the behaviour of a conservative estuarine component (salinity) with that of a component of unknown behaviour. The application of this model to published data shows that in no case has non-conservative behaviour of silicon been demonstrated during estuarine mixing.

27.3.4. CONCEPTS OF TRANSFORMATION AND NEOFORMATION

In the preceding sections the interface reactions taking place when river clays enter the oceans have been considered. These reactions involve exchangeable ions and do not cause any basic structural changes in the clay.

Whether or not authigenic clay mineral formation occurs to a significant extent in the oceans is a question of some controversy, and has an essential bearing on the validity of applying equilibrium concepts, involving mass transfer, to reactions between mineralogical phases and sea water as mechanisms for controlling the composition of sea water itself. In this discussion evidence for such mineral transformation and neoformation is considered. As the topic has been discussed in Chapter 5 and as there is now a voluminous literature (e.g. Sillén, 1966, 1967a, b, c; Garrels, 1965; Mackenzie and Garrels, 1966a, b; Holland, 1965; Kramer, 1965; Helgeson et al., 1969; Helgeson and Mackenzie, 1970; Perry, 1971) on the subject, no attempt will be made to discuss or evaluate Sillén's (1961) hypothesis of a simple equilibrium model for the oceans.

Millot (1970) has used the concepts of *inheritance, transformation* and *neoformation* to account for the source and fate of the clay minerals in the geochemical cycle, and these provide a satisfactory framework in which to discuss the reactivity of minerals in sea water. The oceans inherit detrital minerals which, if stable in sea water, remain unaltered during sedimentation with the result that the corresponding marine sediments reflect the weathering processes operative in the source area of the detrital minerals. Isotopic determinations (Hurley et al., 1959, 1963; Dasch, 1969; Savin and Epstein, 1970) and regional mineralogical data (Biscaye, 1964; Griffin et al., 1968) have shown that these inherited minerals dominate the overall clay mineralogy of marine sediments. However, if the detrital minerals are unstable in sea water, transformations occur and the minerals are modified to reflect their new environment while still retaining a "detrital heritage". *Transformed* clays are thus the result of a structure which has been inherited and subsequently modified by their environment. In addition, neoformations can occur; these are new minerals formed from reactants dissolved in sea water, or are minerals which have been so degraded that they bear little relationship to their original composition. Thus, neoformed clays have little or no "detrital heritage". Transformation can occur either by *degradation* or *aggradation*. Degradation is transformation by subtraction and takes place in an undersaturated environment by leaching of cations and results in the formation of open minerals with variable basal spacings. In contrast, aggradation is the transformation by which degraded minerals produced by weathering are reconstructed under the influence of cation-rich solutions. Degradation-transformation takes place almost exclusively during weathering, whereas aggradation-transformation occurs during sedimentation and diagenesis.

27.3.5. STRUCTURAL CONSEQUENCES OF MAJOR ELEMENT FIXATION

27.3.5.1. *Chlorite and illite*

Both chlorite and illite are common inherited clays in marine sediments. In addition, it has been suggested (Grim *et al.*, 1949; Grim and Johns, 1954; Johns and Grim, 1958; Johns, 1963; Powers, 1954, 1957, 1959; Griffin and Ingram, 1955; Milne and Earley, 1956; Nelson, 1960; Pinsak and Murray, 1960; Gorbunova, 1963; Grim and Loughnan, 1962) that chloritic and illitic phases are presently being formed in marine sediments by aggradation-transformation. The majority of these claims are based on circumstantial evidence and some have little substance. However, phase transformations are so important in ocean chemistry and geochemical budgets that it is essential that the mechanisms of mineral transformation should be evaluated in detail and that the evidence for aggradation-transformation during halmyrolysis should be assessed in terms of these mechanisms.

It has been suggested in Section 27.3.2 that potassium ions which have a low hydration energy are fixed by clay minerals during ion-exchange reactions. This process will be of the greatest significance in areas where there are weathered clays, which have been stripped of potassium, which will then acquire this element in order to upgrade their structures. Consideration of this process has led many workers to suggest that uptake of $K^+$ ions from sea water by clay debris (often montmorillonite) leads to the authigenic growth of illite (e.g. Dietz, 1941; Grim, 1953; Milne and Earley, 1955). However, the process of potassium fixation is not one of structural trans-formation, and Weaver (1958) has shown conclusively that not all smectites are collapsed by $K^+$ ions. It seems clear that the structure of the degraded lattice (Millot's "inheritance") is the main factor controlling aggradation-transformation of montmorillomite to illite. Those montmorillonites which do collapse are often not true montmorillonites (*sensu strictu*), but may be expandable products of the weathering of muscovite and biotite micas. In contrast, those montmorillonites with little or no tendency to collapse have a different origin and are usually formed from unstable volcanic material. In the expanding clay minerals derived from micas there is significant substitution for silicon in tetrahedral sites. This results in the formation of centres of negative lattice charge close to the interlayer and this leads to the strong linkage of potassium to adjacent tetrahedral sheets (Van Olphen, 1963). In contrast, the negative charges of the composite layers of mont-morillorite (*sensu strictu*) are mainly caused by octahedral substitution which results in weaker bonds between unit layers and hence little or no potassium uptake. According to Weaver (1967) most clays adsorb the more abundant ions in preference to potassium, and $K^+$ ions will only be fixed on sites where the lattice charge is higher than $0.7$ per $O_{10}(OH)_2$. For montmorillon-ites few such sites are available, and $Mg^{2+}$ and $Ca^{2+}$ tend to be the predomi-nantly adsorbed ions (Marshall, 1954). In this way extensive tetrahedral substitution promotes lattice contraction following potassium adsorption by clay minerals, and for this reason montmorillonites derived from the weathering of micas are being upgraded in the oceans, whereas montmorillonites (*sensu structu*) are significantly less affected. As has been stressed by Foster (1954) and Yoder and Eugster (1955) a true montmorillonite–illite transformation cannot be achieved "... by stuffing potassium between the layers" since a shift of $Al^{3+}$ from octahedral to tetrahedral sites is required. The formation of illitic phases during halmyrolysis occurs only if the montmorillonite-like clays already have a significant tetrahedral substitution of Al for Si.

In contrast to potassium, the $Mg^{2+}$ ion has a high hydration energy, and selective sorption and fixation of magnesium does not occur during ion-exchange. Hence, the postulated authigenic growth of chlorite cannot be a simple consequence of sorption reactions. Most of the evidence for the formation of chlorite in estuarine and marine environments is based upon X-ray diffraction techniques using the lack of expansion of the characteristic 14 Å (001) reflection as proof of a clay transformation. However, the experimental work of Caillère and Henin (1949) and later investigators (Slaughter and Milne, 1960; Gupta and Malik, 1969; Carstea *et al.*, 1970) has shown conclusively that the stable 14 Å mineral produced by saturating montmorillonite with $Mg^{2+}$ ions is not trioctahedral chlorite since its "brucite" layer decomposes at 400°C. This is similar to the decomposition temperature for brucite (450°C), but considerably lower than that of chlorite (500–700°C). What Caillère and Henin, and others, achieved was the precipitation of $Mg(OH)_2$ in the interlayer position rather than the formation of true chlorite in which the "brucite" layer is attached by much stronger chemical forces. This mineral together with other synthetic products is similar to the magnesium-rich clays which have been claimed to be authigenic chlorites formed by precipitation of octahedrally-coordinated magnesium either as hydroxy-interlayer "islands" (e.g. Grim and Johns, 1954; Powers, 1957) or "atolls" (Frink, 1965), in the interlayer space. Such materials are not truly transformed clays since only their interlayers are altered and their lattices remain detrital (inherited). However, analogy with degradation transformations which take place during weathering suggests that this material, as with the illitic phases discussed above, may be regarded as a type of aggraded clay material, most probably representing an intermediate stage of the transformation which is only completed during deep-burial diagenesis.

Montmorillonite can be formed during weathering by transformation of chlorite, and a reverse mechanism to that proposed for this degradation (e.g. Brown, 1953; Jackson, 1959; Camez, 1962) can be used to describe the upgrading of smectite by magnesium from sea water. This is illustrated in Fig. 27.2 where, for simplicity, the transformation is set out as the formation of one discrete clay mineral from another rather than the formation of interstratified chlorite in the montmorillonite structure. In weathered smectites the interlayer space contains non-exchangeable hydroxy-Al components (Rich, 1968). These act to partially neutralize the charge of the degraded clay lattice, so that when magnesium enters the interlayer space from sea water it is effectively more polar than it would be in exchange positions of well-crystallized clays. This causes the loss of protons (deprotonation) from the interlayer hydration sphere and consequently leads to precipitation of $Mg(OH)_2$ and to the fixation of magnesium in a "brucite" layer, thus,

$$[Al(OH)_2(H_2O)_4]^+ \rightarrow [Al(OH)_2(H_2O)_2.Mg(OH)_2]^+ \rightarrow [Mg_2Al(OH)_6]^+.$$

This reaction is comparable to the deprotonation occurring during polymerization of hydrated aluminium cations to form gibbsite (Hem and Robertson, 1967; Schoen and Robertson, 1970; Elderfield and Hem, 1973). At this stage only partial transformation has occurred resulting in the formation of a material variously termed "Mg-montmorillonite", "Mg-vermiculite", "dioctahedral chlorite", "chloritized montmorillonite", etc. This will be called *dioctahedral chlorite* or *D-chlorite*, and is characterized by the formation of a stable magnesium-rich interlayer without significant alteration of the clay lattice. In order to achieve a full transformation to *T-chlorite* (*trioctahedral chlorite*) the magnesium must penetrate the dioctahedral sheet, entering either empty octahedral sites or forcing $Al^{3+}$ ions out into the tetrahedral sheet. This is an example of the migration of small cations into the octahedral sites of dioctahedral clays (see Section 27.3.1), leading to the development of trioctahedral character. The increased substitution of $Mg^{2+}$ for $Al^{3+}$ in octahedral sites and of $Al^{3+}$ for $Si^{4+}$ in tetrahedral sites increases the net negative charge of the lattice. Consequently, more $Mg^{2+}$ ions pack into interlamellar positions, and hence octahedral sites, resulting in the development of the character of a trioctahedral vermiculite. When this stage occurs the negative lattice charge decreases (since octahedral $Al^{3+}$ ions in trioctahedral vermiculite impart a positive charge whereas octahedral $Mg^{2+}$ ions in dioctahedral montmorillonite impart a negative charge) and so increases the polarity of the increased population of $Mg^{2+}$ ions in the interlayer, which again deprotonates resulting in the formation of a full "brucite" layer. This second deprotonation results from an increase in the magnesium content of the octahedral layer and is analogous to the reverse process (protonation) proposed by Newman (1970) to balance the net loss of layer charge resulting from the decrease in cation occupancy of the octahedral layer on the exchange of potassium from micas. The essential difference between Mg-vermiculite and chlorite is the increased population of $Mg^{2+}$ ions in the interlamellar region of the latter (Mathieson, 1958), and so the complete transformation can be seen as a filling of the interlayer space to form trioctahedral chlorite. In D-chlorite, magnesium is precipitated as "brucite islands" in the interlayer, whereas in T-chlorite the development of trioctahedral character has allowed completion of the full "brucite layer".

Examples of the partial transformation of montmorillonite–chlorite and montmorillonite–illite are to be found among the aggraded clay minerals of the Gulf of Mexico and of the Atlantic coast of the U.S.A. Johns and Grim (1958) suggested that chlorite was being formed in Mississippi river delta sediments and demonstrated that this process was associated with the selective removal of Mg and K from interstitial waters relative to Na. They concluded that magnesium and potassium were being preferentially abstracted from sea water and were entering exchange positions for further fixation to form chlorite and illite. Earlier, Grim and Johns (1954) had shown that this material was dioctahedral and concluded that the "transformation" was related to changes in interlayer cations, and that only minor changes occurred within the stable layer. In a later publication Johns (1963) showed that the increased conversion of montmorillonite to a stable 14Å mineral was associated with an increase in the chloride content of the sediments (Fig. 27.3). Johns suggested that $Cl^-$ ions were being incorporated as magnesium hydroxychloride groups in intermediate layers of chlorite, and argued that his data clearly demonstrated that exchangeable magnesium was being precipitated in the interlayer. By analogy with the experimental study of Biedermann and Chow (1966), who demonstrated that hydrolysis of $Fe^{3+}$ ions in sea water leads to the precipitation of $Fe(OH)_{2.7}Cl_{0.3}$, it is certainly reasonable to expect that magnesium fixed from sea water into degraded montmorillonite would be present as compounds of the type $Mg(OH)_{2-x}Cl_x$.

The transformation observed by Johns and Grim (1958) for the Gulf of Mexico sediments is clearly that of montmorillonite to D-chlorite. To the author's knowledge this is true of all the montmorillonite–chlorite transformations occurring either in sea water or in the surface layers of marine sediments, and there is no evidence that T-chlorite is formed during halmyrolysis. In a study of sediments from the Atlantic coast of the U.S.A., where chlorite-like material was shown to be forming from degraded montmorillonite–illite via a vermiculite stage (Powers, 1957), the "chlorite" in sediments overlain by waters of salinity $1-10\%$ had a thermal stability of less than $200°C$, whereas its stability increased to $450-500°C$ in waters of salinity $10-30\%$. It is possible that this material (D-chlorite) is a precursor of the T-chlorite which could form under favourable conditions (e.g. a long residence time of degraded montmorillonite in the water column), although there is presently no proof of this occurring during halmyrolysis. Griffin (1962) identified traces of chlorite stable to $500-700°C$ in some sediments of the north-east Gulf of Mexico as well as the D-chlorite detected by Johns and Grim (1958). However, major rivers in this area contain traces of metamorphic chlorite, and additional evidence from microfauna in the sediments led Griffin to suggest that

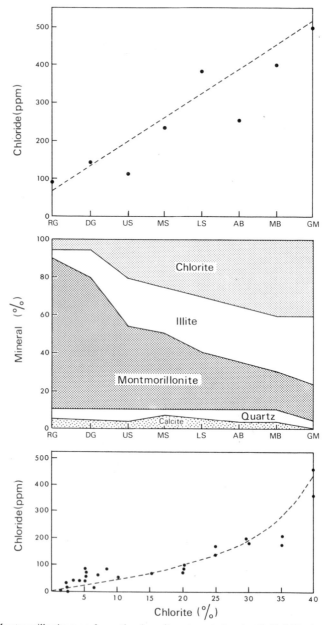

FIG. 27.3. Montmorillonite transformation in sediments entering the Gulf of Mexico. UPPER: Chloride content of the <1 μm fraction; CENTRE: mineralogy of the <1 μm fraction; LOWER: relationship between chloride and chlorite. RG = Guadalupe river, DG = Guadalupe delta, US = upper San Antonio Bay, MS = middle San Antonio Bay, LS = lower San Antonio Bay, AB = Aransas Bay, MB = Mesquite Bay, GM = Gulf of Mexico. (modified with permission from Johns, 1963).

the stable chlorite was detrital and had been displaced by turbidity currents from higher positions close to its source area. This same possibility must be taken into account in the interpretation of partial transformations. For example, Griffin and Goldberg (1963) have interpreted the decrease in concentration of 17Å clay in an east to west traverse across the N. Pacific only partly in terms of fixation of $K^+$ and $Mg^{2+}$ ions and random interlayer collapse, and they have emphasised the detrital influence of the coastal sediments of Canada, Alaska etc. which are high in chlorite and illite and low in montmorillonite.

### 27.3.5.2. Saponite

Another way in which magnesium can be fixed in clay minerals has recently been proposed by Drever (1971, a, b) who suggested that it may replace the iron which has been extracted from marine sediments under anaerobic conditions. i.e.

$$Fe_2 - clay + 3Mg^{2+} + 4S^{2-} \rightarrow Mg_3 - clay + 2FeS_2 + 2e^-$$

S is sulphur derived from the bacterial reduction of sulphate in sea water. In this way a saponite-type composition can be obtained from nontronite, thus,

$$Fe_4[Al_xSi_{8-x}]O_{20}(OH)_4.(\tfrac{1}{2}Ca.Na)_x + 6Mg^{2+} + 8SO_4^{2-} + 15C_{org} + 13H_2O$$

$$\rightarrow Mg_6[Al_xSi_{8-x}]O_{20}(OH)_4(\tfrac{1}{2}Ca.Na)_x + 4FeS_2 + 15HCO_3^- + 11H^+$$

This is not to imply that saponite is actually formed by substitution of $Mg^{2+}$ for the $Fe^{3+}$ in octahedral sites of clay minerals, and the use of a mineral name to describe this and other products of major element fixation (e.g. chlorite and illite) is simply intended to indicate that the composition of the altered clay is moving towards this "structural end member". As Drever has suggested, similar reactions can be postulated for other iron-containing clay minerals and result in a change in chemical composition without marked changes in mineralogy.

Drever examined a series of six cores containing oxic and anoxic zones and found that the non-exchangeable (fixed) magnesium was more abundant in clays from sulphide-rich regions than in clays which were less reducing. The differences were apparently unrelated to mineralogy (Fig. 27.4), and Drever has argued that when the iron in clays (which is unstable with respect to pyrite in anoxic sediments) is released to form a sulphide phase a corresponding amount of magnesium enters the clay from the surrounding interstitial solution.

### 27.3.5.3. Glauconite and chamosite

Glauconite occurs most abundantly in those marine sediments which have

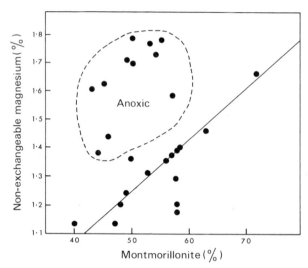

F IG. 27.4. Relationship between non-exchangeable Mg and montmorillonite in the clay fraction of sediments from Banderas Bay, Mexico. (modified with permission from Drever, 1971a).

low accumulation rates, and although it is found in an oxidizing environment it is usually associated with organic matter. In common with many marine phosphate deposits (see Section 27.5.3.1) its presence in near-shore sediments having conditions which favour glauconitisation is not necessarily proof of its Recent formation since detrital glauconite is also common in some continental margin environments (see Pratt, 1963; Bell and Goodell, 1967; Murray and Mackintosh, 1968). The formation of recent glauconite has been reported by several authors, including Burst (1958a), Hulings (1961), Ehlmann et al. (1963), Pratt (1963) and Porrenga (1967).

The term glauconite is used in two senses: (1) as a mineral name for a specific well-ordered, potassium-rich, mica structure; and (2) as a morphological term for green sand-sized grains, pellets, etc., in sediments which may, or may not, contain glauconite. It is most convenient, however, to regard glauconite as a name for a sequence of minerals having a random interstratification of non-expandable 10 Å layers and expandable (montmorillonitic) layers (Table 27.10). According to this classification glauconite (sensu strictu) is material with less than 10% of expandable layers and is an ordered one-layer monoclinic polymorph (1M), containing more iron and less aluminium than illite and with a greater layer charge requiring more interlayer $K^+$ than illite for charge balance. The less-ordered glauconites (1Md) contain between 10% and 20% of expandable layers, and those with >20% of expandable layers are best regarded as mixed layer (glauconite–montmorillonite) clay minerals. The disordered and mixed-layer glauconites (classes 2 and 3 of Table

TABLE 27.10

*Classification of glauconites\**

| Class | Name | % Expandable layers | Description | X-Ray characteristics |
|---|---|---|---|---|
| 1 | 1M glauconite | <10 | Well-ordered, non-swelling high K lattice | Symmetric and sharp diffractions at 10·1, 4·53, 3·3 Å; reflections (112) and (11$\bar{2}$) are always present |
| 2 | 1Md glauconite | 10–20 | Disordered, non-swelling, low K lattice. | Asymmetric basal diffractions broader at base and subdued in intensity; reflections (112) and (11$\bar{2}$) are absent |
| 3 | Interlayered glauconite | >20 | Interlayered clay mineral; extremely disordered. expandable, low K montmorillonite-type lattice | $d$ (001) > 10·15 Å |
| 4 | Mixed mineral (precursor of mineral glauconite) | — | (a) Two or more clay minerals, frequently illite with montmorillonite and illite with chlorite; (b) Mixtures of clay with non-clay minerals. | — |

\* Based on classifications by Burst (1958a, b), Hower (1961) and Bentor and Kastner (1965).

27.10) are typical of low-temperature lattice formation (Yoder and Eugster, 1955; Velde, 1965), and Recent glauconites tend to be of these types (Hower, 1961; Bell and Goodell, 1967; McRae, 1972). This suggests that 1 M glauconite is formed as a result of deep-burial diagenesis via the 1Md polymorph (Hower, 1961), and this is consistent with the transformation reactions resulting in the formation of other types of authigenic clay minerals (see Section 27.3.5.1.).

The present major theory for glauconite formation is the "layer lattice theory" proposed by Burst (1958a, b) and developed by Hower (1961) and Foster (1969). Prior to this, early theories involved the attraction of potassium ions to Fe, Al, or Si hydrous oxide gels (see Murray and Renard, 1891; Collet, 1908) or a vaguely-defined alteration of mud (e.g. Collet, 1908; Takahashi, 1939). Galliher (1935a, b; 1936; 1939) has suggested that all glauconite is formed by transformation of biotite which, although consistent with certain glauconite morphologies such as "booklet" or "accordion" grains (see for example Ballance, 1964; Seed, 1965), is difficult to reconcile with other morphologies and is best regarded as a specific example of the "layer lattice theory". According to this theory the formation of glauconite requires (1) a degraded layer-silicate lattice; (2) a plentiful supply of iron and potassium; and (3) a suitable redox potential. The first stage of formation is the deposition of a degraded 2:1 expandable clay mineral under slightly reducing conditions (Eh 0 to $-200$ mV) either in the bulk sediment or, as is frequently the case, in localized reducing microenvironments such as those found in pellets (e.g. in faecal pellets within foram tests). These conditions allow glauconitization to proceed in preference to pyrite formation. The next stage of formation occurs by sorption of iron and potassium by the degraded clay lattice. The clay must have an initially low lattice charge since clays of high lattice charge sorb potassium rapidly from sea water and collapse to a non-expandable, iron-poor, illitic material (Section 27.3.5.1). In contrast, clays of low lattice charge do not readily sorb potassium but allow iron to diffuse slowly into the octahedral layer where it replaces aluminium and imparts a larger negative layer charge. After this, potassium is more readily sorbed and, with other interlayer cations, neutralizes the layer charge. Potassium is fixed in interlayer positions, collapsing a proportion of the expandable layers, and as glauconitization proceeds there is an increase in lattice charge and hence an increase in the potassium content and a decrease in the proportion of expandable layers. Hence, the various glauconite types (Table 27.10) reflect differences in the extent of layer collapse and of the maturity of the mineral.

Other factors such as temperature and sedimentation rate are also important in their control over the rate and degree of glauconitization. Glauconite is thought to form at the sediment–water interface at low sedimentation rates

(Cloud, 1955), and therefore "mature" glauconite is less likely to develop in regions with a high influx of detritus where rapid sedimentation will arrest the glauconitisation process by burial of developing minerals (McRae, 1972).

The optimum temperature range for glauconite formation has been suggested to be 15–20°C (Takahashi and Yagi, 1929; Porrenga, 1967) on the basis of its present-day formation in warm shelf seas. Braun (1962) has proposed that the formation of glauconite requires a lower temperature than that of chamosite and this is in agreement with the distribution of recent chamosite in marine sediments (Porrenga 1965a, b, 1966, 1967; Giresse, 1965; Von Gaertner and Schellmann, 1965). In recent sediments chamosites are confined to the tropics and Recent chamosite has not been reported in higher latitudes (Porrenga, 1967). This contrasts with glauconite which has a more widespread distribution. Chamosite appears to develop in a similar way to glauconite except that temperatures of > 20°C are considered by Porrenga to be essential for its formation. In the absence of low-temperature synthesis data the relative mutual stability fields of glauconite and chamosite are uncertain. Porrenga has also suggested that chamotization may occur during early diagenesis rather than during synsedimentary stages. The lower potassium and higher ferrous iron contents of chamosite relative to those of glauconite (Table 27.11) are consistent with this suggestion. In addition to these forms, admixed hydrogoethite–chamosite–glauconite has been recognized in pelletal form in sediments of the West African shelf (Emelyanov, 1970). These, and also glauconite pellets from the same area, are of unusual chemical composition (Table 27.11) in that they have high phosphate and organic carbon contents which may indicate intra-sediment formation.

## 27.4. Authigenesis and Supergene Volcanism

### 27.4.1. submarine volcanic phenomena

Boström (1967) has suggested a six-fold classification of the submarine volcanic and metamorphic reactions that can occur in pelagic areas. These are the reactions between: (1) sea water and lava; (2) magmatic emanations and lava; (3) magmatic emanations and sediments; (4) magmatic emanations and sea water; (5) lava and sediments; (6) magma and sediments. Reactions of type (1) are often invoked for the formation of marine authigenic minerals whereby reaction of hot lava and sea water and subsequent submarine weathering result in a characteristic assemblage of palagonite–smectite–zeolite which is found predominantly in regions of basic volcanism. In addition, it has been postulated that certain elements emanate from within, or below, the sediment column and that their reaction with sea water gives

TABLE 27.11

*Chemical composition of Recent glauconite-chamosite-geothite material*

| % | 1 Proto-glauconite (Niger Delta) | 2 Glauconitic sediment (S. African shelf) | 3 Chamosite (Sarawak shelf) | 4 Hydrogoethite-chamosite sediment (mouth of R. Congo) | 5 Goethite pellet (Niger Delta) |
|---|---|---|---|---|---|
| $Fe_2O_3$ | 18·50 | 16·90† | 4·0 | 16·43† | 46·82 |
| FeO | 3·00 | 3·60 | 16·94 | 1·20 | <1 |
| $K_2O$ | 2·74 | 4·25 | 0·5 | 0·47 | 0·58 |
| $P_2O_5$ | 0·24 | 0·76 | — | 2·62 | 2·64 |
| Org. C | 0·5 | | — | | 0·5 |
| Volatiles | 8·55 | 5·38 | 9·3 | 4·71 | 12·15 |

* Partial chemical analyses from data of Porrenga (1967) for samples 1, 3 and 5, and Emelyanov (1970) for sample 2 and 4.

† Total iron as $Fe_2O_3$.

rise to chemical precipitates such as sulphates (barite, celestobarite), oxides, (goethite, manganite) and silicates (sepiolite, palygorskite). This process involves reactions of types (2)–(4), and types (5) and (6) may conveniently be linked to this group since these five reactions constitute aspects of the diagenetic, metamorphic and metasomatic processes that take place as a result of submarine volcanism. Inasmuch as the ultimate stage of these reactions often involves hydrogenous precipitation, such processes will be discussed in the context of their link with authigenic mineral formation on the sea floor. Hydrogenous sedimentation also occurs as a result of emanative systems where there may be no volcanic or magmatic contribution to the composition of the ultimate authigenic minerals. In these systems the role of "volcanism" is merely to provide a heat source and the resulting emanation is geothermal rather than hydrothermal (see Sections 27.4.3.1). It is convenient to consider these geothermal systems at the same time as the hydrothermal ones (where the emanating "solution" is at least partly magmatic—American Geological Institute, 1962) and thus two general types of volcanic-linked reactions should be recognized as being important in the formation of the hydrogenous material in marine sediments: (1) reactions between sea water and lava; (2) reactions involving hydrothermal and geothermal solutions. It has long been realized that these reactions are linked to authigenic mineral formation; the former being suggested by Murray and Renard (1891), and the latter by von Gumbel (1878) and Murray and Irvine (1894).

### 27.4.2. REACTIONS BETWEEN SEA WATER AND LAVA

#### 27.4.2.1. *Structure and devitrification of volcanic glass*

The initial reaction between lava and sea water is one of quenching, resulting in the formation of volcanic glass and, if the initial physico-chemical state of the melt is appropriate, sometimes of high temperature silicate crystallites. The glassy phases found in marine sediments are not all similar but relate compositionally to the volcanic material from which they were derived. Similarly, the formation of authigenic minerals by the submarine weathering of volcanic glass depends on the composition and mode of formation of these glassy phases. Because of this, it is necessary to know something of the structure of molten lavas and volcanic glass, and of the mechanisms of glass devitrification, in order to assess when and where halmyrolysis of submarine volcanics is likely to occur.

Present views on glass and molten lava structures are largely based on the theories of Zachariasen (1932) and on confirmatory X-ray work by Warren and co-workers (see Warren and Briscoe, 1938). According to the Zachariasen–Warren theory, $SiO_4$ tetrahedra in a silicate glass form three-dimensional networks linked by forces similar to those in the corresponding crystal. This leads to the concept of "network-formers", which are the oxides forming the glass skeleton, and "network-modifiers" which are the metal oxides which contribute the large interstitial cations. Singly-bonded oxygens are referred to as "non-bridging oxygens". The relative amounts of network-formers and network-modifiers present govern the viscosity of the system. In silica-rich systems the strong association of $SiO_4$ groups leads to a high viscosity, whereas in metal-rich systems the $SiO_4$ network is disrupted by large modifying cations, thus producing a lower viscosity. This leads to a structural model (Scarfe, 1972) for molten lavas and glasses. The poly-component rock melt contains oxygen ions, network-forming polyanions and network-modifying cations (Table 27.12), and the higher the proportion

TABLE 27.12

*Classification of ions in silicate rocks\**

| Cation | Coord. no. | Type† |
|---|---|---|
| $P^{5+}$ $Si^{4+}$ $Al^{3+}$ | 4 | NWF |
| $Ti^{4+}$ $Al^{3+}$ $Fe^{3+}$ | 6 (or 4) | I |
| $Fe^{2+}$ $Mn^{2+}$ $Mg^{2+}$ | 6 | u NWM |
| $Ca^{2+}$ $Na^+$ $K^+$ | 6 | NWM |

\* Based on data by Scarfe (1972) after Dietzel (1942).
† NWF = Network-forming ion.
  NWM = Network-modifying ion.
  uNWM = usually network-modifying ion.
  I = intermediate ion.

of network-formers the higher the viscosity. Hence, acidic lavas are more viscous than are basaltic lavas having the same temperature and water content. The viscosity of the melt is a major factor influencing the rate of hydration of volcanic glass, and this, together with compositional factors, has an important bearing on the palagonitization process.

Grauer and Hamilton (1950) have described the rate of crystallization in silicate melts in terms of the constraint of viscosity on the thermodynamic process, and Marshall (1961) has used this concept to explain the devitrification process. Since glasses are thermodynamically unstable, free energy is the driving force of crystallization so that as the temperature falls below the liquidus temperature the glass structure changes from that of a disordered liquid to that of an ordered (supercooled) liquid. In the absence of a constraint, the system would cross an energy barrier related to the formation and growth of nuclei and fall to its lowest free energy state which is that of a crystallite (Figure 27.5—Left). However, a constraint on this thermodynamic

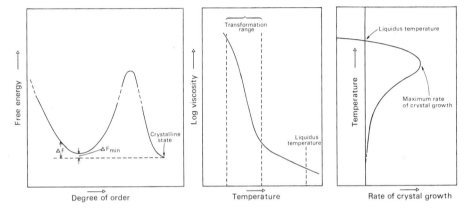

FIG. 27.5. Schematic representation of glass devitrification. LEFT: variation of degree of order (or viscosity) with free energy for a glass at a temperature below the transformation range; CENTRE: viscosity change of glass as a function of temperature; RIGHT: rate of crystal growth in glasses as a function of temperature. (With permission from Marshall, 1961).

process is the rapid increase in the viscosity of the glass with decreasing temperature (Fig. 27.5—Centre) which prevents relaxation of the system to complete equilibrium. The combination of these two processes controls the rate of crystal growth (Fig. 27.5—Right) which reaches a maximum generally within 100°C of the liquidus temperature, and then decreases exponentially thus imparting an apparent stability to the supercooled glassy state. The first stage of devitrification is the diffusion of water molecules into the glass, and Marshall has calculated the activation energy of the diffusion process to be about 30 kcal mol$^{-1}$, which agrees with experimental values (Scholze and Mulfinger, 1959; Hawkins, 1961). Using this value he has estimated the time-temperature relationship for devitrification of a perlitic glass. His results (Fig. 27.6) show that devitrification is extremely slow at low temperatures, whereas at higher temperatures the predominance of the thermodynamic driving force over the constraint of viscosity results in rapid hydration. By extrapolating the data to higher temperatures Bonatti (1965) has suggested that devitrification taking many millions of years at 20°C may be achieved at 500°C in the order of hours.

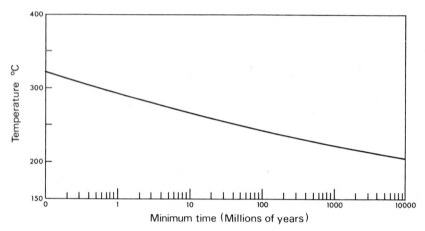

F𝖨𝗀. 27.6. Minimum time required for thermal reconstruction of natural glass as a function of temperature. Modified, with permission, from Marshall (1961).

It is therefore apparent that hydration occurs more effectively at high temperatures when the glass viscosity is low. In addition, hydration is more effective in glass formed from a basic (low viscosity) melt than in that from an acidic (high viscosity) melt. Marshall (1961) has suggested that the smaller amounts of Si–O and Al–O bonds (network formers) in basic melts will increase the rate of devitrification in glasses formed from them by about 20% as compared to acidic glasses. Although this increase produces a negligible effect at low temperatures, it is probably significant during hydration at high temperatures. Another factor affecting the rate of hydration is the rate of diffusion of water into the glass; this is greater in hydrous glasses than in dry ones (Haller, 1963), and for this reason hydration at high temperatures will increase any subsequent low temperature hydration above that predicted by Marshall's model. Because of the strong temperature dependence of hydration, Bonatti (1965) has suggested that two genetically different types of volcanic glass are present on the sea floor: (1) glass formed by lava–sea water interaction during a submarine volcanic eruption; and (2) glass formed during a subaerial eruption (a lithogenous sediment component), or formed during a submarine eruption when the lava was prevented from direct interaction with sea water (as for example the lava inside a glass crust of the first type). Volcanic glass of type (1) contains up to 20% of water if formed from a basic lava, and will decompose on the sea floor at low temperatures much more rapidly than type (2) (e.g. sideromelane which is often found within the cores of lava fragments) which has a lower water content. The effects of hydration of basalts and basaltic glass appear to be reflected in the compositions of the interstitial solutions of marine sediments.

Bischoff and Ku (1970) have observed that interstitial water chlorinities are much more variable in sediments from the region of the mid-Atlantic ridge than in those of adjacent non-ridge areas in which they are remarkably constant. They attribute these differences to a combination of silicate hydrolysis, which removes water and hence increases chlorinities, and the addition of hydrothermal solutions which decreases chlorinities.

Theoretically, crystallisation should follow hydration, and so it is possible that the products of glass devitrification include high temperature crystallites. Diffusion of water molecules into the glass structure breaks Si—O—Si, Si—O—Al and Al—O—Al bonds by attachment of hydroxyls, and after this, relatively little energy is required to orientate Si or Al tetrahedra to form a crystalline structure (Marshall, 1961). Whether or not this will occur depends on the temperature, the rate of cooling, the viscosity and the original water content of the initial lava. Experimental studies (Hawkins and Roy, 1963) have shown that smectites and zeolites can form at high temperatures by lava–water interaction. In agreement with the concepts discussed above, Hawkins and Roy found that silica-rich materials were less altered than were those of intermediate and basic compositions. In addition, the presence of free MgO (presumably acting as a network modifier) was observed to be important, and rocks containing less than 2–3% MgO did not alter to montmorillonite.

### 27.4.2.2. *Alteration of acidic and basic lavas*

Consideration of the physico-chemical state of melts and glasses has led to predictions that volcanic glasses derived from acidic and basic lavas will alter in different ways. It is appropriate, therefore, before discussing the nature of the alteration products to see if such differences do occur. In an assessment of the volcanic contribution to South Pacific sediments Peterson and Goldberg (1962) examined the distribution of quartz and feldspars over this region, and their results together with those from the work of Peterson and Griffin (1964) allow a comparison to be made between the alteration products of the two types of glass.

Peterson and Goldberg found that over most of the South Pacific calcic plagioclase was the predominant feldspar type indicative of a region of basic volcanism. In contrast, on the East Pacific Rise between 40° and 50° S high temperature alkali varieties were the dominant feldspars, and together with quartz, delineated a province of acidic vulcanism. The volcanic glass associated with the calcic plagioclase region was hydrated, dark reddish brown in colour and often altered. It is this type of hydrated basic glass which is usually referred to as palagonite. In contrast, colourless acid glass was associated with the quartz-alkali feldspar volcanic region. Because the alteration of volcanic debris to montmorillonite has been well documented in the literature since Murray and Renard (1891) first reported it, Peterson and Griffin (1964) examined the relationship between the predominant feldspar the type of glass and the abundance of montmorillonite to establish whether the clay mineralogy of the sediment is controlled by the dominant type of volcanism. Their results, which are shown in Fig. 27.7, demon-

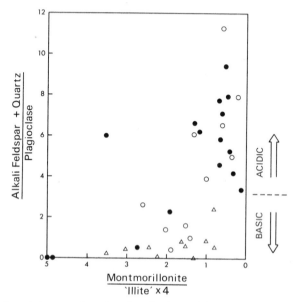

FIG. 27.7. Relationship between clay minerals and volcanic debris in Southeast Pacific sediments. Solid circles = no glass identified; open circles = presence of glass ($\eta < 1\cdot54$), triangles = presence of glass ($\eta > 1\cdot54$). Modified from Peterson and Griffin (1964). $\eta$ = refractive index. The clay ratio, including the factor of 4, is used to represent the enrichment of montmorillonite in the sediment.

strate that basic volcanic debris is associated with the presence of abundant montmorillonite in the sediment and that acidic volcanic debris is found with relatively small amounts of montmorillonite. Clearly therefore, as predicted theoretically, it is the devitrification of basic volcanic glass (pala-gonite) which results in authigenic mineral formation whereas acid volcanic glass is apparently less affected. However, not all palagonite is equally altered. From a study of South Pacific deep-sea basalts Bonatti (1967) suggested two different ways in which the emplacement of lava on the sea floor can result in different degrees of palagonitization: (1) by quiet effusion, when only a thin crust of glass is formed on immediate contact with sea water, in this instance the lava cools below this crust and so is not significantly altered; (2) by explosive reaction, causing thermal shattering and pulverization of the lava on contact with sea water. The large surface area of the resultant volcano-clastic deposit ("hyaloclastite") favours hydration and extensive palagonitization of the lava while it is still hot. In contrast, non-explosive lava flows are relatively unaffected by sea water (Fig. 27.8). Bonatti has suggested that viscosity is one factor determining the mode of eruption of the lava. The lavas of the non-explosive type were probably very fluid at the time of eruption compared with the more viscous melts which reacted with explosive shattering. This has led to the suggestion (Bonatti and Arrhenius, 1970) that relatively viscous lavas of intermediate (andesite-trachyte) composition should be more extensively altered than basaltic lava, which is incompatible with the devitrification mechanism discussed in Section 27.4.2.1. However, as Bonatti (1967) has clearly shown (by the absence of first generation phenocrysts in the non-explosive lava indicating liquid ejection on the ocean floor) the viscosity differences between the two lava types were related to temperature and not to composition, and in fact the hyaloclastic lavas were more basic than the non-explosive lavas which contain less network-forming cations and more network-modifying cations. Shattering occurs when cooling is rapid at temperatures in the annealing range (Marshall, 1961), and so it is possible that the viscosity gradient in lower temperature basaltic lavas is still significant at annealing temperatures when the anisotropic stresses caused by differential contraction will favour hyaloclastite formation. On compositional grounds extensive devitrification would not be expected for rhyolitic lavas, and there is a virtual absence of acid hyaloclastites (Pichler, 1965).

### 27.4.2.3. Smectites and Zeolites

The authigenic mineral assemblage resulting from lava–sea water interaction is predominantly palagonite–smectite–zeolite. Palagonite is formed by hydration during cooling of lavas and its low temperature transformation leads to the formation of smectites and zeolites. In addition, smectites and certain varieties of zeolite may be formed at high temperatures during the palagonitization process itself. In this latter instance, these minerals are more the products of hydrothermal reaction than they are of a halmyrolytic process. The mechanisms of crystallization are not yet understood, but the evidence of an association with palagonite is exceptionally clear. For example, Bonatti (1963) has given photomicrographic evidence of various stages of the alteration process from palagonite grains without apparent zeolite growth, through the formation of zeolite nuclei, to the full development of twinned crystals of phillipsite at the expense of palagonite. Similarly, Nayudu (1964) has described the Horizon nodule as having a nucleus predominantly composed of phillipsite which is growing out of palagonite. Other authors (e.g. Morgenstein, 1967) have observed the direct alteration of palagonite into

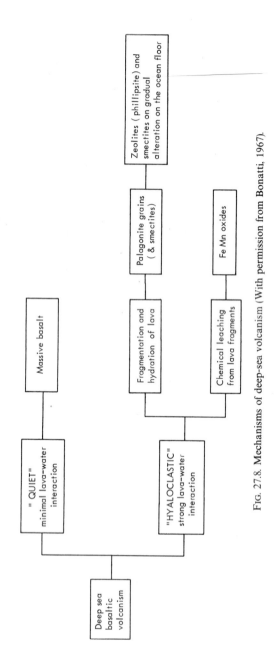

FIG. 27.8. Mechanisms of deep-sea volcanism (With permission from Bonatti, 1967).

montmorillonite and goethite. The regional distribution of montmorillonite and phillipsite in Pacific Ocean sediments and the coincidence of these minerals with a province of basic volcanism implies an origin from igneous silicate phases.

Not all zeolites in marine sediments have such an obvious volcanic origin. Arrhenius (1963) has reported microcrystalline zeolite within dissolving siliceous skeletal material and has suggested that high concentrations of biogenous silica in interstitial solutions will favour zeolitization. Three types of zeolites are therefore apparently observed in marine sediments: (1) high-temperature volcanic ones, formed by the interaction between sea water or interstitial water and hot basaltic lava during its intrusion; (2) low-temperature volcanic types, formed from decomposing igneous silicates and volcanic glass and ash; and (3) non-volcanic ones, favoured by high concentrations of the constituent elements in interstitial solution. Hawkins and Roy (1963) have demonstrated that the zeolite formed by lava–water interaction at high temperatures is analcime and that it is stable above 250°C below which phillipsite is the stable zeolite. For this reason, the phillipsite in South Pacific sediments is apparently of type (2) since its range of stability falls well below the temperatures at which crystal growth is maximal (see Section 27.4.2.1). The low-temperature formation of phillipsite is also compatible with the phase relationships between analcime and phillipsite since, according to Coombs $et$ $al.$ (1959), entropy considerations suggest that the least hydrated phase would be the high temperature one. For these two zeolites,

$$(0.5 \text{ Ca, Na, K})[\text{AlSi}_{1.67}\text{O}_{5.33}]2\text{H}_2\text{O}$$
$$\text{phillipsite}$$

$$+ \text{Na}^+ + 0.33\text{SiO}_2 \rightarrow \text{Na}[\text{AlSi}_2\text{O}_6]\text{H}_2\text{O} + (0.5 \text{ Ca, Na, K})^+ + \text{H}_2\text{O}$$
$$\text{analcime}$$

the large entropy of free water suggests that the above reaction must have a positive entropy change implying that phillipsite becomes more stable at lower temperatures relative to analcime. Similarly, for the heulandite–clinoptilolite group,

$$(0.5 \text{ Ca, Na})[\text{AlSi}_5\text{O}_{12}]4\text{H}_2\text{O}$$
$$\text{clinoptilolite}$$

$$\rightarrow (0.5 \text{ Ca, Na})[\text{AlSi}_{3.5}\text{O}_9]3\text{H}_2\text{O} + 1.5\text{SiO}_2 + \text{H}_2\text{O}$$
$$\text{haulandite}$$

clinoptilolite should be the stable low-temperature phase. In fact, the heulandite-clinoptilolite zeolites are rarely found in present day surface marine sediments. The zeolites identified by Biscaye (1964) in Atlantic Ocean sediments (clinoptilolite, heulandite, phillipsite) were not of Recent age, and there appears to be little zeolite in present day Atlantic sediments as compared with those of the South Pacific where the phillipsite concentration of the sediment exceeds 50% over extensive areas. (Recent results from the Deep Sea Drilling Project have shown phillipsite and other zeolites to be a widespread feature of deeper sections of the sediment column in the Atlantic and other major oceans). However, Hathaway and Sachs (1965) and Bonatti and Joensuu (1968) have identified clinoptilolite (the predicted low-temperature phase) in association with sepiolite, quartz and montmorillonite (see Section 27.4.3.5), and a halmyrolytic transformation of volcanic ash is the suggested origin in each case. The association of this zeolite with quartz is in agreement with the work of Coombs *et al.* (1959) who have shown that phillipsite is stable in a Si-deficient environment, whereas heulandite (and, by inference, clinoptilolite) can coexist with quartz (Fig. 27.9). Clinoptilolite has also been identified in diatomaceous Antarctic sediments in which there is an excess of silica in solution from the dissolution of opaline organisms (Goodell, 1965). Furthermore, phase rule considerations allow five phases to coexist at equilibrium (at any given arbitrary temperature and pressure)

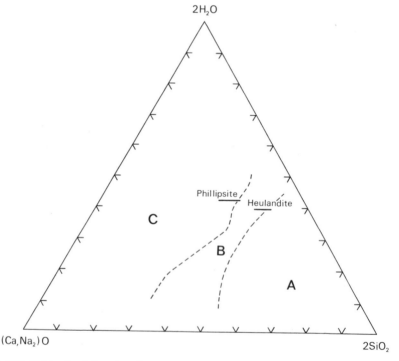

FIG. 27.9. Fields of stability of zeolites. A: field of phases favoured by supersaturation in silica; B: field of phases which can commonly co-exist with quartz (or opal); C: field of phases favoured by a Si-deficient environment. (Modified with permission from Coombs *et al.* (1959)).

with the system

$$Na_2O.Al_2O_3 - CaO.Al_2O_3 - MgO.Al_2O_3 - SiO_2 - H_2O,$$

and therefore phases can coexist with an aqueous solution. For this reason it is not necessary to postulate metastability for any one phase.

Although phillipsite appears to be quantitatively the most important zeolite in marine sediments, it has been found that other members of the phillipsite–harmotome group are widespread (Arrhenius, 1963; see also Morgenstein, 1967). Consideration of the relative stabilities of phillipsite and harmotome

$$0.5Ba[AlSi_3O_8]3H_2O + (0.5\ Ca,\ Na,\ K)^+$$
harmotome

$$\rightarrow (0.5\ Ca,\ Na,\ K)[AlSi_{1.67}O_{5.33}]2H_2O + 0.5Ba^{2+} + 1.33SiO_2 + H_2O$$
phillipsite

suggests that harmotome is in fact the stable member of this group at lower temperatures. However, Arrhenius and Bonatti (1965) have stated that the formation of harmotome is limited to the early stages of volcanic glass–sea water interaction, and that the slower-growing phillipsite engulfs harmotome nuclei (and also smectite phases). Although entropy considerations suggest that harmotome should be the stable phase, its replacement by phillipsite is probably related to the partition of Na, K, Ca and Ba between sea water and zeolite. Although distribution coefficient data are not available it seems unlikely that the high $(0.5\ Ca,\ Na,\ K)^+/Ba^{2+}$ ratio of sea water would result in a Ba-rich zeolite during slow crystal growth on the sea floor unless its water content was significantly greater than that of its Ca, Na, K-rich counterpart.

Evidence for the slow growth of phillipsite is convincing. In the South Pacific, large and abundant phillipsite crystals are present where the sedimentation rate is low, whereas on topographic highs, where biogenous carbonates increase the sedimentation rate, the concentrations of phillipsite are lower and the crystals are smaller, even where the non-carbonate fraction is dominated by palagonite and other volcanic debris (Bonatti, 1963). This is compatible with the greater abundance of phillipsite in the South Pacific (average sedimentation rate $0.45$ mm $10^{-3}$ yr) as compared with the North Pacific ($1.5$ mm $10^{-3}$ yr) and the virtual absence of phillipsite from Recent Atlantic Ocean surface sediments ($1.85$ mm $10^{-3}$ yr). In addition, the high concentrations of thorium (up to 2200 ppm), barium, zinc and other trace elements in sea-floor zeolites (Arrhenius, 1963; Bonatti, 1963) is suggestive of extensive contact with sea water. A recent experimental study by Mariner and Surdam (1970) also points to the slow precipitation of marine phillipsite as well as offering some insight into the mechanisms of its formation.

G

The more siliceous zeolites are the most common alteration products of siliceous glass in lower pH environments, and the less siliceous ones are found in higher pH environments. In view of this, the increasing solubility of silica with pH cannot be used to explain these variations, and Mariner and Surdam have suggested that the observed correlation between pH and zeolite mineralogy is due to the relative differences in the ease of dissolution of Al and Si from the glass. Although silica solubility increases with pH, in alkaline solutions the solubility of alumina increases more rapidly. For this reason the Si/Al ratio of the solution in equilibrium with an aluminosilicate glass will decrease with increasing pH (Fig. 27.10—Left). Mariner and Surdam have also suggested that zeolites formed at low temperatures crystallize from a poorly-ordered gel, the Si/Al ratio of which is controlled by the Si/Al ratio of the solution, and they have proposed that the gel results from the reaction between uncharged silicic acid and the aluminate ion to form Al-silicate chains, thus:

$$
\underset{\underset{\text{OH}}{|}}{\overset{\overset{\text{OH}}{|}}{\text{OH}-\text{Si}-\text{OH}}} + \left(\underset{\underset{\text{OH}}{|}}{\overset{\overset{\text{OH}}{|}}{\text{OH}-\text{Al}-\text{OH}}}\right)^{-} \rightarrow \left(\underset{\underset{\text{OH}}{|}\;\underset{\text{OH}}{|}}{\overset{\overset{\text{OH}}{|}\;\overset{\text{OH}}{|}}{\text{OH}-\text{Si}-\text{O}-\text{Al}-\text{OH}}}\right)^{-} + H_2O.
$$

These chains group around hydrated cations in order to attain electrical neutrality, and this leads to the formation of cage-like polyhedral zeolite units. Mariner and Surdam were concerned with the formation of zeolite in saline alkaline lakes, but the mechanism which they proposed can be applied to the marine environment. In addition, their hypothesis does not necessarily rely on a volcanic source for the zeolite components, but may be applied generally to all low-temperature (hydrogenous) marine zeolites. In their experiments Mariner and Surdam found that phillipsite crystallized from gels formed over the pH range 9·1 to 10·5, whereas no gels were obtained in the range 7 to 9 in 70 days (Fig. 27.10—Right). This is probably because of the slow rate of solution of alumina in the latter pH range and is indicative of a slow growth rate for marine phillipsites. The pH of phillipsite crystallization in marine sediments is, of course, not necessarily that in sea water since volcanic zeolites form within palagonite, and non-volcanic zeolites form from pore solutions of sediments. Comparison of the composition of marine zeolites with Mariner and Surdam's experimental data would suggest that the more siliceous clinoptilolite should be the common zeolite at the normal pH of sea water, thus emphasizing the likely formation of phillipsite from a concentrated alkaline solution between, or inside, palagonite grains. Indeed, Deffeyes (1959) has suggested that the formation of zeolites from volcanic

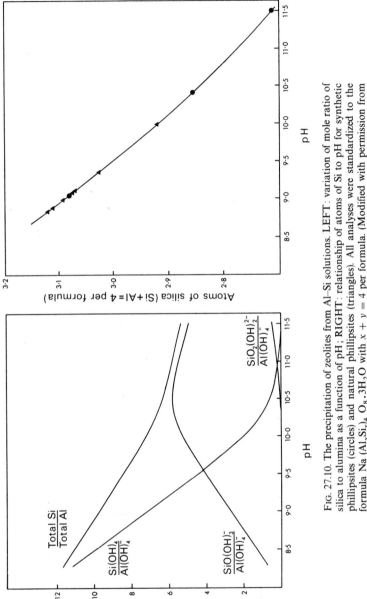

FIG. 27.10. The precipitation of zeolites from Al–Si solutions. LEFT: variation of mole ratio of silica to alumina as a function of pH; RIGHT: relationship of atoms of Si to pH for synthetic phillipsites (circles) and natural phillipsites (triangles). All analyses were standardized to the formula Na (Al$_x$Si$_{y/4}$ O$_8$·3H$_2$O with $x + y = 4$ per formula. (Modified with permission from Mariner and Surdam 1970).

glass is feasible only because glass hydrolysis results in a sufficiently alkaline pH for zeolitization.

Further research is clearly needed before the observed distribution of marine zeolites can be fully understood. For example, the work of Hawkins and Roy (1963) suggests that analcime is a high-temperature zeolite, metastable with respect to phillipsite at low temperatures. However, it is present in Recent sediments of the Gulf of Naples (Muller, 1961) together with newly formed opal, quartz and clay minerals (Section 27.4.2.5) and is probably formed at low temperatures. Similarly, analcime has been reported by Norin (1953) as an alteration product of volcanic ash. Muller (1967) has suggested that analcime and not phillipsite is formed because the parent glass had a trachyte–leucite chemistry in contrast to the basaltic composition of the Pacific palagonite. In addition, Eugster and Jones (1968) have described aluminosilicate gels formed by alkali trachyte rocks which artificially crystallized to analcime. However, huge deposits of analcime are found in Upper Jurassic and Lower Cretaceous deposits of the Central Congo Basin (Vanderstappen and Verbeek, 1964) in the absence of even traces of volcanic material.

The development of smectites from basic volcanic material is similarly a process which needs further clarification. There appears to be no doubt that montmorillonite can form directly from palagonite (e.g. Arrhenius and Bonatti, 1963; Morgenstein, 1967), and that it is a common constituent of sediments containing volcanic ash or glass. However, there is some evidence that smectites are metastable with respect to phillipsite (Arrhenius and Bonatti, 1965) or iron chlorite (Rex, 1967) under sea-floor conditions. Both high-temperature and low-temperature formations have been attributed to these minerals. The premise that certain smectites are formed hydrothermally in the contact zone between hot lava and sea water is primarily based on the extrapolation of hydrothermal synthesis data (Hauser and Reynolds, 1939; Hawkins and Roy, 1963)—See Section 27.4.2.1. A critical parameter is obviously the length of time the lava remains at temperatures at which crystal growth is maximal (see Fig. 27.5). These conditions are more likely to occur when: (1) the lava is ejected at high temperature; (2) the lava or glass is rich in volatiles, and consequently will have a high heat content compared with "drier" glasses; (3) the glass or lava is ejected in association with hydrothermal solutions. Because of the poor crystallinity and limited expandibility of most marine montmorillonites Arrhenius (1963) has suggested that they form by in situ decomposition of volcanic glass rather than by high temperature crystallization. Matthews (1962) has described submarine weathering of yellow fibro-palagonite to montmorillonite which "certainly occurred at about 0°C." In contrast, Peterson and Griffin (1964) have described the rapid formation of montmorillonite from basic volcanic material and have suggested that the reaction might occur while the material is hot. Similarly, Muller (1967) has ascribed the origin of the nontronite in Pacific sediments to direct contact between hot lava and sea water.

Experimental studies (see for example Caillère *et al.*, 1957; Pedro, 1961; Siffert, 1962) indicate that montmorillonite-type phases can be readily formed at low temperatures, although a somewhat more strongly alkaline solution than sea water is required. The development of smectites inside palagonite grains therefore seems feasible. However, Matthews (1962) has reported that montmorillonite forms by transformation of the palagonite skin on the surface of lava flows under conditions at which normal sea water pH levels prevail. The most important factor determining whether or not montmorillonite will precipitate is the availability of sufficient magnesium (Deer *et al.*, 1966), although very high magnesium concentrations lead to the formation of sepiolite (Millot, 1970; see also Section 27.4.3.5). Before questions such as this can be resolved with certainty the relationships between marine zeolites, smectites and other authigenic silicates must be determined.

### 27.4.2.4. *Iron and manganese oxides*

The alteration products of lava–sea water reactions are frequently associated with deposits of oxides of iron and manganese. These oxides occur in association with basalts and basaltic glass both as exterior features (e.g. crusts of goethite and manganese oxide on the surfaces of lava flows, manganese nodules with volcanic glass interiors, hydroxy–iron interlayers in smectites, limonite, goethite and ferro-manganese cement etc.), and interior features (e.g. segregations of goethite inside palagonite grains and at grain boundaries, dendritic manganese structures, opaque ferro-manganese oxide particles and crystal needles and rods of black opaque manganese oxide inside glass, etc.). Bonatti (1967) has observed that such oxides are more usually found in association with hyaloclastic lavas than they are with non-explosive lava flows (see Section 27.4.2.2). He has suggested that manganese and ferrides are leached from lavas at high temperatures so that the more comminuted lava resulting from hyaloclastic volcanism suffers extensive chemical leaching, whereas lavas formed by quiet effusion do not. This view has been questioned on many grounds, particularly because of the large quantities of basalt needed to achieve the required mass balance and the absence of manganese-deficient decomposition residues (see Chapter 28 for a detailed discussion of authigenic ferro-manganese oxide minerals). Since analyses of such iron and manganese oxides give equivocal evidence for their origin it is more appropriate to see if changes can be observed in the composition of ocean-floor basalts during low-temperature weathering (Table 27.13). The difficulty in interpreting some of these results in the context of halmyrolysis is that, for some of the rocks studied, chemical

TABLE 27.13

*Compositional changes caused by submarine weathering of ocean-floor basalts*

| Source* | Basalt increase (+) | Basalt decrease (−) |
|---|---|---|
| 1 | + H₂O + Fe₂O₃ | − Mg                           − Fe |
| 2 | + Fe + K | − Mg − Ca − Na |
| 3 |  | − Ca − Na |
| 4 |  | − Ca − SiO₂ |
| 5 | + K + Rb + Cs |  |
| 6 | + H₂O + Fe₂O₃ |  |
| 7 | + K + Rb | − Ca                 − Sr |
| 8 | + K + Rb |  |
| 9 | + H₂O + Fe + K        + Ti + Mn + P + Na | − Mg − Ca − Si |
| 10 | + H₂O + K | − Mg − Ca − Na − Si        − Fe − Mn |

* data from: 1. Engel *et al.* (1965); 2. Nicholls (1963); 3. Moore (1966); 4. Hart and Nalwalk (1970); 5. Hart (1969); 6. Miyashiro *et al.* (1969); 7. Cann (1969, 1970); 8. Philpotts *et al.* (1969); 9. Hart (1970); 10. Hart (1973).

differences will reflect metamorphic alteration as well as low-temperature weathering. For example, the loss of silica and magnesium to sea water from the basalts may be due to the alteration of olivine and clinopyroxene to smectites and chlorites, although some of the silica lost is probably part of the "excess silica" formed by the devitrification of volcanic glass (Section 27.4.2.5). Calcium may be lost by conversion of anorthite to smectite, although greenschist facies metamorphism leads to a depletion of calcium sites in rocks (Cann, 1970). The increase in potassium may reflect the replacement of plagioclase by muscovite or orthoclase (see Section 27.4.2.6), and could also account for part of the calcium loss.

The most comprehensive data on the weathering of sea floor basalts has been provided by Hart (1970, 1973) who has calculated the temporal basalt weathering exchange with sea water (Table 27.14). He used two different

TABLE 27.14

*Basalt weathering exchange with sea water\**

| Method | 1 | 2 |
|---|---|---|
| constant: | $Al_2O_3$ (assumed) | $Al_2O_3$<br>$TiO_2$ |
| gained from sea water | 1·6 $TiO_2$<br>3·7 $H_2O$<br>0·9 $K^+$<br>3·3 $Fe^{2+}$<br>0·06 $Mn^{2+}$<br>0·21 $Na^+$ | 1·98 $H_2O$<br>0·45 $K^+$ |
| lost to sea water: | 7·4 $SiO_2$<br>2·1 $Ca^{2+}$<br>2·3 $Mg^{2+}$ | 0·45 $Fe^{2+}$<br>0·006 $Mn^{2+}$<br>0·25 $Na^+$<br>2·2 $SiO_2$<br>1·42 $Ca^{2+}$<br>0·92 $Mg^{2+}$ |

* Data from Hart (1970, 1973). Methods: (1) by analysis of surface samples, (2) using seismic velocities (see text for details); units are $10^{-9}$ g cm$^{-3}$ yr$^{-1}$.

methods: (1) Comparison of the compositions of progressively-altered basalts as a function of their distance from mid-ocean ridge spreading centres (Hart, 1970); (2) Comparison of the compositions of basalts of known seismic velocity on the assumption that the rate of change of seismic velocity as a function of age represents an integrated value for their rate of alteration (Hart, 1973). These methods give good agreement for most of the major elements in the rocks with the exception of iron, manganese and sodium. Hart's opinion is that the rates calculated using the seismic data technique should be the more reliable ones, a conclusion which is supported by a microprobe study by Melson and Thompson (1972) who showed that the clay alteration products of basalts are depleted in manganese and sodium relative to the initial glass. Because of this, the halmyrolysis of basalt may provide a source of iron and manganese in sea water. The association of both iron and manganese oxides with hyaloclastic lavas may be a consequence of the more significant hydration, and larger surface area, of this type of basalt compared with those of the non-explosive type. This would enhance the amount, and possibly the rate, of manganese precipitation from sea water onto the lava surfaces (Elderfield et al., 1972).

### 27.4.2.5. Silica

Quartz is a common component of sediments in regions of acid vulcanism (Section 26.4.2.2). However, this material is not a low-temperature sediment component. Cristobalite is known to be a common product of the devitrification of silica glass (Rieck and Stevels, 1951) but, although there are many reports of geologically older silica deposits, e.g. the Upper Jurassic–Middle Eocene North Atlantic deep-sea cherts (Ewing et al. 1969; Peterson et al., 1970; Calvert, 1971) in which the "volcanic association" silica–zeolite–smectite–(sepiolite) is found, hydrogenous $SiO_2$ minerals have only rarely been reported in Recent marine sediments.

Up to 10% quartz, opal and chalcedony is present in sediments of the Gulf of Naples, and Muller (1961) has described this material as a product of Recent low-temperature neoformation resulting from decomposition of volcanic glass. Because the alteration products of volcanic glass (zeolites and clay minerals) are always poorer in silica than is the original glass, Muller (1967) has argued that large quantities of silica must be released during the transformation process. If this takes place very rapidly Muller has suggested that all the silica liberated will not be able to dissolve in the sea water and will be partially fixed in the sediment. This seems an unlikely hypothesis unless the excess silica is somehow prevented from being mixed with normal sea water as it would otherwise surely be dispersed. Rex (1967) has observed that vesicles are preserved in a Pacific Ocean tuff breccia as opaline spherules which originally formed as linings of the gas vacuoles in the basaltic glass. He has suggested that one of the first stages of glass alteration involves the diffusion of silica into these vesicles where thin shells of opal are deposited. The precipitation of silica in a closed micro-environment, and its later transference to the sediment on further alteration of the glass, offers a reasonable mechanism for the preservation of this excess silica. This is analogous to the formation of the majority of Recent volcanic silica deposits. These result when hot spring waters debouch into restricted environments. On cooling these waters become saturated with respect to amorphous silica because they have not been significantly diluted by outside waters which are undersaturated in silica. Relative to normal sea water, silica is significantly enriched in interstitial solutions of most marine sediments (Siever et al., 1965; Fanning and Schink, 1969; Bischoff and Ku, 1970), but its concentration rarely approaches the equilibrium solubility of about 60 ppm for amorphous silica in low-temperature bottom waters.

### 27.4.2.6. Authigenic feldspars

Low-temperature feldspathization is not of quantitative significance in the sea-floor environment. Authigenic orthoclase has been reported by Mellis (1952, 1959) and Matthews (1962) as being formed by replacement of calcic plagioclase. If the orthoclase is a low-temperature phase it probably re-crystallized under highly alkaline conditions as it is metastable with respect to zeolites below pH 9 at low temperatures (Arrhenius, 1963). However, Matthews has suggested that feldspathization may have occurred at moderate temperature (less than 200°C) and this pH constraint may not necessarily be applicable.

Minor amounts of other alteration products of submarine volcanism are occasionally found in marine sediments. The various descriptions (e.g. Matthews, 1962; Bonatti, 1967) of altered submarine lavas are good pointers to the types of product which are usually released to the sediment and to those which are not easily available for sedimentation. Matthews, for example, has described the alteration of the glassy interior of a lava flow to chlorophaeite, which is replaced by chlorite, montmorillonite, limonite and leucoxene. This material is much less accessible than are the altered surfaces of lavas, and only rarely will it be found in the associated sediment. For this reason montmorillonite, but not chlorite, is found in sediments near the sites of basic volcanic activity. Hydrogenous carbonates and hormites may be formed as a result of lava–sea water interaction. However, they may also be precipitated as a result of hydrothermal injection and for this reason they are discussed in Section 27.4.3.

### 27.4.3. REACTIONS INVOLVING HYDROTHERMAL AND GEOTHERMAL SOLUTIONS

### 27.4.3.1. The origins and compositions of thermal solutions

Early suggestions (e.g. Von Gumbel, 1878) that submarine hot springs are the source of certain elements found in marine sediments have been strengthened by recent discoveries of authigenic mineral deposits above active divergent plate margins such as those of the East Pacific Rise (Arrhenius, 1952; Skornyakova, 1964; Boström and Peterson, 1966, 1969) and the Red

Sea (Degens and Ross, 1969). These two regions provide an interesting comparison because, although both deposits are thought to form by authigenic precipitation resulting from the upward percolation and subsequent discharge onto the sea floor of subterranean solutions, different sources and types of thermal emanation have been proposed to explain the chemistry and mineralogy of the resultant precipitates. The Red Sea deposits have formed from geothermal brines with apparently little, or no, contribution from volcanic or magmatic sources (Bischoff, 1969; Craig, 1969), whereas the East Pacific Rise deposits appear to have formed from hydrothermal solutions with contributions from both volcanic and magmatic sources (Boström et al., 1971). Since hydrogenous reactions occur after discharge of the thermal solutions into sea water, the postulated sources of such solutions are of interest in this context only in respect to their bearing on the composition of the hydrogenous sediment, and for this reason will be discussed only briefly.

Since there is considerable confusion in the literature as to the meanings of the terms *geothermal* and *hydrothermal* solutions a simple distinction will be made for the purpose of this discussion. *Geothermal solutions* will be considered to be those formed by thermal—but not chemical—transfer from a geothermal heat source, whereas both thermal and chemical transfer are involved in the formation of *hydrothermal solutions*. In the schematic model for the origin of a hydrogenous deposit from a geothermal solution shown in Fig. 27.11 (Upper) convective flow is induced by a heat source. This causes the inflow of water, which sinks through a permeable sedimentary basin, is heated at depth, and as a result rises and is discharged as a brine. Authigenic minerals are then precipitated from the brine either by reaction with sea water, or by cooling which causes the continuous release of metals from their chloride complexes. The geothermal solution may become mineralized by extraction of metals from the sediments or rocks it encounters, or it may be largely hot super-saline sea water. The thermal solutions associated with the hydrogenous deposits of hydrothermal origin may also have some contribution from volcanic and magmatic sources. The schematic model for this type of deposit (Fig. 27.11—Lower) indicates that there may be a "primary" contribution, as a result of expulsion of a hydrothermal fluid during degassing of the mantle and/or during crystallization of magma. In addition, there may be "secondary" contributions arising from metamorphism and hydrothermal leaching of basalt during its slow cooling by removal of elements which are fractionated initially into residual siliceous fluids.

The chemistry of a geothermal solution is determined by the sediment or rock through which it percolates. For example, the brines which discharge into the axial deeps of the Red Sea are probably composed of metal–chloride

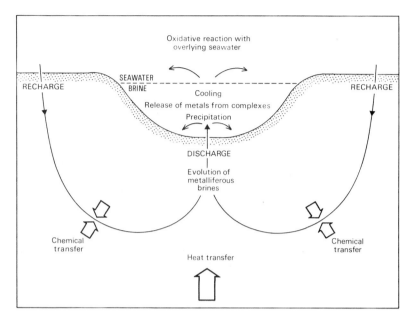

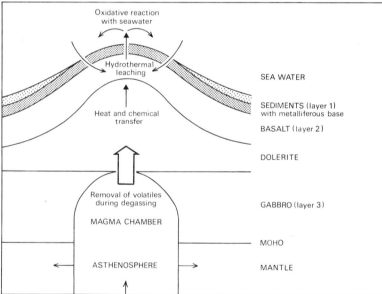

FIG. 27.11. Schematic models for the generation of metalliferous sediments in regions of high heat flow. UPPER: the origin of a hydrogenous deposit from a geothermal solution (the "Red Sea model"). LOWER: the origin of a hydrogenous deposit from a hydrothermal solution (the "East Pacific Rise model").

and metal–bisulphide complexes of low Eh and near-neutral pH (Bischoff, 1969); the chloride salts being derived from evaporite beds and the heavy metals from shale bands, both of which occur in the path of the brine's "convective cell." The chemistry of hydrothermal solutions is somewhat more predictable since the processes involved in their formation can be evaluated. Degassing of the mantle during primary differentiation of pyrolite may release volatiles such as Hg, As, Cd and B (Boström et al., 1971) which are among the "excess elements" of Horn and Adams' (1966) geochemical balance. A comparison of the compositions of submarine silicic and basaltic rocks suggests that Mg, Ca, Co, Cr, Cu and Ni may be released during the formation of acidic differentiates; primary and retrograde greenschist meta-morphism may release Ca, K and Fe (Deffeyes, 1970; Hart, 1973); hydro-thermal leaching may cause mobilization of elements such as Fe, Cu and Ni which form residual liquids and immiscible sulphide components in slowly-cooling basalts (Corliss, 1971; see also Elderfield (1975) for a review of these volcanogenic processes).

### 27.4.3.2. Iron and manganese minerals

Hydrous oxide minerals of iron and manganese are present in ridge sedi-ments from all the major oceans, and this association has naturally led to suggestions (see for example: Skornyakova, 1964; Boström and Peterson, 1969; Boström et al., 1969, 1971; Horowitz, 1970) that there is a genetic link between authigenic ferro-manganese oxide precipitation and ridge volcanism. These suggestions are strengthened by the direct observation of precipitation of iron and manganese hydroxides from hydrothermal solutions above an active submarine volcano (Zelenov, 1964), by the enrichment of iron and manganese in marine sediments near hot springs above active subduction zones (Honnorez, 1969; Baas Becking Laboratory, Ann. Rept., 1971) and by the high concentrations of manganese and iron found in sea water immedi-ately prior to a volcanic eruption (Elderfield, 1972b). Of all the ridge sedi-ments studied those most likely to have a hydrothermal source are found on the East Pacific Rise. These sediments are significantly richer in iron than are other ridge sediments and also contain high concentrations of trace elements, including the volatiles boron, arsenic, cadmium and mercury (Section 27.4.3.1). There is no doubt that submarine volcanism significantly affects the composition of sediments from the South Pacific Ocean (see Section 27.4.2.2), and isotopic evidence for the East Pacific Rise sediments (Bender et al., 1971; Dasch et al., 1971) indicates that volcanic processes have influenced the sediment chemistry. On these, and other grounds, it has been suggested (Boström et al., 1971) that the authigenic mineral deposits of the East Pacific Rise have precipitated from hydrothermal solutions which have been released at the ridge crest and which contain the enriching metals, some of which originate as carbonate-rich solutions formed during degassing of the mantle, and some as a result of shallow hydrothermal leaching of basalt. However, there is no conclusive evidence that all ferro-manganese ridge deposits were formed exclusively by precipitation from hydrothermal solutions. Other, non-volcanic, theories (Wedepohl, 1960; Turekian, 1967, 1966—see Section 27.3.3.2) have been proposed to explain the composition of the hydrogenous phases in mid-Atlantic sediments which are significantly different from those of the East Pacific Rise. Cobalt and nickel, which are enriched in mid-ocean sediments from the Atlantic (Turekian and Imbrie, 1967), are depleted in sediments from the crest of the East Pacific Rise compared to those from its flanks (Boström and Peterson, 1969). Further-more, the Atlantic ridge sediments are not enriched with hydrogenous iron oxides (Chester and Messiha-Hanna, 1970) to anything like the extent to which they are in those from the East Pacific Rise. Although volcanic and/or hydrothermal processes are obviously important in ocean-ridge sedimenta-tion it is clear that some caution must be exercised in making ocean-wide generalizations based on evidence from one area.

The formation of authigenic iron and manganese minerals from geothermal brines is known to have occurred in the Red Sea deposits (Bischoff, 1969) in which iron montmorillonite and iron hydroxides, oxides and sulphides, and manganese oxides and carbonates have been detected. The sequence of formation of the iron-bearing minerals is illustrated in Fig. 27.12. Both iron and manganese originate from the brine; however, their enrichment in the brine as compared to normal sea water is unusually high and the ultimate origin of these elements is uncertain (Craig, 1969). There is a possibility of some magmatic contribution (Emery et al., 1969; Brooks et al., 1969), as well as of solution from halite–sylvite zones of evaporite and salt dome deposits (Craig, 1969).

### 27.4.3.3. Barite and celestobarite

Authigenic barite is not a common component of sediments in most oceanic regions, and its concentration is rarely more than one per cent of the carbonate-free sediment. Biogenous barite is found in sediments beneath areas of high organic productivity; it is formed apparently during bacterial oxidation of particulate organic matter (Chow and Goldberg, 1960; Arrhenius, 1963; Turekian, 1968). High concentrations (up to 10% of the carbonate-free sediment) of barite are present in the sediments of the East Pacific Rise. This barite contains about 5 mole % celestite in solid solution and its similar distribution pattern to that of hydrogenous iron oxides may indicate an hydrogenous origin related to hydrothermal injection at the crest of the Rise (Arrhenius and Bonatti, 1965; Boström and Peterson, 1966; Boström et al., 1971). In addition, Dunham and Hanor (1967) have found

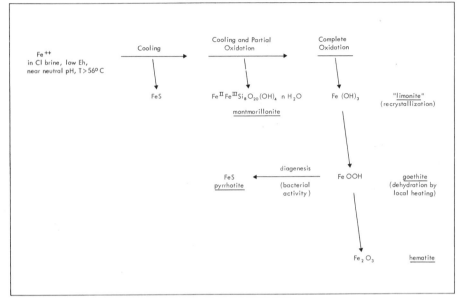

FIG. 27.12. Pathway of formation of hydrogenous iron-bearing minerals from the Red Sea brine (With permission from Bischoff, 1969).

high barite concentrations along the continental extension of the East Pacific Rise in western North America, and Hanor and Brass (1968) have observed high levels of barium in subsurface Tertiary deposits about 2200 km from the Rise. The location of these are consistent with estimates of spreading rates in the South Pacific (Bonatti and Arrhenius, 1970).

### 27.4.3.4. Carbonate minerals

Biogenous carbonates are a common component of marine sediments, and they usually occur as unconsolidated oozes which are concentrated below regions of high productivity and beneath waters shallower than the carbonate compensation depth. Lithified carbonates and recrystallized limestones have been occasionally identified in marine sediments (see for example Murray and Lee, 1909; Hamilton and Re, 1959; Fischer and Garrison, 1967; Thompson et al. 1968); and inorganically precipitated crystals of calcite, aragonite and dolomite have been found in the deep-sea environment (Bonatti, 1966; Thompson, 1972). The general recognition of these sediment types has led to a variety of theories for their origin which often emphasise the role of submarine volcanism. These volcanic-linked theories are of two main types: (1) precipitation from hydrothermal solutions, either directly from carbonate-rich solutions (Bonatti, 1966; Thompson, 1972), or indirectly by processes such as the re-precipitation of biogenous carbonates previously dissolved either by magmatic acidic volatiles, or as a consequence of the raising of the temperature of the bottom waters at the site of a submarine eruption (Bonatti, 1966); (2) precipitation from carbonate-saturated interstitial solutions formed as a result of basalt weathering (Murata and Erd, 1964; Bischoff and Ku, 1970; Thompson, 1972).

The crystalline aragonite described by Thompson (1972) is a convincing example of a product of the first of these mechanisms. Aragonite is commonly found in association with serpentinized peridotite, both as veins running through the rock and as surface precipitates. Such aragonites are enriched in Cr, Cu, Fe, Mn, Ti, Ni and Zr relative to those of shallow-water origin. Thompson has suggested that the aragonite precipitates from cold carbonate-rich solutions and since serpentinization may be caused by the interaction of the peridotite with volatiles from degassing of the mantle (Thompson and Melson, 1970), the carbonate similarly may have a primary origin. The interstitial waters of a serpentinizing peridotite are probably enriched in magnesium (Thompson, 1972), and the inhibiting effect of Mg ions on calcite precipitation (Bischoff and Fyfe, 1968—see Section 27.5.2) may explain the formation of the metastable polymorph. The presence of lithified carbonates is often related to the submarine weathering of basalt. The close association of lithified carbonates and basement rock which has been recognized by many workers (see e.g. Thompson, 1972) suggests that lithification takes place by the interaction of biogenous carbonate sediment with interstitial waters from which calcium carbonate precipitates following supersaturation caused by hydrolysis of basalt. Bischoff and Ku (1970) have suggested that such supersaturation is caused by the release of hydroxyl ions to the interstitial solutions of sediments during low-temperature weathering of submarine igneous rocks. Thompson (1972) has concluded that the resultant increase in pH is probably sufficient to cause lithification of some foraminiferal oozes, and that recrystallized limestones possibly represent an end stage of the lithification process.

It is clear that the formation of such inorganic and lithified carbonates is not yet understood. Bartlett and Greggs (1969) consider that the complexity of the lithification phenomenon is exemplified by the recognition of some lithified carbonates in which adjacent lithified and non-lithified layers differ in age by more than 30000 years. According to these authors "the development of a lithified carbonate rock from the soft, nonlithified ooze formed on the sea bottom must involve not one, but many, periods of lithification interrupted by periods of solution. The processes involved, lithification, solution and related chemical changes, must be random and fluctuating in intensity, duration and location".

### 27.4.3.5. *Hormites*

Although not commonly reported in marine sediments palygorskite and sepiolite (the hormites) are of particular interest because they represent a clear example of the neoformation of clay minerals in the marine environment. Occurrences of palygorskite are known in lacustrine and lagoonal sediments (see e.g. Kerr, 1937; Heystek and Schmidt, 1953), and in shallow-water marine sediments in which it is believed to form by the reaction of lithogenous clays with Mg-rich solutions (Muller, 1961). Both palygorskite and sepiolite have been found in deep-sea deposits (Hathaway and Sachs, 1965; Heezen *et al.*, 1965; Bonatti and Joensuu, 1968; Peterson *et al.*, 1970; Bowles *et al.*, 1971) usually on, or near, volcanic topographic features, and their association with smectites and volcanic ash has led some workers to conclude that they have been formed from this volcanic parent material. However, the low-temperature synthesis of sepiolite by Wollast *et al.* (1968), and the reasoning of Bowles *et al.* (1971), clearly indicate that hormites can also form by primary precipitation.

Bowles and his-co-workers have assessed the three possible origins of the hormite minerals. Using the terminology defined in Section 27.2 these are: (1) lithogenous, transported to, and deposited in, the deep-sea environment; (2) hydrogenous (halmyrolysate), formed by alteration of some parent material; (3) hydrogenous (precipitate), formed from their constituent elements dissolved in sea water. Since a lithogenous origin is most unlikely, most workers believe that these minerals have formed in the oceans by hydrogenous reactions. Bonatti and Joensuu (1968) have suggested that palygorskite is produced by the reaction of magnesium-rich hydrothermal solutions with montmorillonite, which is itself formed hydrogenously (see Section 27.4.4) by the alteration of volcanic glass according to the generalized scheme:

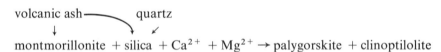

$$\text{montmorillonite} + \text{silica} + Ca^{2+} + Mg^{2+} \rightarrow \text{palygorskite} + \text{clinoptilolite}$$

Further, they have predicted that "authigenic Mg-silicates of the palygorskite–sepiolite group will be found frequently in deep-sea sediments associated with active ridges and fracture zones". Hathaway and Sachs (1965) have suggested a similar mechanism

$$\text{silica} + Mg^{2+} \rightarrow \text{sepiolite}$$

volcanic
ash

$$\text{montmorillonite} + \text{quartz} + \text{clinoptilolite}$$

The postulated mechanisms of formation from volcanic ash are based primarily on mineral associations which, in fact, are not always apparent. For example: Bowles *et al.* (1971) found no evidence of ash, smectites, etc., in the palygorskite-rich materials which they examined which were associated with quartz, calcite and dolomite. The equilibrium saturation diagram for the system $CaO—MgO—CO_2—SiO_2—H_2O$ (Fig. 27.13—Left) predicts that sepiolite may coprecipitate with quartz, quartz–calcite or calcite–dolomite. Of the five samples examined by Bowles and his co-workers three samples contained the assemblage palygorskite–quartz, one contained palygorskite–quartz–calcite, and one palygorskite–quartz–dolomite. According to Fig. 27.13—Left, the association sepiolite–quartz–dolomite cannot exist and Bowles *et al.* have suggested that the palygorskite–quartz–dolomite assemblage exists metastably, because of the extreme temperature gradients resulting from interaction of hot hydrothermal solutions with cold bottom waters (see Fig. 27.13—Right). Thus, analogy with known phase relationships

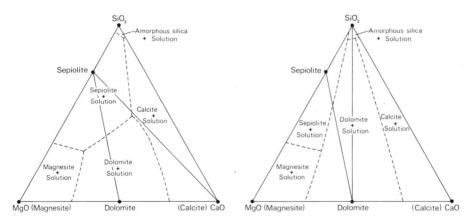

FIG. 27.13. Sepiolite stability relationships. LEFT: equilibrium saturation diagram for the system $CaO—MgO—CO_2—SiO_2—H_2O$ at 25°C and 1 atm; approximate positions of the primary field boundaries are shown by dashed lines. RIGHT: equilibrium saturation diagram modified to show the metastable extension of the pertinent boundary curves required to achieve the coexistence of sepiolite and quartz with dolomite (Bowles *et al.*, 1971).

for sepiolite, together with an absence of volcanic parent material, led Bowles *et al.* to suggest that deep-sea palygorskite may form as a chemical precipitate by the direct reaction of Mg-rich hydrothermal solutions with sea water.

The low-temperature synthesis of sepiolite was performed by Wollast *et al.* (1968) simply by adding dissolved silica to sea water at pH 8. The product which they obtained was compatible with the reaction:

$$8\,Mg^{2+} + 12\,SiO_{2(aq)} + 22\,H_2O \rightarrow (MgO)_8(SiO_2)_{12}(H_2O)_{14} + 8H^+.$$

This would give a precipitate with a $SiO_2/MgO$ ratio of 3:2 which is identical with that deduced from the theoretical composition of sepiolite $Si_{12}Mg_8O_{30}$ $(OH)_4(H_2O)_4 8H_2O$. Wollast and his co-workers used the free energy associated with this theoretical composition in a discussion of the influence of magnesium, silica and pH on the sepiolite–sea water equilibrium. Their activity diagram for the system $MgO—SiO_2—H_2O$ is illustrated in Fig. 27.14, together with some additional data for natural water compositions.

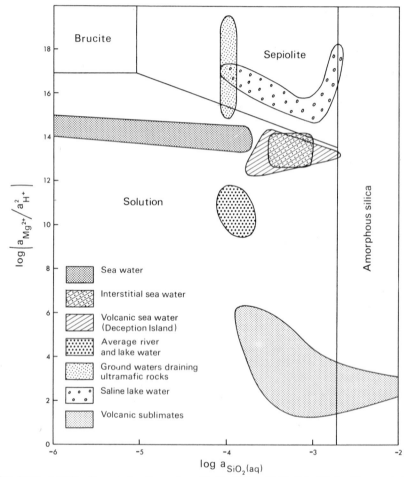

FIG. 27.14. Activity diagram for the system $MgO—SiO_2—H_2O$. (Modified with permission from Wollast *et al.* 1968).

As suggested by Bowles *et al.* (1971) the addition of Mg-rich solutions to sea water will cause sepiolite to precipitate. Wollast *et al.* concluded that sepiolite does not form, even metastably, when sea water is enriched with silica if the pH is "relatively low". In fact, it can be predicted from Fig. 27.14 that sepiolite should precipitate from sea water by the addition of silica provided that the pH is greater than 7·6; at a normal sea water pH of 8·3, sepiolite precipitation would require a concentration of 13·7 ppm $SiO_2$. The range of interstitial water compositions overlaps the sepiolite saturation boundary, and precipitation is likely in those sediments in which the dissolution of siliceous organisms, alteration of volcanic ash, etc., results in high interstitial silica concentrations. However, most sea waters have lower silica concentrations than these, and are undersaturated with respect to sepiolite. Some waters in volcanic regions may become enriched in silica, and therefore sepiolite may be precipitated locally if the normal sea water pH levels are maintained. If not, the waters become saturated with respect to amorphous silica unless they are also enriched with magnesium. For example the values for, "volcanic" sea waters from Deception Island (Elderfield, 1972b) fall within the silica saturation boundary rather than within that of sepiolite; this is in agreement with the observed precipitated phase. Similarly, results from chemical analyses of sublimates from volcanic fumaroles (White and Waring, 1963) lie well away from the sepiolite stability field, mainly because of their low pH. In contrast, some waters draining ultramafic rocks are supersaturated with respect to sepiolite, and it is conceivable that chemical weathering of mafic minerals could result in the precipitation of sepiolite. Likewise, serpentinization of femics should release some Mg and Si to solution (Thayer, 1966) and cause precipitation of palygorskite (Bonatti and Joensuu, 1968).

Some uncertainties are inherent in predictions based on determined thermodynamic functions since Christ *et al.* (1973) have shown that the activity-product constant of poorly-crystallized natural sepiolite is considerably more positive than that of well-crystallized sepiolite and approx. 1000 times lower than that of the synthetic product obtained by Wollast *et al.* Christ *et al.* have suggested that both the high surface energy of poorly-crystallized sepiolite, and the possibility of the system being a non-equilibrium one, will influence hormite solubility, and hence the extent to which its presence in marine sediments is indicative of a particular range of pore water composition. However, the present evidence seems to indicate that sepiolite and, by implication palygorskite, can form by direct precipitation in the oceans. The association of most deep-sea hormite occurrences with tectonic features on the sea floor suggests that precipitation is most probably caused by the reaction of sea water with magnesium-rich geothermal, or hydrothermal, solutions emanating from fractures and fissures. However, precipitation is theoretically possible whenever sufficient excess of magnesium and/or silica is present in sea water under alkaline conditions. In addition, palygorskite and sepiolite may be formed by alteration of parent material, probably montmorillonite, by Mg-rich solutions during halmyrolysis.


## 27.5. PRECIPITATION OF SUPERGENE MATERIAL IN MARINE SEDIMENTS

### 27.5.1. SITES AND MECHANISMS OF PRECIPITATION

The hydrogenous supergene minerals in marine sediments were originally derived from the overlying water column. This group of minerals consists of non-volcanic precipitates, but does not include any halmyrolysates or any precipitates formed as a result of volcanic or magmatic processes. It is difficult to distinguish between a halmyrolysate and a precipitate when the role of the pre-existing sediment component in the formation of hydrogenous minerals is uncertain. Some of the processes described in this section (e.g. phosphate precipitation) rely on pre-existing sediment components as sources of elements. However, these pre-existing components are biogenous and so the hydrogenous phase may be classified as a precipitate rather than as a halmyrolysate (see Section 27.2).

The main types of precipitation processes have been described briefly in Section 27.1. Because it is necessary to exceed the solubility product some sort of closed environment is obviously favourable to the formation of a supergene mineral. This might be the intra-sediment environment, in which elements released from decomposing biogenous material are concentrated in pore solutions, or a stratified water column which tends to trap elements which diffuse from the sediment, or a partially-closed basin in which evaporation is maximal. If precipitation occurs under open ocean conditions then the precipitated elements must have undergone some change in speciation, (e.g. as a result of pH or redox changes), or have encountered changes in environment (such as those which occur when river waters enter the oceans). Alternatively, kinetic factors may have prevented earlier precipitation.


### 27.5.2. EVAPORATION OF SEA WATER

The first, and most abundant, salts formed by the evaporation of sea water are calcium carbonate, calcium sulphate and sodium chloride. The formation of these salts is followed by the complex precipitation of more soluble chlorides and sulphates of sodium, magnesium and potassium. The pathway of crystallization which follows halite precipitation is illustrated in Fig. 27.15, and follows the sequence bloedite–epsomite–kainite–hexahydrite to a theoretical composition on complete evaporation comprising halite, calcium salts and kieserite, carnallite and bischofite. Since calcium carbonate is the first mineral to precipitate from evaporating sea water it is quantitatively the most important of the evaporite minerals found in Recent shallow

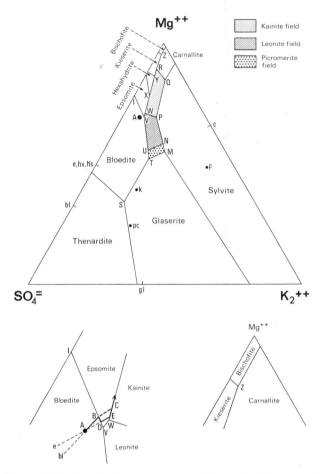

FIG. 27.15. The crystallization of salts from sea water following initial precipitation of halite. UPPER: Jänacke diagram for system $NaCl$—$KCl$—$MgCl_2$—$Na_2SO_4$—$H_2O$ at 25°C; Salt points: c = carnallite, bl = bloedite, gl = glaserite, e = epsomite, hx = hexahydrite, ks = kieserite, k = kainite, p = picromerite; A is the composition of seawater. LOWER LEFT: enlargement to show part of path of crystallization of sea water; LOWER RIGHT: enlargement to show details at the Mg corner. Possible pathways of crystallization at 25°C are ABCXYRZ and ABDEXYRZ. Modified with permission from Krauskopf (1967) after Braitsch (1962).

marine sediments. Gypsum and anhydrite are sometimes found in Recent environments and halite crusts may precipitate on intertidal flats, although the latter mineral is ephemeral and is dissolved when the surface is flooded by sea water, washed with rain or covered in dew. The only evaporite minerals which may be considered as hydrogenous components of marine sediments are those which can coexist with sea water, and for this reason the ephemeral and the more exotic evaporite salts are not discussed here. These topics have been reviewed by Braitsch (1962), Stewart (1963), Borchert (1965) and Krauskopf (1967).

Although aragonite is the first mineral to precipitate on the evaporation of sea water it can be predicted from simple thermodynamic considerations that dolomite ought in fact to be formed. Further, if $CaCO_3$ were to be precipitated, it should in fact be calcite which is less soluble than aragonite. Factors influencing the stability of the calcium carbonate minerals in sea water have been reviewed recently by Bathurst (1971) who has emphasized the important role played by magnesium ions in sea water in controlling the nature of the carbonate precipitate. As Bathurst has pointed out, the free energy change for the reaction

$$2CaCO_{3(arag)} + Mg^{2+} \rightarrow CaMg(CO_3)_2 + Ca^{2+}$$

is only $-3.99\ kcal\,mol^{-1}$, and so the predicted dolomitization of aragonite may easily be reversed. Fyfe and Bischoff (1965) have emphasised the statistical requirements necessary for growth of the highly-ordered dolomite crystal, and Lippmann (1960) and Garrels et al. (1960), among others, have recognised the significant endothermic dehydration necessary to release the $Mg^{2+}$ ion from its water dipole so that it can associate with carbonate ions to form dolomite. Lippmann has invoked the greater hydration energy of $Mg^{2+}$ compared with that of $Ca^{2+}$ to explain the inhibiting effect of $Mg^{2+}$ ions on calcite growth, and Bathurst, has shown by a simple calculation, that this factor (endothermic dehydration) acts against the free energy difference between aragonite and magnesian calcite (exothermic lattice building) to the extent that precipitation of aragonite in preference to Mg calcite is favoured by about $10\ kcal\,mol^{-1}$; whereas simple solubility considerations predict that the precipitation of Mg calcite is favoured by about $2\ kcal\,mol^{-1}$.

Aragonite mud is present, often in vast quantities, in sediments of warm, shallow, marginal seas such as those of the Bahamas–Florida platform and the Trucial coast of the Persian Gulf. Since the detection of inorganically precipitated aragonite is indirect, being based on circumstantial chemical evidence, it is often difficult to distinguish it from physiologically precipitated aragonite. This difficulty is typified by studies of Bahamian aragonite mud. Calcium carbonate is present in the sediment as aragonite algal needles, faunal skeletons, oolitic coats, grapestone matrix etc. (Bathurst,

1971), and as the growth of physiological aragonite can be easily observed, evidence that some may have formed inorganically is largely based on water chemistry. According to Cloud (1962) only 25% of the sediment is biogenic, and a large proportion of the sediment must therefore have been formed by inorganic precipitation. Some data from Smith's (1940) classic study of the Bahama Bank are illustrated diagrammatically in Fig. 27.16. A combination of evaporation over the whole area and restricted mixing with ocean water gives rise to a salinity gradient over the Bank (Fig. 27.16—Left),

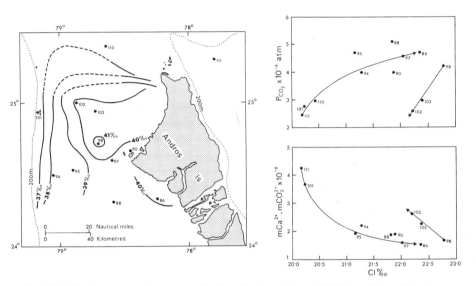

FIG. 27.16. Carbonate precipitation on the Great Bahama Bank. LEFT: isohalines off Andros Island; UPPER RIGHT: relationship between $P_{CO_2}$ and chlorinity; LOWER RIGHT: relationship between $m_{Ca^{2+}} \cdot m_{CO_3^{2-}}$ product and chlorinity. (Modified with permission from Bathurst 1971; after Smith, 1940).

and the increase in salinity away from the ocean is associated with a corresponding rise in $P_{CO_2}$ (Fig. 27.16—Upper right) and a fall in the product $m\,Ca^{2+}\;m\,CO_3^{2-}$ (Fig. 27.16—Lower right) as aragonite is precipitated. In a later study, Broecker and Takahashi (1966) showed that the Bank water is supersaturated, with aragonite by as much as 15%, and that the rate of calcium carbonate precipitation is proportional to the degree of supersaturation. This is important circumstantial evidence for the inorganic precipitation mechanism because this relationship is not found when precipitation occurs physiologically, but is if aragonite is being formed inorganically (Bathurst, 1971).

Further evidence for the inorganic precipitation of aragonite can be seen from studies of lagoonal muds of the Abu Dhabi region of the Persian Gulf. Aragonite is present as small needles and grains which contain *ca.* 9400 ppm of strontium (Kinsman 1966, 1969). This level is similar to that in inorganic-ally precipitated aragonite (8200 ppm) which Kinsman predicted on the basis of the experimentally determined partition coefficient at 29°C. In contrast, the local molluscan aragonite has much lower strontium concentrations (1500–2000 ppm), and although local corals contain *ca.* 7755 ppm Sr they are too scarce to be a source of the large amounts of fine-grained aragonite mud. Unfortunately, the strontium partition method can only give an indication of whether an aragonite is organic or inorganic in origin. Endolithic algae are very active in the Abu Dhabi region, and it is possible that aragonite precipitation in the lagoonal muds is physiologically controlled (Kendall and Skipworth, 1969), or that it occurs inorganically and is subsequently modified by local microbiological factors (Bathurst, 1971).

Gypsum starts to precipitate from seawater at salinity 117‰ and, like dolomite, tends to form in ephemeral lagoons, lakes and tidal flats in which flood waters evaporate. The relationship between authigenic mineral precipitation in lagoonal sediments and that in ephemeral waters can be illustrated in the Abu Dhabi region where halite, aragonite, gypsum, anhydrite, dolomite, celestite, magnesite and huntite are found in the *sabkha* which are a band of salt flats inland from algal flats. The patterns of authigenic mineral precipitation can be recognized from the chemistry of the brines. In lagoonal waters having a Mg:Ca atomic ratio of *ca.* 5·5 aragonite is precipitated, whereas on the algal flats this ratio increases to about 11 and gypsum is precipitated. At such high Mg:Ca ratios aragonite is no longer stable and an exchange reaction takes place between the magnesium ions in the brine and the calcium of aragonite leading to dolomitization in the sabkha environment and also to the precipitation of anhydrite.

### 27.5.3. PRECIPITATION OF MICRONUTRIENT ELEMENTS

#### 27.5.3.1. *Phosphates*

Most of the phosphate in marine sediments, occurs as discrete nodules, concretions or phosphorite rock, in which it is present as carbonate fluor-apatite. In addition, phosphate may be found as a sorbed phase, particularly on iron-rich sediments. Sedimentary apatites have the general formula $Ca_{10-a-b} Na_a Mg_b (PO_4)_{6-x} (CO_3)_x F_y, F_2$, where $x$ ranges from near zero to 1·5, $y \sim 0.4x$ and $b \sim 0.4a$ (McClellan and Lehr, 1969). There is a voluminous literature on phosphate minerals including recent reviews of both marine phosphate deposits (Tooms *et al.*, 1969), and the influence of physico-chemical factors on apatite precipitation (Gulbrandsen, 1969). Tooms *et al.* have discussed the mechanisms of phosphate formation in terms of: (1) chemical precipitation; (2) inorganic replacement; and (3) organic activity. However, precipitation brought about by organisms is often difficult to distinguish from that produced by mechanisms (1) and (2). To these three mechanisms some authors (e.g. Mansfield, 1940) also add vulcanism which they think may assist apatite precipitation.

The chemical precipitation theory is in fact a polygenetic one. Kazakov (1937, 1938) was the first to relate phosphate precipitation to the upwelling of sea water enriched with phosphate from the dissolution of particulate organic matter. Upwelling is accompanied by a decrease in $CO_2$ partial pressure and by an increase in the pH of sea water which result in it becoming supersaturated with respect to carbonate fluorapatite. As a result of this, it would be expected that phosphate precipitation would occur along the eastern continental margins of the Atlantic and Pacific where a combination of trade-wind stress and the Coriolis effect causes upwelling at the shelf edges. Phosphates are indeed found in such regions, but the majority of these deposits are not geologically Recent and consequently they cannot be cited as evidence for the upwelling hypothesis. However, this hypothesis is apparently supported by two reported occurrences of Recent phosphates in sediments from the eastern continental margins of the South Atlantic and South Pacific, i.e. off South West Africa (Baturin, 1969, 1970; Baturin *et al.*, 1970a; Senin, 1970) and Peru (Baturin *et al.*, 1970b; Shishkina, 1971) respectively. These are both regions of intense upwelling and of extremely high primary productivity. The latter results in mass mortality of marine life, but the present evidence suggests that authigenic phosphate formation occurs within the sediment and is not caused by phosphate saturation in the nutrient-rich sea water. According to Baturin (1970) almost all of the phosphate in upwelling waters is utilized by phytoplankton, and the high organic productivity of these regions is quite consistent with the rate at which phosphate is supplied to them. Baturin has suggested that phosphate is precipitated diagenetically as a result of the following sequence. (1) The breakdown of phosphate-rich organic remains; (2) phosphate enrichment in interstitial waters resulting in their saturation with respect to calcium phosphate; (3) the precipitation of phosphatic gels and (4) the consolidation of the gels by dehydration and the loss of organic matter and silica.

In addition to the formation of carbonate fluorapatite by primary precipitation there is good reason to believe that a hydrogenous phosphate phase (cf. apatite) may be present in some sediments as a result of the direct removal of phosphate ions from sea water. It is well known that phosphate ions are sorbed by hydrous iron and aluminium oxides and by clay minerals (e.g. Van Olphen, 1963, Bache, 1964; Muljadi *et al.*, 1967; Pissarides *et al.*, 1968). This sorption reaction has often been represented in terms of the precipitation of iron and/or aluminium phosphates, such as (Al, Fe) $(H_2PO_4)_n$ $(OH)_{3-n}$ (Bache, 1964), which would suggest that phosphate tends to be associated with sediments as a hydrogenous precipitate or as an adsorbed phase. Goldschmidt (1954) has suggested that the high phosphate contents (1–2% $P_2O_5$) of some clay and zeolite-rich deep-sea sediments represent iron phosphate precipitation, and Berner (1973) has recently found $P_2O_5$ levels of up to 4% in the iron-rich volcanogenic sediments of the East Pacific Rise. Berner's data indicate a good correlation between iron and phosphorus with an average $Fe_2O_3/P_2O_5$ ratio of 4; this was interpreted in terms of phosphate uptake as a chemisorbed layer on fine-grained ferric oxides, and/or a discrete poorly ordered precipitate of ferric phosphate on the surface of the iron oxides. In view of the high iron to phosphorus ratio, the latter explanation seems more reasonable, and if this is the case then about 30% of the iron is present as amorphous ferric phosphate. Alternatively, the Fe-P correlation may indicate a combination of adsorption of phosphate on ferric oxide and iron-rich clay minerals, such as nontronite.

### 27.5.3.2. *Silica and silicates*
Non-volcanic precipitates of silica and silicates occur in some marine sediments. For convenience, this mode of formation has been discussed in Section 27.4, together with silica (27.4.2.5), zeolites (27.4.2.3) and hormites (27.4.3.5) of volcanically-linked origin.

### 27.5.4. OXIDATION AND REDUCTION

### 27.5.4.1. *Iron and manganese oxides*
The importance of iron and manganese oxide minerals, and their influence over the geochemistry of marine sediments, are attested by the voluminous literature on manganese nodules. The mineralogy, geochemistry and genesis of manganese nodules are described in Chapter 28 of this volume, and this topic will not be discussed here. Some aspects of the geochemistry of the ferro-manganese oxide phases in marine sediments are discussed in Section 27.3.3.2, and volcanogenic iron and manganese oxides are briefly described in Sections 27.4.2.4 and 27.4.3.2. In addition to these types, iron and manganese oxides are also precipitated in the oceans in the absence of a detrital core, rock surface or sediment. The much quoted study by Zelenov (1964) describes the precipitation of Fe and Mn oxides from metalliferous solutions emanating from the submarine Banu Wuhu volcano off Indonesia. Oxide precipitation in the absence of volcanism has been reported for stratified sea waters at the oxygen zero boundary; for example Spencer and

Brewer (1971) have described the precipitation of manganese in the waters of the Black Sea at the boundary between the oxygenated surface waters and the $H_2S$-containing deep waters as a result of its upward advective and diffusive transfer to the oxidised zone. The resulting hydrogenous phase has only a limited residence time in the oceans since the manganese oxide particles sink into the sulphide zone and are redissolved. In contrast, if the oxygen zero boundary occurs at, or below, the water-sediment interface (as it does when reducing sediments are overlain by an oxygenated unstratified water column) manganese oxides may precipitate as a crust on the sediment surface or become concentrated at horizons in the sediment itself (see e.g. Gorham and Swaine, 1965; Manheim, 1965; Li et al., 1969; Calvert and Price, 1970, 1972). All three modes of formation have been observed by Taylor (personal communication 1973) for a Norwegian fjordal environment.

### 27.5.4.2. Sulphides

Pyrite is almost invariably present in organic-rich marine sediments in which it is precipitated from the interstitial water as a result of bio-geochemical processes. The iron is supplied by detrital iron minerals which are unstable in anaerobic environments and which, as a result of both bacterial and inorganic processes, release ferrous iron to the pore waters. In the latter it reacts with dissolved $H_2S$ formed by the bacterial reduction of sulphate, and to a lesser extent by the decomposition of organic sulphur compounds. It then precipitates initially as metastable iron sulphides which subsequently transform to pyrite. Pyrite formation has been studied in detail in the Black Sea (see e.g. Ostroumov, 1953; Ostroumov et al., 1961; Volkov, 1961; Emelyanov and Shimkus, 1962), the Gulf of California (see e.g. Emery and Rittenberg, 1952; Kaplan et al., 1963; Berner, 1964a), the Wadden Sea, Netherlands (Von Straaten, 1954) and the Wash, eastern England (Love, 1967). The mechanisms of sulphide precipitation have been studied by, among others, Zobell and Rittenberg (1948), Harrison and Thode (1958a, b), Berner (1963; 1964a, c; 1967a, b; 1969; 1970; 1971) and Rickard (1969).

Berner (1971) has listed three principal factors which limit the precipitation of iron sulphide minerals in sediments. These are: (1) the concentration and reactivity of iron compounds; (2) the availability of dissolved sulphate; (3) the concentration of metabolisable organic matter. Usually, it is the reactivity of detrital iron minerals, rather than their concentration, which influences the extent of pyrite formation. Ferruginous coatings on detrital grains and iron-rich clay minerals are much more reactive than are coarse-grained iron minerals, and in addition their fine grain size aids their breakdown. It is evident that this factor might limit pyrite formation in carbonate muds and other non-lithogenous sediments, but organic iron compounds may provide an alternative source of the element. Sulphate diffuses from overlying sea water and is readily advailable. For these reasons Berner has concluded that the most important factor in pyrite formation is the concentration of available organic matter. The organic matter is utilised by sulphate-reducing bacteria to produce hydrogen sulphide. This reacts with ferrous iron to form sulphides such as mackinawite ($Fe_3S_4$) and greigite ($Fe_{1+x}S$) which are thought to react with elemental sulphur to form pyrite, as formalised by the reactions (see also Chapter 33).

$$8 \, FeOOH + CH_4 + 15CO_2 + 2H_2O \rightarrow 8Fe^{2+} + 16HCO_3^-$$
$$SO_4^{2-} + 2CH_2O \rightarrow H_2S + 2HCO_3^-$$
$$Fe^{2+} + H_2S \rightarrow FeS + 2H^+$$
$$FeS + S^0 \rightarrow FeS_2$$

Hence, pyrite is formed during early diagenesis, often within a few cm of the sediment surface. Sulphide precipitation can also occur at or above the sediment–water interface. In the Black Sea, authigenic manganese oxides precipitate above the oxygen zero boundary (Section 27.5.4.1) and sulphides of copper, zinc and iron precipitate below it (see Chapter 16; Spencer and Brewer, 1971). Since the deep waters of the Black Sea contain high concentrations of hydrogen sulphide (up to $7.3 \, ml \, l^{-1}$) it is probable that pyrite formation is controlled by the supply and distribution of detrital iron minerals rather than by the content of metabolisable organic material in the sediments (Ostroumov et al., 1961; Berner, 1971).

## REFERENCES

American Geological Institute (1962). "Dictionary of Geological Terms", 545 pp. Dolphin Books, New York.

Arrhenius, G. (1952). Rep. Swed. Deep Sea Exped. 5, Goteborg.

Arrhenius, G. (1954). Bull. Geol. Soc. Amer. 65, 1228.

Arrhenius, G. (1963). In "The Sea" (M. N. Hill, ed.), Vol. 3, pp. 655–727. Inter-science, New York.

Arrhenius, G. and Bonatti, E. (1965). In "Progress in Oceanography" (M. Sears, ed.), Vol. 3, pp. 7–22. Pergamon, New York.

Baas Becking Geological Laboratory (1971). "Annual Report", 17 pp. Canberra, Australia.

Bache, B. W. (1964). Soil Sci. 15, 110.

Ballance, P. F. (1964). J. Sed. Petrol. 34, 91.

Bartlett, G. A. and Greggs, R. G. (1969). Science, N.Y. 166, 740.

Bathurst, R. G. C. (1971). Carbonate Sediments and their Diagenesis. In "Developments in Sedimentology", Vol. 12, 620 pp. Elsevier, Amsterdam.

Baturin, G. N. (1969). Dokl. Akad. Nauk SSSR, 189, 1359.

Baturin, G. N. (1970). In "The Geology of the East Atlantic Continental Margin" (F. M. Delany, ed.), Vol. 1. Rep. No. 70/13 Inst. Geol. Sci., pp. 87–97.

Baturin, G. N., Kochenov, A. V. and Petelin, V. P. (1970). *Lithology and Mineral Resources,* 2 (quoted in Baturin, 1970).
Bell, D. L. and Goodell, H. G. (1967). *Sedimentology,* 9, 169.
Bender, M., Broecker, W., Gornitz, V., Middel, U., Kay, R. and Sun, S. (1971). *Earth Planet. Sci. Letters,* 12, 425.
Bentor, Y. K. and Kastner, M. (1965). *J. Sed. Petrol.* 35, 155.
Berner, R. A. (1963). *Geochim. Cosmochim. Acta,* 27, 563.
Berner, R. A. (1964a). *J. Geol.* 72, 293.
Berner, R. A. (1964b). *Geochim. Cosmochim. Acta,* 28, 1497.
Berner, R. A. (1964c). *Mar. Geol.* 1, 117.
Berner, R. A. (1967a) *Amer. J. Sci.* 265, 773.
Berner, R. A. (1967b). *In* "Estuaries" (G. Lauff, ed.), pp. 268–272, American Association for the Advancement of Science.
Berner, R. A. (1969). *Amer. J. Sci.* 267, 19.
Berner, R. A. (1970). *Amer. J. Sci.* 268, 1.
Berner, R. A. (1971). "Principles of Chemical Sedimentology", 240 pp. McGraw-Hill, New York.
Berner, R. A. (1973). *Earth Planet. Sci. Letters,* 18, 77.
Biedermann, G. and Chow, T. J. (1966). *Acta Chem. Scand.* 20, 1376.
Bien, G. S., Contois, D. E. and Thomas, W. H. (1958). *Geochim. Cosmochim. Acta,* 14, 35.
Biscaye, P. E. (1964). *Yale Univ. Dept. Geol. Geochem. Tech. Rep.* 8, 86 pp.
Bischoff, J. L. (1969). *In*: "Hot Brines and Recent Heavy Metal Deposits in the Red Sea" (E. T. Degens and D. A. Ross, eds), pp. 368–401. Springer-Verlag, Berlin.
Bischoff, J. L. and Fyfe, W. S. (1968). *Amer. J. Sci.* 266, 65.
Bischoff, J. L. and Ku, T. (1970). *J. Sed. Petrol.* 40, 960.
Bolt, G. H., Sumner, M. E. and Kamphort, A. (1963). *Soil Sci. Soc. Am. Proc.* 27, 294.
Bonatti, E. (1963). *Trans. N.Y. Acad. Sci.* (2) 25, 938.
Bonatti, E. (1965). *Bull. volc.* 28, 257.
Bonatti, E. (1966). *Science,* 153, 534.
Bonatti, E. (1967). *In* "Researches in Geochemistry" (P. H. Abelson, ed.), Vol. 2, pp. 453–491. Wiley, New York.
Bonatti, E. and Arrhenius, G. (1970). *In* "The Sea" (A. E. Maxwell, ed.) Vol. 4(1), pp. 445–464. Wiley, New York.
Bonatti, E. and Joensuu, O. (1968). *Amer. Mineral.* 53, 975.
Borchert, H. (1965). *In* "Chemical Oceanography" (J. P. Riley and G. Skirrow, eds), Vol. 2, pp. 205–276. Academic Press, London and New York.
Boström, K. (1967). *In* "Researches in Geochemistry" (P. H. Abelson, ed.), Vol. 2, pp. 421–452. Wiley, New York.
Boström, K. and Peterson, M. N. A. (1966). *Econ. Geol.* 61, 1258.
Boström, K. and Peterson, M. N. A. (1969). *Marine Geol.* 7, 427.
Boström, K., Peterson, M. N. A., Joensuu, O. and Fisher, D. E. (1969). *J. Geophys. Res.* 74, 3261.
Boström, K., Farquarson, B., and Eyl, W. (1971). *Chem. Geol.* 10, 189.
Bowles, F. A., Angino, E. A., Hosterman, J. W., and Gale, O. K. (1971). *Earth Planet Sci. Letters,* 11, 324.
Braitsch, O. (1962). *Mineral Petrog. Mitt.* 3, 232.
Braun, H. (1962). *Z. Erzbergb. Metalhutten wes.* 15, 613.
Broecker, W. S. and Takahashi, T. (1966). *J. Geophys. Res.* 71, 1575.
Burst, J. F. (1958a). *Bull. Am. Assoc. Pet. Geol.* 42, 310.

Burst, J. (1958b). *Amer. Mineral.* **43**, 481.

Burton, J. D. and Liss, P. S. (1968). *Nature, Lond.* **14**, 35.

Burton, J. D., Liss, P. S. and Venugoplan, U. K. (1970). *J. Cons. perm. int. Explor. Mer.* **33**, 134.

Brooks, R. R., Kaplan, I. R., and Peterson, M. N. A. (1969). *In* "Hot Brines and Recent Heavy Metal Deposits in the Red Sea" (E. T. Degens and D. A. Ross, eds), pp. 180–203. Springer-Verlag, Berlin.

Brown, G. (1953). *Clay Minerals Bull.* **2**, 64.

Caillere, S. and Henin, S. (1949). *Mineral Mag.* **28**, 612.

Caillere, S., Hening, S., and Esquevin, J. (1957). *Bull. Grpe. Fr. Arg.* **9**, 67.

Calvert, S. E. (1968). *Nature, Lond.* **219**, 919.

Calvert, S. E. (1971). *Contr. Mineral. Petrol.* **33**, 273.

Calvert, S. E. and Price, N. B. (1970). *Contr. Mineral. Petrol.* **29**, 215.

Calvert, S. E. and Price, N. B. (1972). *Earth Planet Sci. Letters,* **16**, 245.

Calvet, R. and Prost, R. (1971). *Clays Clay Minerals,* **19**, 175.

Camez, T. (1962). *These Sci., Strasbourg et Mem. Serv. Carte Geol. Als. Lor.* **20**, 90 pp.

Cann, J. R. (1969). *J. Petrol.* **10**, 1.

Cann, J. R. (1970). *Earth Planet. Sci. Letters,* **10**, 7.

Carroll, D. and Starkey, H. C. (1960). *In* "Clays and Clay Minerals, Proc. 7th Natl. Conf." (A. Swineford, ed.), pp. 80–101. Pergamon, New York.

Carstea, D. D., Harward, M. E. and Knox, E. G. (1970). *Clays Clay Minerals,* **18**, 213.

Chester, R. (1971). *In*: "Introduction to Marine Chemistry" (J. P. Riley and R. Chester), pp. 283–421. Academic Press, London and New York.

Chester, R. and Hughes, M. J. (1967). *Chem. Geol.* **2**, 249.

Chester, R. and Johnson, L. R. (1971). *Nature, Lond.* **231**, 176.

Chester, R. and Messiha-Hanna, R. G. (1970). *Geochim. Cosmochim. Acta,* **34**, 1121.

Chow, T. J. and Goldberg, E. D. (1960). *Geochim. Cosmochim. Acta,* **20**, 192.

Christ, C. L., Hostetler, P. B. and Siebert, R. M. (1973). *Amer. J. Sci.* **273**, 65.

Cloud, P. E. (1955). *Bull. Am. Assoc. Pet. Geol.* **39**, 484.

Cloud, P. E. (1962). *Geochim. Cosmochim. Acta,* **26**, 867.

Collet, L. W. (1908). "Les depots marins", 320 pp. Doin, Paris.

Coombs, D. S., Ellis, A. J., Fyfe, W. S. and Taylor, A. M. (1959). *Geochim. Cosmochim. Acta,* **17**, 53.

Corliss, J. B. (1971). *J. Geophys. Res.* **76**, 8128.

Couch, E. L. and Grim, R. E. (1968). *Clays Clay Minerals,* **16**, 249.

Craig, H. (1969). *In* "Hot Brines and Recent Heavy Metal Deposits in the Red Sea" (E. T. Degens and D. A. Ross, eds.), pp. 208–242. Springer-Verlag, Berlin.

Dasch, E. J. (1969). *Geochim. Cosmochim. Acta,* **33**, 1521.

Dasch, E., Dymond, J. and Heath, G. (1971). *Earth Planet. Sci. Letters,* **13**, 175.

Deer, W. A., Howie, R. A. and Zussman, J. (1966). "An Introduction to the Rock-Forming Minerals", 528 pp. Longmans, London.

Deffeyes, K. S. (1959). *J. Sed. Petrol.* **29**, 602.

Deffeyes, K. S. (1970). *In* "The Megatectonics of Continents and Oceans" (C. H. Johnson and B. C. Smith, eds.), pp. 194–222. Rudgers Univ. Press, N.J.

Degens, E. T. and Ross, D. A. editors (1969). "Hot Brines and Recent Heavy Metal Deposits in the Red Sea", 600 pp. Springer-Verlag, Berlin.

Degens, E. T., Williams, E. G. and Keith, M. L. (1957). *Bull. Am. Assoc. Pet. Geol.* **41**, 2427.

Degens, E. T., Williams, E. G. and Keith, M. L. (1958). *Bull. Am. Assoc. Pet. Geol.* **42**, 981.

Dewis, F. J., Levinson, A. A. and Bayliss, P. (1972). *Geochim. Cosmochim. Acta*, **36**, 1359.
Dietz, R. S. (1941). "Clay Minerals in Recent Marine Sediments", Ph.D. thesis, Univ. Illinois.
Dietzel, A. (1942). *Z. Elektrochem.* **48**, 194.
Drever, J. I. (1971a). *Science* **172**, 1334.
Drever, J. I. (1971b). *J. Sed. Petrol.* **41**, 982.
Dunham, A. C. and Hanor, J. S. (1967). *Econ. Geol.* **62**, 82.
Ehlmann, A. J., Hulings, N. C. and Glover, E. D. (1963). *J. Sed. Petrol.* **33**, 87.
Eisenman, G. (1962a). *Biophys. J. Suppl.* **2**, 259.
Eisenman, G. (1962b). *In* "Membrane Transport and Metabolism" (A. Kleinzeller and A. Kotyk, eds), pp. 163–179. Academic Press, New York and London.
Elderfield, H. (1972a). *Nature, Lond.* **237**, 110.
Elderfield, H. (1972b). *Marine Geol.* **13**, 1.
Elderfield, H. (1975). *In*: "Marine Manganese Deposits" (G. P. Glasby, ed.). Elsevier, *in press*.
Elderfield, H. and Hem, J. D. (1973). *Mineral Mag.* **39**, 89.
Elderfield, H., Gass, I. G., Hammond, A. and Bear, L. M. (1972). *Sedimentology*, **19**, 1.
Emelyanov, E. M. (1970). *In* "The Geology of the East Atlantic Continental Margin" (F. M. Delany, ed.), Vol. 4. *Rep. no. 70/16 Inst. geol. Sci.* pp. 97–103.
Emelyanov, E. M. and Shimkus, K. M. (1962). *Okeanologiya*, **2**, 1040.
Emery, K. O. and Rittenberg, S. C. (1952). *Bull. Am. Assoc. Pet. Geol.* **36**, 735.
Emery, K. O., Hunt, J. M. and Hays, E. E. (1969). *In* "Hot Brines and Recent Heavy Metal Deposits in the Red Sea" (E. T. Degens and D. A. Ross, eds), pp. 557–571. Springer-Verlag, Berlin.
Engel, A. E. J., Engel, C. G. and Havens, R. G. (1965). *Geol. Soc. Am. Bull.* **76**, 719.
Eugster, H. P. and Jones, B. F. (1968). *Science*, **161**, 160.
Ewing, M., Worzel, J. L., Beall, A. O., Berggren, W. A., Bukry, D., Burk, C. A., Fisher, A. G. and Pessagno, E. A. (1969). "Initial Reports of the Deep Sea Drilling Project", Vol. 1, 672 pp. Washington, D.C.
Fanning, K. A. and Schink, D. R. (1969). *Limnol. Oceanogr.* **14**, 59.
Fischer, A. G. and Garrison, R. E. (1967). *J. Geol.* **75**, 488.
Fleet, M. E. L. (1965). *Clay Minerals*, **6**, 3.
Foster, M. D. (1954). *In* "Clays and Clay Minerals, Proc. 2nd Natl. Conf." (A. Swineford and N. Plummer, eds), pp. 386–397. National Academy of Science. Publ. 327. Washington, D.C.
Foster, M. D. (1969). *U.S. Geol. Surv. Prof. Paper* **614F**, 17 pp.
Frederickson, A. F. and Reynolds, R. C. (1960). *In* "Clays and Clay Minerals, Proc. 7th Nat. Conf." (A. Swineford, ed.), pp. 203–213. Pergamon, New York.
Frink, C. R. (1965). *Soil. Sci. Soc. Am. Proc.* **29**, 379.
Galliher, E. W. (1935a). *Bull. Am. Assoc. Pet. Geol.* **19**, 1569.
Galliher, E. W. (1935b). *Geol. Soc. Am. Bull.* **46**, 1341.
Galliher, E. W. (1936). *Proc. Geol. Soc. Am.* 345.
Galliher, E. W. (1939). *In*: "Recent Marine Sediments" (P. D. Trask, ed.), pp. 513–515. American Association of Petrochemical Geology, Tulsa, Okla.
Garrels, R. M. (1965). *Science*, **148**, 69.
Giresse, P. (1965). *C.R. Acad. Sci., Paris*, **260**, 5597.
Glaeser, R. and Mering, J. (1967). *C.R. hebd. Seanc. Acad. Sci., Paris* **265**, 833.
Goldberg, E. D. and Arrhenius, G. (1958). *Geochim. Cosmochim. Acta*, **13**, 153.
Goldschmidt, V. M. (1954). "Geochemistry", 730 pp. Clarendon Press, Oxford.

Goldschmidt, V. M. and Peters, C. (1932). *Nachr. Ges. Wiss. Gottingen, Math. Physik. III*, 402; *IV*, 528.

Goodell, H. G. (1965). *In* Contrib 11, Sed. Res. Lab., Dept. of Geol., Florida State University.

Gorbunova, Z. N. (1963). *Litologia y Poleznye Iskopamye*, **1**, 28.

Gorham, E. and Swaine, D. J. (1965). *Limnol. Oceanogr.* **10**, 268.

Grauer, O. H. and Hamilton, E. H. (1950). *J. Res. U.S. Natl. Bur. Std.* **44**, 495.

Greene-Kelly, R. (1955). *Mineral. Mag.* **30**, 604.

Gregor, B. (1968). *Nature, Lond.* **219**, 360.

Gregor, H. P. (1948). *J. Amer. Chem. Soc.* **70**, 1293.

Griffin, G. M. (1962). *Geol. Soc. Am. Bull.* **73**, 737.

Griffin, G. M. and Ingram, R. L. (1955). *J. Sed. Petrol.* **25**, 194.

Griffin, J. J. and Goldberg, E. D. (1963). *In* "The Sea" (M. N. Hill, ed.), Vol. 3, pp. 728–741. Interscience, New York.

Griffin, J. J., Windom, H. and Goldberg, E. D. (1968). *Deep-Sea Res.* **15**, 433.

Grim, R. E. (1953). "Clay Mineralogy", 384 pp. MacGraw-Hill, New York.

Grim, R. E. and Johns, W. D. (1954). *In* "Clays and Clay Minerals, Proc. 2nd Natl. Conf." (A. Swineford and N. Plummer, eds), pp. 81–103. National Academy of Science Publ. 327, Washington, D.C.

Grim, R. E. and Loughnan, F. C. (1962). *J. Sed. Petrol.* **32**, 240.

Grim, R. E., Dietz, R. S. and Bradley, W. F. (1949). *Geol. Soc. Am. Bull.* **60**, 1785.

Gross, M. G. and Nelson, J. L. (1966). *Science*, **154**, 879.

Gulbrandsen, R. A. (1969). *Econ. Geol.* **64**, 365.

Gupta, G. C. and Malik, W. U. (1969). *Clays Clay Minerals*, **17**, 331.

Haller, W. (1963). *Phys. Chem. of Glasses*, **4**, 217.

Hamilton, E. L. and Rex, R. W. (1959). *U.S. Geol. Surv. Prof. Pap.* **260-W**. pp. 785–798. Washington, D.C.

Hanor, J. S. and Brass, G. W. (1968). *In* "Program of annual mtg., Geol. Soc. Amer.", p. 126 (abstract).

Harder, H. (1970). *Sediment. Geol.* **4**, 153.

Harrison, A. G. and Thode, H. G. (1958a). *Bull. Am. Assoc. Pet. Geol.* **42**, 2642.

Harrison, A. G. and Thode, H. G. (1958b). *Trans. Faraday Soc.* **54**, 84.

Harriss, R. C. (1966). *Nature, Lond.* **212**, 275.

Hart, R. (1970). *Earth Planet. Sci. Letters*, **9**, 269.

Hart, R. (1973). *Can. J. Earth Sci.* **10**, 799.

Hart, S. H. (1969). *Earth Planet. Sci. Letters*, **6**, 295.

Hart, S. R. and Nalwalk, A. J. (1970). *Geochim. Cosmochim. Acta*, **34**, 145.

Hathaway, J. C. and Sachs, P. L. (1965). *Amer. Mineral*, **50**, 852.

Hauser, E. A. and Reynolds, H. H. (1939). *Amer. Mineral*, **24**, 590.

Hawkins, D. B. (1961). "Experimental Hydrothermal Studies Bearing on Rocks Weathering and Clay Mineral Formation". University microfilms, Ann Arbor, Michigan, Order No. 61-67851.

Hawkins, D. B. and Roy, R. (1963). *Geochim. Cosmochim. Acta*, **27**, 1047.

Heezen, B. C., Nesteroff, W. D., Oberlin, A., and Sabatier, G. (1965). *C.R. Acad. Sci. Paris*, **260**, 5819.

Helgeson, H. C. and Mackenzie, F. T. (1970). *Deep-Sea Res.* **17**, 877.

Helgeson, H. C., Garrels, R. M. and Mackenzie, F. T. (1969). *Geochim. Cosmochim. Acta*, **33**, 455.

Hem, J. D. and Robertson, C. E. (1967). *U.S. Geol. Surv. Water-Supply Pap.* **1827A**, 55 pp.

Heystek, H. and Schmidt, E. R. (1953). *Trans. Geol. Soc. S. Afr.* **56**, 99.

Hoffmann, V. and Klemen, R. (1950). *Z. Anorg. Allg. Chem.* **262**, 95.

Holland, H. D. (1965). *Proc. Nat. Acad. Sci. U.S.* **53**, 1173.

Honnorez, J. (1969). *Mineral Dep.* **4**, 114.

Horn, M. K. and Adams, J. A. S. (1966). *Geochim. Cosmochim. Acta,* **30**, 279.

Horowitz, A. (1970). *Marine Geol.* **9**, 241.

Hower, J. (1961). *Amer. Mineral.* **46**, 313.

Hulings, N. C. (1961). *In* "First National Coastal and Shallow Water Research Conf." pp. 488–491. Texas Christian University.

Hurley, P. M., Hart, S. R., Pinson, W. H. and Fairbairn, H. W. (1959), *Geol. Soc. Am. Bull.* **70**, 1622.

Hurley, P. M., Heezen, B. C., Pinson, W. H. and Fairbairn, H. W. (1963). *Geochim. Cosmochim. Acta,* **27**, 393.

Jackson, M. L. (1959). *Clays Clay Minerals,* **6**, 133.

Jackson, M. L. (1963). *Clays Clay Minerals,* **11**, 29.

Johns, W. D. (1963). *Fortschr. Geol. Rheinland Westfalen,* **10**, 215.

Johns, W. D. and Grim, R. E. (1958). *J. Sed. Petrol.* **28**, 186.

Johnson, V., Cutshall, N. and Osterberg, C. (1967). *Water Resources Res.* **3**, 99.

Kaplan, I. R., Emery, K. O. and Rittenberg, S. C. (1963). *Geochim. Cosmochim. Acta,* **27**, 297.

Kazakov, A. V. (1937). *U.S.S.R. Sci. Inst. Fertilisers and Insectfungicides Trans.* **142**, 95.

Kazakov, A. V. (1938). *Sovetskaya Geologiya,* **8**, 33.

Keller, W. D. (1963). *In* "Clays and Clay Minerals. Proc. 11th Natl. Conf." (W. F. Bradley, ed.), pp. 136–137. Pergamon, Oxford.

Kendall, C. G. St. C. and Skipworth, P. A. D'E. (1969). *Bull. Am. Assoc. Pet. Geol.* **53**, 841.

Kerr, P. F. (1937). *Amer. Mineral.* **22**, 534.

Kharkar, D. P., Turekian, K. K. and Bertine, K. K. (1968). *Geochim. Cosmochim. Acta,* **32**, 285.

Kinsman, D. J. J. (1966). *In* "Second Symposium on Salt" (J. L. Rau, ed.), Vol. 1, pp. 302–326. N. Ohio Geological Society, Cleveland, Ohio.

Kinsman, D. J. J. (1969). *J. Sed. Petrol.* **39**, 486.

Kramer, J. R. (1965). *Geochim. Cosmochim. Acta,* **29**, 921.

Krauskopf, K. B. (1967). "Introduction to Geochemistry", 721 pp. McGraw-Hill, New York.

Landergren, S. (1945). *Arkiv. Kemi. Mineral. Geol.* **19A**, (No. 26), 1.

Landergren, S. (1958). *Geol. Fören. Stockholm Forh.* **80**, 104.

Landergren, S. and Carvajal, M. C. (1969). *Arkiv. Mineral. Geol.* **5**, 13.

Lerman, A. (1966). *Sedimentology,* **6**, 267.

Levinson, A. A. and Ludwick, J. C. (1966). *Geochim. Cosmochim. Acta,* **30**, 855.

Li, Y. H., Bischoff, J. and Mathieu, G. (1969). *Earth Planet. Sci. Letters,* **7**, 265.

Ling, G. N. (1962). "A Physical Theory of the Living State", Chap. 4. Blaisdell, New York.

Lippmann, F. (1960). *Fortschr. Mineral,* **38**, 156.

Liss, P. S. and Spencer, C. P. (1970). *Geochim. Cosmochim. Acta,* **34**, 1073.

Love, L. G. (1967). *Sedimentology,* **9**, 327.

Mackenzie, F. T. and Garrels, R. M. (1965). *Science,* **150**, 57.

Mackenzie, F. T. and Garrels, R. M. (1966a). *J. Sed. Petrol.* **36**, 1075.

Mackenzie, F. T. and Garrels, R. M. (1966b). *Amer. J. Sci.* **264**, 507.

Mackenzie, F. T., Garrels, R. M., Bricker, O. P. and Bickley, F. (1967). *Science*, **155**, 1404.
Maeda, H. (1952). *Publ. Seto Mar. Biol. Lab.* **2**, 249.
Manheim, F. T. (1965). *In* "Symposium on Marine Geochemistry". University of Rhode Island, Occ. Publ. no. 3, 217.
Mansfield, G. R. (1940). *Amer. J. Sci.* **238**, 863.
Mariner, R. H. and Surdam, R. C. (1970). *Science* **170**, 977.
Marshall, C. E. (1954). *In* "Clays and Clay Minerals, Proc. 2nd Natl. Conf." (A. Swineford and N. Plummer, eds), pp. 364–385. National Academy of Science Publ. 327, Washington, D.C.
Marshall, R. R. (1961). *Geol. Soc. Amer. Bull.* **72**, 1493.
Mathieson, A. McL. (1958). *Amer. Mineral.* **43**, 216.
Matthews, D. H. (1962). *Nature, Lond.* **194**, 368.
McClellan, G. H. and Lehr, J. R. (1969). *Amer. Mineral.* **54**, 1374.
McRae, S. G. (1972). *Earth Sci. Rev.* **8**, 397.
Mellis, O. (1952). *Nature, Lond.* **169**, 624.
Mellis, O. (1959). *Meddel. Oceanog. Inst. Goteborg. Ser. B.* **8** (6), 1.
Melson, W. G. and Thompson, G. (1972). *Geol. Soc. Am. Mem.* (cited in Hart, 1973).
Millot, G. (1970). "Geology of Clays", 429 pp. Springer-Verlag, New York.
Milne, I. H. and Earley, J. W. (1955). *Bull. Am. Assoc. Pet. Geol.* **42**, 328.
Miyashiro, A. F., Shido, F. and Ewing, M. (1969). *Contr. Mineral. Petrol.* **23**, 38.
Moore, J. G. (1966). *U.S. Geol. Surv. Prof. Pap.* **550-D**, 163.
Morgan, J. J. and Stumm, W. (1964). *J. Colloid. Sci.* **19**, 347.
Morgenstein, M. (1967). *Sedimentology,* **9**, 105.
Mortland, M. M. and Ellis, B. G. (1959). *Soil Sci. Soc. Am. Proc.* **23**, 363.
Muljadi, D., Posner, A. M. and Quirk, J. P. (1967). *Soil Sci.* **17**, 212.
Muller, G. (1961). *Beitr. Mineral. Petrog.* **8**, 1.
Muller, G. (1967). *In* "Diagenesis in Sediments" (G. Larsen and G. V. Chilingar, ed), *Developments in Sedimentology* **8**, pp. 127–177. Elsevier, Amsterdam.
Murata, K. J. and Erd, R. C. (1964). *J. Sed. Petrol.* **34**, 633.
Murray, J. and Irvine, R. (1894). *Trans. Roy. Soc. Edinburgh,* **37**, 721.
Murray, J. and Lee, G. V. (1909). *Mem. Mus. comp. Zool. Harv.* **38**, 7.
Murray, J. and Renard, A. L. (1891). "Report on Deep-Sea Deposits Based on Specimens Collected During the Voyage of H.M.S. *Challenger*", 525 pp. H.M. Stationery Office, London.
Murray, J. W. and Mackintosh, E. E. (1968). *Can. J. Earth Sci.* **5**, 243.
Nayudu, Y. R. (1964). *Bull. Volcan.* **27**, 391.
Nelson, B. W. (1960). *In* "Clays and Clay Minerals, Proc. 7th Natl. Conf." (A. Swineford, ed.), pp. 135–142. Pergamon, New York.
Newman, A. C. D. (1970). *Clay Minerals,* **8**, 361.
Nicholls, G. D. (1963). *Sci. Progr.* **51**, 12.
Norin, E. (1953). *Bull. Geol. Soc. Inst. Univ. Upsala,* **34**, 279.
Ostroumov, E. A. (1953). *Akad. Nauk SSSR, Inst. Okeonologii Trudy,* **7**, 70.
Ostroumov, E. A., Volkov, I. I. and Fomina, L. C. (1961). *Akad. Nauk SSSR, Inst. Okeanologii Trudy,* **50**, 93.
Pedro, G. (1961). *Clay Min. Bull.* **4**, 266.
Perry, E. A. (1971). *Deep-Sea Res.* **18**, 921.
Perry, E. A. (1972). *Amer. J. Sci.* **272**, 150.
Peterson, M. N. A. and Goldberg, E. D. (1962). *J. Geophys. Res.* **67**, 3477.
Peterson, M. N. A. and Griffin, J. J. (1964). *J. Marine Res.* **22**, 13.

Peterson, M. N. A., Edgar, N. T., Cita, M., Gratner, S., Goll, R., Nigrini, C. and von der Borch, C. (1970). "Initial Reports of the Deep Sea Drilling Project", Vol. 2, 501 pp. Washington, D.C.

Philpotts, J. A., Schnetzler, C. C. and Harts, S. R. (1969). *Earth Planet. Sci. Letters*, 7, 293.

Pichler, H. (1965). *Bull. Volc.* 28, 293.

Pinsak, A. P. and Murray, H. H. (1960). *In* "Clays and Clay Minerals, Proc. 7th Natl. Conf." (A. Swineford, ed.), pp. 162–178. Pergamon, New York.

Pissarides, A., Stewart, J. W. B. and Rennie, D. A. (1968). *Can. J. Soil Sci.* 48, 151.

Porrenga, D. H. (1965a). *In*: "Clays and Clay Minerals, Progr. 14th Natl. Conf." (S. W. Bailey, ed.), pp. 25–26. Pergamon, Oxford.

Porrenga, D. H. (1965b) *Geol. Mijnb.* 44, 400.

Porrenga, D. H. (1965a). *In* "Clays and Clay Minerals, Proc. 15th Natl. Conf." (S. W. Bailey, ed.), pp. 221–233. Pergamon, Oxford.

Porrenga, D. H. (1967). "Clay Mineralogy and Geochemistry of Recent Marine Sediments in Tropics Areas", 145 pp. Druk, Stolk-Dordt.

Potts, R. H. (1959). "Cationic and Structural Changes in Missouri River Clays when Treated with Ocean Water". M.S. thesis, Univ. Missouri, Colombia, Missouri.

Powers, M. C. (1954). *In* "Clays and Clay Minerals, Proc. 2nd Natl. Conf." (A. Swineford and N. Plummer, eds), pp. 68–80. National Academy of Science Publ. 327, Washington, D.C.

Powers, M. C. (1957). *J. Sed. Petrol.* 27, 355.

Powers, M. C. (1959). *In* "Clays and Clay Minerals, Proc. 6th Natl. Conf." (A. Swineford, ed.), pp. 309–326. Pergamon, London.

Pratt, W. L. (1963). *In* "Essays in Marine Geology in Honor of K. O. Emery" (R. L. Miller, ed.), pp. 97–119. University of South California Press, Los Angeles.

Rex, R. W. (1967). *In* "Clays and Clay Minerals, Proc. 15th Natl. Conf." (S. W. Bailey, ed.), pp. 195–203. Pergamon, Oxford.

Rich, C. I. (1964). *Soil Sci.* 98, 100.

Rich, C. I. (1968). *Clays Clay Minerals*, 16, 15.

Rickard, D. T. (1969). *Stockholm Contr. Geol.* 20, 67.

Rieck, Cl. D. and Stevels, M. J. (1951). *J. Soc. Glass Tech.* 35, 284.

Russell, K. L. (1970). *Geochim. Cosmichim. Acta,* 34, 893.

Savin, S. M. and Epstein, S. (1970). *Geochim. Cosmochim. Acta,* 34, 43.

Sawhney, B. L. (1972). *Clays Clay Minerals*, 20, 93.

Sayre, W. W., Guy, H. P. and Chamberlain, A. R. (1963). *U.S. Geol. Surv. Prof. Pap.* 433-A.

Scarfe, C. M. (1972). "Viscosity and Related Properties of Basic Magmas". Ph.D. thesis, 220 pp., Univ. of Leeds.

Schink, D. R. (1967). *Geochim. Cosmochim. Acta,* 31, 987.

Schoen, R. and Robertson, C. E. (1970). *Amer. Mineral.* 55, 43.

Scholze, H. and Mulfinger, H. P. (1959). *Glastechn. Ber.* 32, 381.

Seed, D. P. (1965). *Amer. Mineral.* 50, 1097.

Senin, Y. M. (1970). *Lithology and Mineral Resources,* 1, 11.

Shaw, D. M. and Bugry, R. (1966). *Can. J. Earth. Sci.* 3, 49.

Shiskina, O. V. (1971). *Dokl. Akad. Nauk SSR,* 201, 263.

Siever, R. (1968). *Earth Planet. Sci. Letters,* 5, 106.

Siever, R., Beck, K. C. and Berner, R. A. (1965). *J. Geol.* 73, 39.

Siffert, B. (1962). *Mem. Serv. Carte Geol. Als. Lor.* No. 21.

Sillén, L. G. (1961). *In* "Oceanography" (M. Sears, ed.), *Amer. Assoc. Adv. Sci.* **67**, 549.

Sillén, L. G. (1966). *Tellus,* **18**, 198.

Sillén, L. G. (1967a). *In* Equilibrium Concepts in Natural Water Systems" (R. F. Gould, ed.), pp. 57–69. American Chemical Society, Washington, D.C.

Sillén, L. G. (1967b). *Chemistry in Britain,* **3**, 291.

Sillén, L. G. (1967c). *Science,* **156**, 1189.

Skornyakova, I. S. (1964). *Internat. Geol. Rev.* **7**, 2161.

Slaughter, M. and Milne, I. H. (1960). *In* "Clays and Clay Minerals, Proc. 7th Natl. Conf." (A. Swineford, ed.), pp. 114–124. Pergamon, New York.

Smith, C. L. (1940). *J. Marine Res.* **3**, 1.

Sonnen, M. B. (1965). "Zinc Adsorption by Sediments in a Saline Environment", M.S. thesis (Civil Engineering studies, Sanitary Engineering series No. 24), 42 pp, Univ. Illinois.

Sosman, R. B. (1965). "The Phases of Silica", 388 pp. Rudgers University Press, New Brunswick.

Spencer, D. W. and Brewer, P. G. (1971). *J. Geophys. Res.* **76**, 5877.

Stefansson, U. and Richards, F. A. (1963). *Limnol. Oceanogr.* **8**, 394.

Stewart, F. H. (1963). *In* "Data of Geochemistry" (M. Fleischer, ed.), Chap. Y, 52 pp. *U.S. Geol. Surv. Prof. Pap. 440 Y.*

Stumm, W. and Morgan, J. J. (1970). "Aquatic Chemistry", 583 pp. Wiley, New York.

Takahashi, J. (1939). *In* "Recent Marine Sediments" (P. D. Trask, ed.), pp. 503–512. American Association of Petrochemical Geology, Tulsa, Okla.

Takahashi, J. and Yagi, T. (1929). *Econ. Geol.* **24**, 838.

Tettenhorst, R. (1962). *Amer. Mineral.* **47**, 769.

Thayer, T. P. (1966). *Amer. Mineral.* **51**, 685.

Thompson, G. (1972). *Geochim. Cosmochim. Acta,* **36**, 1237.

Thompson, G. and Melson, W. G. (1970). *Earth Planet. Sci. Letters,* **8**, 61.

Thompson, G., Brown, V. T., Melson, W. G. and Cifelli, R. (1968). *J. Sed. Petrol.* **38**, 1035.

Tiller, K. G., Hodgson, J. F. and Peech, M. (1963). *Soil Sci.* **95**, 392.

Tooms, J. S., Summerhayes, C. P. and Cronan, D. S. (1969). Oceanogr. Mar. Biol. Ann. Rev. **7**, 49.

Turekian, K. K. (1967). *In* "Progress in Oceanography" (M. Sears, ed.), Vol. 4, pp. 227–244. Pergamon, New York.

Turekian, K. K. (1968). *Geochim. Cosmochim. Acta,* **32**, 603.

Turekian, K. K. (1971). *In* "Impingement of Man on the Oceans" (D. W. Hood, ed.), pp. 9–73. Wiley, New York.

Turekian, K. K. and Imbrie, J. (1967). *Earth Planet. Sci. Letters,* **1**, 161.

Vanderstappen, R. and Verbeek, T. (1964). *Ann. Musee Roy. Congo Belge, Ser. Sci. Geol.* **47**. (quoted in Muller, 1967).

Van Olphen, H. (1963). "An Introduction to Clay Colloid Chemistry", 301 pp. Interscience, New York.

Von Straaten, L. M. J. U. (1954). *Leidse Geol. Medel.* **19**, 1.

Vasil'ev, V. P. (1960). *Russian J. Phys. Chem.* **34**, 840.

Velde, B. (1965). *Amer. Mineral.* **50**, 436.

Volkov, I. I. (1961). *Tr. Inst. Okeanol., Akad. Nauk SSSR,* **50**, 68.

Von Gaertner, H. R. and Schellmann, W. (1965). *Min. Petr. Mitt.* **III**, (10), 349.

Von Gumbel, G. (1878). *Sitz. Berichte K. Bayer. Akad. d. Wiss. München, Mathen-Physik. Klasse.* pp. 189–209.

Wahlberg, J. S. and Fishman, M. J. (1962). *U.S. Geol. Surv. Bull.* **1140-A**, 30 pp.

Warren, B. E. and Briscoe, J. (1938). *J. Amer. Ceram. Soc.* **21**, 259.

Weaver, C. E. (1958). *Amer. Mineral.* **43**, 839.

Weaver, C. E. (1967). *Geochim. Cosmochim. Acta,* **31**, 2181.

Wedepohl, K. (1960). *Geochim. Cosmochim. Acta,* **18**, 200.

White, D. E. and Waring, G. A. (1963). *In* "Data of Geochemistry" (M. Fleischer, ed.), Chap. K, 27 pp. *U.S. Geol. Surv. Prof. Pap.* **440-K**.

Wollast, R., Mackenzie, F. T. and Bricker, O. P. (1968). *Amer. Mineral.* **53**, 1645.

Yoder, H. S. and Eugster, H. P. (1955). *Geochim. Cosmochim. Acta,* **8**, 225.

Zachariasen, W. H. (1932). *J. Amer. Chem. Soc.* **54**, 3841.

Zelenov, K. K. (1964). *Dokl. Akad. Nauk SSSR, Earth Sci. Sect.* **155**, 91.

Zobell, C. E. and Rittenberg, S. C. (1948). *J. Mar. Res.* **7**, 602.

Chapter 28

# Manganese Nodules and other Ferro-manganese Oxide Deposits

D. S. CRONAN

*Department of Geology, University of Ottawa,
Ottawa 2, Canada\**

## 28.1. INTRODUCTION

Manganese nodules and other ferro-manganese oxide deposits are precipitates of hydrous manganese and iron oxides which are abundant in all the oceans of the world, in shallow marine environments and in many temperate lakes. They were first discovered in the deep ocean during the circumnavigation of the globe by H.M.S. *Challenger* (1873–76), and were described by Murray (1876) and, in much more detail, by Murray and Renard (1891). During the past decade and a half their economic value has been recognized, and this has stimulated sampling operations which have provided much of the material used in the detailed investigations summarized in this chapter.

The literature on manganese nodules is very extensive. In many respects the original descriptions by Murray and Renard (1891) are a model of their kind, and together with the work of Agassiz (1906), they provide much of the early information on the subject. These investigations were followed by a period of

---

\* Present address, Geology Dept, Imperial College of Science and Technology, London, SW7.

little activity, but interest revived during the early nineteen fifties (Goldberg, 1954). One of the first attempts to collect together a body of analytical data on manganese nodules was that of Menard *et al.* (1964) who assembled a large number of analyses of Pacific nodules from various sources. About the same time, Mero (1965) showed that there are extensive regional variations in the composition of nodules throughout the Pacific Ocean. Since 1965, work on nodules has become more diversified. Manheim (1965) and Price (1967) have investigated shallow water occurrences, and work on the deep-sea varieties has been continued by, among others, Bender *et al.* (1966), Barnes (1967), Cronan (1967, 1972a, c), Cronan and Tooms (1967a, b, 1968, 1969), Price and Calvert (1970), Skornyakova and Andrushchenko (1968, 1970) and Glasby (1970, 1972). Reviews have been published by Mero (1965), Manheim (1965), Price (1967), Tooms *et al.* (1969) and Skornyakova and Andrushchenko (1970).

## 28.2. DISTRIBUTION

Manganese nodules vary considerably in abundance throughout the World Ocean. Large areas of the sea floor contain them in high concentrations, but within these regions their distribution is often very patchy. Bottom photography and underwater television have shown them to be quite variable in size, shape and distribution over short distances (Kaufman and Siapno, 1972). For example, areas where fifty per cent or more of the sea floor is covered with nodules grade into areas of near zero coverage within a few hundreds of metres (Fig. 28.1). The factors determining this variability are complex.

In the Pacific Ocean, nodules are most abundant in a broad area between about 6°N and 20°N extending from approximately 120°W to 160°E (Skornyakova and Andrushchenko, 1970; Ewing *et al.*, 1971). Locally, high concentrations of nodules cover as much as 75 per cent, or more, of the sea floor in this region (Fig. 28.2). Sedimentation rates are lower in this area than in the equatorial zone to the south (Hays *et al.*, 1969; Opdyke and Foster, 1970) and the principal sediment types are siliceous ooze and red clay. In the South Pacific, nodule distribution is more variable. Menard and Shipek (1958) have estimated that a quarter to a half of the sea floor in parts of this area is covered with nodules, and Ewing *et al.* (1971) have found high concentrations in a province comprising the Manihiki Plateau, Society Islands, Tahiti and the Tuamotu Archipelago. Other areas of the Pacific where nodules are abundant include the region between the East Pacific Rise and the New Zealand Plateau, and in the vicinity of Tasmania (Ewing *et al.*, 1971). Furthermore, in the absence of nodules, ferro-manganese oxide encrustations coat many exposed rock surfaces throughout the Pacific, and are especially abundant in

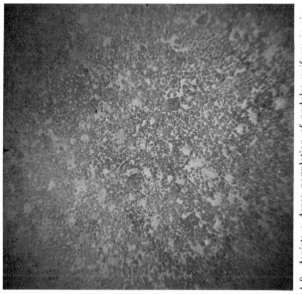

1-8, depicts a dense population of nodules uniform in size, shape and distribution.

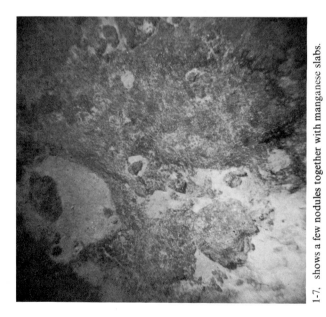

1-7, shows a few nodules together with manganese slabs.

FIG. 28.1. A series of photographs which illustrates the variability of the morphology and distribution of nodules over a small area. The six pictures were taken during a traverse of approximately 2·8 km in the Pacific Ocean by the Deepsea Ventures Inc. ship *Prospector*.

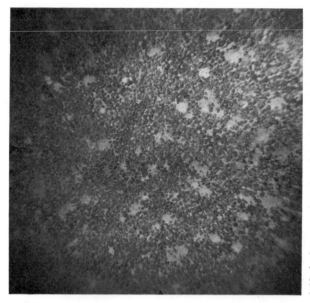

1-10, depicts a continuation of the dense nodule distribution in 1-8 and 1-9.

Fig. 28.1

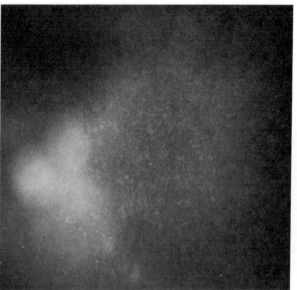

1-9, exhibits a nodule distribution similar to that shown in 1-8, but taken from a greater distance above the sea floor.

1-11, makes a sharp contrast with 1-8, 1-9 and 1-10, and shows a field of manganese encrusted boulders from which nodules are almost absent.

1-12, exhibits a return to the dense nodule distribution of 1-8, 1-9 and 1-10.

Fig. 28.1

All pictures were taken at 10·2 minute intervals, except between 1-10 and 1-11. The interval between these was 20·4 minutes, owing to a rapid increase in the roughness of the terrain which necessitated the hoisting of the camera out of range of the sea floor.

Notes by courtesy of W. D. Siapno. Photographs by courtesy of R. Kaufman and W. D. Siapno of Deepsea Ventures Inc., Gloucester Point, Virginia.

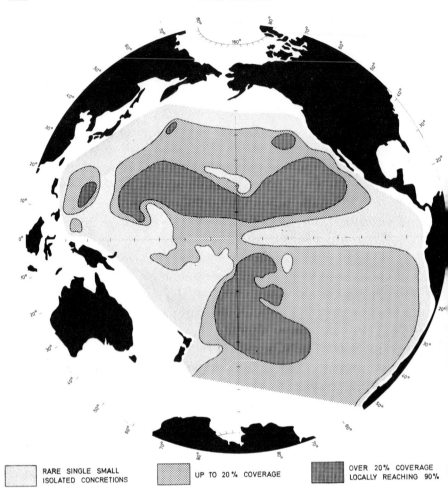

FIG. 28.2. Map showing the distribution of manganese nodules on the floor of the Pacific Ocean (after Skornyakova and Andrushchenko, 1970).

elevated volcanic areas, such as the Mid-Pacific Mountains and the Southern Borderland Seamount Province (Cronan and Tooms, 1969).

The abundance and spatial density of nodules in the Atlantic Ocean is lower than in the Pacific. Local high concentrations occur in areas such as the Blake Plateau and the Scotia Sea, but these are of restricted areal extent. According to Ewing *et al.* (1971), nodules in the Atlantic are most abundant in the areas of clay deposition between the abyssal plains bordering the continents and the Mid-Atlantic Ridge. However, encrustations are found on most exposed volcanic rocks away from the median valley of the Mid-Atlantic Ridge, and are often more abundant than manganese nodules. Thick ferro-manganese oxide coatings have been recovered both from the region of the Ridge at 45° N, and also from the King's Trough area on its eastern flank. In the former, the encrustations thicken away from the median valley, and Aumento (1969) considered this to result from longer growth periods associated with the increasing age of the sea floor away from the Ridge crest as a consequence of sea floor spreading. The age of the sea floor may therefore be a factor in determining the abundance and distribution of ferro-manganese oxide deposits.

There is a paucity of data on the distribution of nodules in the Indian Ocean. However, Bezrukov (1962, 1963) has found high concentrations in parts of the southern Indian Ocean, and Laughton (1967) and Glasby (1970) have reported locally high concentrations on the flanks of the Carlsberg Ridge. Isolated occurrences have been noted in the Somali Basin by Cronan and Tooms (1967b), and Willis (1970) has described nodules from off the coast of South Africa. Recent studies by Bezrukov and Andrushchenko (1972) have indicated that the highest nodule concentrations in the Indian Ocean are in the basins far from land, as is found in the Pacific.

One of the most important factors in determining the distribution of nodules in the World Ocean is the rate of accumulation of their associated sediments. With certain exceptions, high sedimentation rates are thought to inhibit nodule growth, perhaps by burying potential nuclei, or even the nodules themselves, before they can grow to an appreciable size. For example, according to Ewing et al. (1971) the limits of the large nodule province in the North Pacific (Fig. 28.2) are determined by increases in sedimentation rates at the periphery of the North Pacific Basin due to turbidity current and hemipelagic deposition, coupled with an increase in biogenic sedimentation associated with the equatorial zone of high productivity to the south. Horn et al. (1972a) have found that major nodule provinces in all oceans occur where rates of sedimentation are lowest, generally in areas of a red clay or siliceous ooze accumulation. However, high sedimentation rates do not always inhibit nodule growth as nodules are locally abundant in some areas of rapid sedimentation as a result of the diagenetic remobilisation of manganese in buried sediments, followed by its upward migration and reprecipitation at the sediment–water interface (Section 28.5.3.1). One such area occurs off Baja California (Cronan and Tooms, 1969), where reducing conditions occur at shallow depths beneath the sediment–water interface.

Another important factor in determining the world-wide distribution of nodules which is related, in part, to sedimentation rates, is the distribution of potential nodule nuclei. Almost all nodules have a nucleus of some sort, usually consisting of volcanic material such as pumice, palagonite or other altered volcanic rock. Such nuclei are more abundant in volcanic than in non-volcanic areas, and this may partly account for the high concentrations of nodules often found in areas of submarine volcanic activity (see Section 28.6). The virtual absence of nodules in areas of tubidity current deposition, such as abyssal plains, may result from the burial of potential nucleating agents by the turbidity flows.

Fewer data are available on sub-surface nodules than on their surface counterparts. Nevertheless, they are known to be common in the upper few metres of pelagic sediments. Cronan and Tooms (1967a) found that nodules were more or less equally abundant within the upper two metres of the sediment of more than one hundred Pacific cores as they were in the surface layer (Fig. 28.3). However, Horn et al. (1972b) have found from a study of approximately fifty other cores from all the oceans that nodules in the top 3 m of sediment are only about thirty five per cent as abundant as at the surface. These estimates are, of course, too limited to give a statistically valid estimate of the concentration of buried nodules in the World Ocean, but, nevertheless they do serve to indicate their order of magnitude.

Until recently, knowledge of buried nodules was confined to the upper few metres of pelagic sediments. However, the work of the JOIDES Deep Sea Drilling Project has shown that they also occur at considerably greater depths. Many of the nodules found in DSDP cores have probably slumped from the surface as a result of the drilling process (McManus et al., 1970). However, one buried nodule, which occured at a depth of 131 m in DSDP 162 from the north eastern Pacific, is thought, on the basis of its occurrence and the nature of its surrounding sediment, to have been formed in-situ (Section 28.5.3.2.). The possible formation of nodules at depth within the sediment column has considerable implications in attempts to explain their overall distribution on the ocean floor. Their preferential concentration at the sediment surface has been variously ascribed to burrowing organisms (Mero, 1965), upward diffusion of reduced manganese (Lynn and Bonatti, 1965), or seismic activity. However, an alternative possibility is that it may in some areas, in part, result from non-deposition of sediment causing a lack of burial of nodules or from localized sediment erosion leading to concentrations of previously buried nodules at the surface (Cronan and Tooms, 1967a). The nodules could originally have formed in situ at depth, or could more likely have been buried after formation at the surface. Differential exposure and burial of nodules by erosion and deposition of sediments, coupled with factors such as the availability of suitable nuclei, topographic controls, and sedimentation patterns (Moore and Heath, 1966), might partly account for the high local variability in nodule distribution mentioned at the outset of this section.

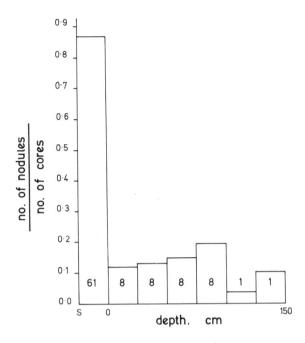

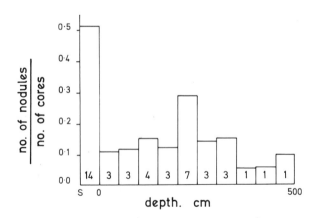

FIG. 28.3. Sub-surface nodule distribution in Pacific sediments. The numbers at the foot of each of the columns refer to the actual number of nodules found over each depth interval (from *Deep-Sea Res.*, **14**, 1967, by courtesy of Pergamon Press Ltd.).

1A

1B

2A

2B

Fig. 28.4. Photographs illustrating two typical morphologies of manganese nodules (A is top view and B is side view). Nodule 1 is a nearly spherical deep-sea nodule from the Carlsberg Ridge. Nodule 2 is a discoidal near-shore concretion which has grown around a large pebble from off British Columbia.

## 28.3. MORPHOLOGY AND INTERNAL STRUCTURE

Most deep-sea manganese nodules tend to be spherical to oblate in form (Fig. 28.4), but irregular shapes and agglomerates are not uncommon. Often, nodules from any particular locality are morphologically similar, although in some instances more than one morphological population has been found to occur within a limited area (Cronan and Tooms, 1967b; Kaufman and Siapno, 1972).

In a study of ferro-manganese oxide deposits from a large number of localities in the South Pacific and South Atlantic Oceans, Goodell et al. (1970) divided the concretions into four morphological varieties on the basis of the nature of their nuclei, the thickness of their ferro-manganese oxide phases, and their external shape. *Nodules* were roughly spherical and were nearly always associated with evidence of current movements, suggesting that they may have attained their shape as a result of overturning due to winnowing of the underlying sediment. *Crusts* were the most abundant variety of ferro-manganese deposit and ranged in thickness from a few millimetres to tens of centimetres. *Botryoidals* consisted of oval or flattened concretions with several nuclei which resulted in the development of grape-like clusters. *Agglomerates* consisted of irregular masses, composed of pebbles or granules usually of glacial origin, encrusted or cemented together by a thin ferro-manganese oxide coating. These morphological types were shown to have a degree of geographical dependency; agglomerates, for example, were absent north of 50° S. Grant (1967) suggested that the morphology of these concretions was dependent on a number of factors, including the nature and proximity of the source of the elements which they contain, and the rate of concretion formation. To these might be added the shape and nature of the nucleus.

Shallow marine and continental margin nodules often have morphologies different from those found in the deep ocean (Fig. 28.4). A wide range of shapes occur in near-shore environments, but there is a tendency for flattened, saucer- or disc-like concretions to be more abundant than in deep-sea areas. Such nodules having these near-shore morphologies are also common in lakes. According to Manheim (1965), such flattened concretions reflect the derivation of manganese, and associated elements, by upward diffusion through the interstitial waters of the underlying sediments rather than by direct precipitation from the overlying sea water. In the nodules formed by this process the sides of the nuclei nearest the source acquires the thickest oxide layers. The thin coating on the tops of the nuclei probably represents the fraction precipitated from sea water. The usual absence of a coating on the underside of the nucleus could be due to its burial in poorly oxidizing, or reducing, sediments. These conclusions are

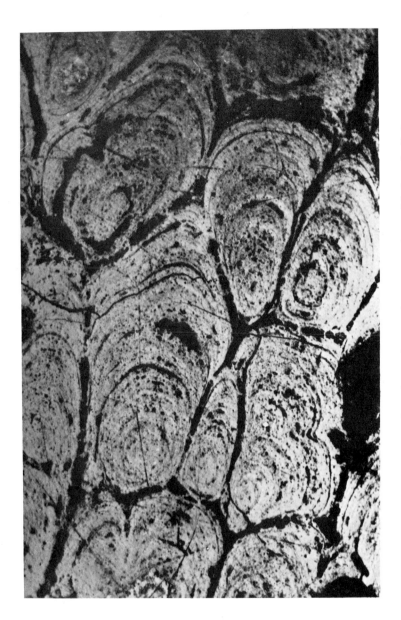

FIG. 28.5. Photograph of radially elongated segregation structures in a nodule from the Carlsberg Ridge (from *Deep-Sea Res.*, **15**, 1968, by Courtesy of Pergamon Press Ltd.).

supported by the findings of Winterhalter (1966) who described ring shaped concretions from the Baltic Sea which were convex on their upper surfaces, and flat or concave on their lower surfaces as a result of partial corrosion. In contrast, Morgenstein (1972) has described mushroom shaped concretions collected from shallow water in the Hawaiian Archipelago which grow on indurated ferruginous nuclei, the buried portions of which are free of oxide deposits. He concluded that these concretions grow at the sediment–water interface as a result of the precipitation of elements solely from the overlying seawater.

The internal structures of ferro-manganese oxide deposits are exceedingly variable and complex. Andrushchenko and Skornyakova (1969) have described at least five distinct structures in the ferro-manganese oxide phases of Pacific deep-sea nodules, and have noted that a variety of more complex structures also exist. The five principal structures are: (a) concentrically banded collomorphic structures, which were the most common variety encountered, and which were thought to have formed by the precipitation of colloidal iron and manganese oxides; (b) collomorphic globular structures, thought to have resulted from the coagulation of a gel, with subsequent dehydration of the iron and manganese oxides; (c) parallel laminar and shelly laminar structures, which were thought to have formed as a result of the precipitation of ferro-manganese oxides on relatively unaltered nuclei; (d) dendrites, thought to have resulted from the chemical weathering of volcanic rocks; and, (e) cataclastic structures, related to the break-up of nodules on the sea floor, most probably as a result of shrinkage and/or fracturing by movement.

Structures similar to these have been described in nodules from a variety of locations (see, for example: Wiseman, 1937; Arrhenius, 1963; Mero, 1965; Sorem, 1967; Cronan and Tooms, 1968; Glasby, 1970). Cronan and Tooms (1968) found the collomorphic globular structure (Fig. 28.5) to be the most common in a suite of nodules from the Carlsberg Ridge. They rejected the possibility that the globules might be aggregates of previously formed micro-nodules and concluded that the migration of manganese, and associated elements, to centres of nucleation could provide a mechanism for globule formation. Evidence for post-depositional recrystallization of manganese in nodules is provided by the presence of recrystallized dendrites of todorokite in some nodules (Sorem, 1972). Glasby (1970) has noted that continental margin nodules lack the concentrically banded structures found in most deep-sea concretions, and has concluded that this is a result of the growth rates being higher in near-shore environments. Most near-shore nodules are relatively young geologically, and their lack of structures similar to those found in deep-sea concretions would support the contention that such structures may, at least in part, be a result of ageing.

## 28.4. MINERALOGY

The mineralogy of the manganese phases present in ferro-manganese oxide deposits has been the subject of some confusion. Buser and Grütter (1956) found what they considered to be three principal manganese minerals in nodules. They termed them 7Å manganite, $\delta$-$MnO_2$ and 10Å manganite respectively. The first two of these appeared to be identical respectively to the synthetic 7Å manganous manganite and $\delta$-$MnO_2$ of Buser et al. (1954). The 10Å manganite, which in fact has a basal spacing of 9·7Å, was thought to be a modification of the 7Å manganite. Other workers (Straczek et al., 1960; Levinson, 1962; Manheim, 1965; Cronan and Tooms, 1969) adopted a different approach and compared the powder patterns of manganese nodules with those of known manganese minerals. They concluded that the principal patterns obtained by X-ray diffraction analysis of the nodules could be assigned to the previously described mixed $Mn^{2+}$–$Mn^{4+}$ oxides *todorokite* (Yoshimura, 1934) and *birnessite* (Jones and Milne, 1956). By the same technique, Andrushchenko and Skornyakova (1969) reported the presence of todorokite and birnessite in Pacific nodules. Other ferro-manganese minerals that were present included psilomelane, pyrolusite and woodruffite, which were often intimately intergrown with each other. In addition, rancieite and nsutite have been reported by Sorem and Gunn (1967) and Manheim (1965).

From a comparison of the published powder patterns of the minerals suggested to be present in nodules (Table 28.1), Cronan and Tooms (1969) concluded that the phases which some workers had been referring to as 10Å and 7Å manganite, were those which others were calling todorokite and birnessite respectively. Subsequently, Giovanoli et al. (1971) from a study of synthetic phases have distinguished two groups of manganites. These were termed the buserite and birnessite families; neither of them has a well-defined single composition, and they can probably be considered the equivalents of the 10Å manganite and 7Å manganite of Buser and Grütter (1956).

Another alternative has been put forward by Burns and Brown (1972) who considered that nodules, when initially formed, contain only one distinct authigenic ferro-manganese oxide phase, i.e. 10Å manganite. $\delta$-$MnO_2$ was thought to result from modification of the 10Å manganite by partial oxidation of the $Mn^{2+}$ and loss of structural water. Furthermore, Brown (1972) considered that the 7Å line in the X-ray diffraction pattern of nodules is simply the 101 plane of 10 Å manganite. However, this would seem to be unlikely in view of the not infrequent occurrence of nodules showing the 7·1 Å line but not the 9·7 Å line and vice versa.

No attempt will be made to resolve these differences in nomenclature.

TABLE 28.1

X-ray powder patterns of selected manganese minerals and nodules (from Cronan, 1967)

| 1 Todorokite→ dÅ | I | 2 Birnessite dÅ | I | 3 Challenger Sta. 297 dÅ | I | 4 MV-65-1-41 dÅ | I | 5 Loch Fyne dÅ | I | 6 MP 43 a dÅ | I | 7 10Å manganite dÅ | 8 δ-MnO$_2$ dÅ |
|---|---|---|---|---|---|---|---|---|---|---|---|---|---|
| 9·56–9·65 | 100 | | | 9·56 | 100 | | | 9·66 | 100 | | | 9·7 | |
| 6·98–7·20 | 15 | 7·27 | S | 7·10 | 55 | 7·18 | 100 | | | | | | |
| 4·76–4·81 | 80 | | | 4·81 | 45 | | | 4·81 | 75 | | | 4·8 | |
| 4·42–4·45 | 10 | | | | | | | 4·48 | 5 | | | | |
| | | 3·60 | W | | | 3·57 | 30 | 3·34 | 5 | | | 3·25 | |
| 3·40 | 5 | | | | | | | | | | | | |
| 3·19–3·20 | 30 | | | 3·19 | 40 | | | | | | | | |
| 3·10–3·11 | 10 | | | 3·11 | 10 | | | | | | | | |
| 2·45–2·46 | 25 | 2·44 | M | 2·44 | 40 | 2·44 | 20 | 2·45 | 15 | 2·43 | b | 2·46 | 2·43 |
| 2·39–2·40 | 45 | | | 2·39 | 30 | | | 2·40 | 20 | | | 2·39 | |
| 2·33–2·36 | 15 | | | 2·35 | 20 | | | 2·35 | 10 | | | | |
| 2·21–2·23 | 30 | | | | | | | 2·23 | 20 | | | | |
| 2·13–2·16 | 10 | | | | | | | 2·13 | 15 | | | 2·18 | |
| 1·98–2·00 | 10 | | | 1·97 | 10 | | | 1·97 | 25 | | | | |
| 1·92–1·93 | 10 | | | | | | | 1·916 | 5 | | | | |
| 1·83 | 5 | | | | | | | | | | | | |
| 1·78 | 10 | | | | | | | | | | | | |
| 1·73–1·75 | 10 | | | | | | | | | | | | |
| 1·68 | 5 | | | 1·67 | 20 | | | | | | | | |
| 1·53–1·56 | 10 | | | 1·53 | 15 | | | 1·53 | 5 | | | | |
| 1·49 | 30 | | | | | | | | | | | | |
| 1·419–1·43 | 30 | 1·412 | M | 1·42 | 40 | 1·41 | 15 | 1·41 | 25 | 1·41 | b | 1·42 | 1·41 |
| 1·38–1·40 | 15 | | | 1·40 | 30 | | | 1·40 | 25 | | | | |
| 1·33 | 5 | | | | | | | | | | | | |

1. Todorokite: Range of d spacings from Straczek et al. (1960), Frondel et al. (1960) and Levinson (1960).
2. Birnessite: Data from Jones and Milne (1956).
3. Challenger 297 (37° 29' S, 83° 07' W) containing both todorokite and birnessite.
4. MV-65-1-41 (24° 34' N, 113° 28' W) containing birnessite.
5. Loch Fyne nodule, containing todorokite.
6. MP 43 a (13° 00' N, 165° 00' E), containing δ-MnO$_2$.
7. 10 Å manganite: Data from Buser and Grütter (1956).
8. δMnO$_2$: Data from Buser and Grütter (1956).

It is sufficient to say here that X-ray diffraction data from a large number of nodules (Barnes, 1967; Andrushchenko and Skornyakova, 1969; Cronan and Tooms, 1969) indicate that they contain at least two distinct abundant manganese phases; one with a strong 9·7Å line and the other with a broad band near 2·44Å. These phases may be intergrown with each other and with other manganese minerals. In comparison, the 7·1Å line occurs infrequently. In this chapter the two principal phases will normally be termed *todorokite* and $\delta$-$MnO_2$ respectively, but the terminology of previous authors will also be used in certain cases when their work is discussed.

According to Buser (1959), the principal differences between the phases in manganese nodules are their degrees of oxidation and hydration. Buser and Grütter (1956) first suggested that there was a structural relationship between 10Å manganite, 7Å manganite and $\delta$-$MnO_2$, in that their degree of disorder and oxidation increased from the 10Å manganite through 7Å manganite to the $\delta$-$MnO_2$. Similarly, there is evidence for oxidation differences between todorokite and $\delta$-$MnO_2$. The O:Mn ratios of todorokite vary from 1·74 to 1·87 (Straczek et al., 1960; Frondel et al., 1960) in contrast to a ratio of up to 1·99 for synthetic $\delta$-$MnO_2$ (Bricker, 1965). Oxidation differences between the minerals in ferro-manganese nodules, and the possible need for stabilization of their structures by cation substitution (Buser and Grütter, 1956), have important implications in the chemistry and genesis of ferro-manganese oxide deposits, and these will be discussed subsequently.

Data on the mineralogical composition of nodules in relation to their environments of deposition indicate that there are regional variations in nodule mineralogy throughout the World Ocean (Barnes, 1967; Cronan and Tooms, 1969; Glasby, 1972). For example, if the areas marginal to the continents are excepted, todorokite-rich nodules in the Pacific occur in regions of greater average depths than do those rich in $\delta$-$MnO_2$ (Cronan and Tooms, 1969). However, the former also occur in relatively shallow inshore areas both in the Pacific and elsewhere (Cronan, 1967; Grill et al., 1968; Calvert and Price, 1970; Glasby, 1972). In contrast, nodules containing $\delta$-$MnO_2$ as their principal phase are widespread and most abundant in elevated submarine volcanic mountain areas such as the Mid-Pacific mountains, the Southern Borderland Seamount Province, and the island groups of the South Pacific. It is evident from the occurrence of todorokite in both shallow water inshore nodules, and in those from the deep ocean, that variations in nodule mineralogy cannot result from the effects of parameters which vary directly with depth, such as pressure, although the mineralogy may vary in a general way with depth in pelagic areas. In view of the probable oxidation differences between the two principal minerals in nodules, the degree of oxidation (redox potential) of the environment of deposition may play an important role in determining nodule mineralogy. Owing to its lower degree of oxidation, todorokite may be more likely to form in less oxidizing marine environments than $\delta$-$MnO_2$, in agreement with its known occurrences, such as those in continental margin sediments which often have a lower Eh than do those of the open ocean (Baas Becking *et al.*, 1960). In contrast, the tops of seamounts, where $\delta$-$MnO_2$ is particularly abundant, are among the most highly oxidizing environments in the ocean (Mero, 1965). Deep water pelagic sediments may have intermediate redox potentials and nodules from those areas sometimes contain a mixture of todorokite and $\delta MnO_2$ (Barnes, 1967).

The mineralogy of nodules from the Atlantic and Indian Oceans is less well known than is that of their Pacific counterparts. However, Bezrukov and Andrushchenko (1972) have analysed nodules from a variety of Indian Ocean localities, and have concluded that the principal minerals present are vernadite (a hydrated colloidal manganese dioxide; possibly a precursor of todorokite; Strunz, 1970), and psilomelane. Todorokite and $\delta$-$MnO_2$ were thought to occur in subordinate amounts, and this led the authors to comment that this was an apparent discrepancy between their results and those of Cronan and Tooms (1969) who considered $\delta$-$MnO_2$, in particular, to be an important mineral in Indian Ocean nodules. Comparison of the powder patterns of the phases involved can help to resolve this difference. The pattern for psilomelane given by Bezrukov and Andrushchenko shows two principal lines, one between 2·43 and 2·45Å and the other between 1·41 and 1·43Å which are almost identical in wavelength to the two lines reported for $\delta$-$MnO_2$ by Cronan and Tooms (1969). It is possible, therefore, that the principal difference between these phases is one of terminology rather than of mineralogy. If this is so it illustrates the necessity for establishing a clearly defined terminology for the minerals in nodules to reduce the possibility of further confusion.

The mineralogical composition of a nodule appears to exert an important influence on its chemical composition. Various authors (Barnes, 1967; Cronan and Tooms, 1969; Okada and Shima, 1970) have found from bulk chemical analyses that Ni and Cu are enhanced in nodules rich in todorokite, whereas those containing $\delta$-$MnO_2$ as their principal mineral phase are enriched in Co and Pb; these findings are supported by electron probe data. For example, Sano and Matsubara (1970) found that Ni and Cu were enriched in the todorokite rich phase of a Pacific nodule, and Burns and Fuerstenau (1966) have pointed out that Ni, Cu, Zn, K and Mg are enriched in the 10Å manganite (todorokite-rich) layers of the nodules which they examined.

The causes of the partition of elements between the two principal mineralogical phases of nodules have not been fully established. Goodell *et al.*

(1970) considered them to be largely a matter of colloidal and solution chemistry. Alternatively, Buser (1959) and Burns and Fuerstenau (1966) have suggested that divalent manganese in the disordered layer of 10 Å manganite (todorokite) might be replaceable by other metal ions leading to a stable phase containing Ni, Cu, Zn and other elements. Similarly, Cronan and Tooms (1969) suggested that the enrichment of Ni and Cu in todorokite-rich nodules from the Pacific and Indian Oceans might be a result of the ability of these metals to substitute as divalent ions for the $Mn^{2+}$ in todorokite. The general association of Co and Pb with $\delta$-$MnO_2$-rich nodules may result from the replacement of $Mn^{4+}$ in $\delta$-$MnO_2$ by $Co^{3+}$ and $Pb^{4+}$ (Goldberg, 1961; Sillén, 1961), or their substitution for $Fe^{3+}$ in the iron oxide phases of the nodules, such as FeOOH (Goldberg, 1963; Burns, 1965). In any event, a prerequisite for their enrichment would seem to be their oxidation from soluble $Co^{2+}$ and $Pb^{2+}$ ions to higher oxidation states under the conditions associated with $\delta$-$MnO_2$ formation.

Some todorokite-rich nodules from continental margin areas are not enriched in Ni and Cu, in contrast to the deep-sea varieties (Grill et al., 1968; Cronan and Tooms, 1969; Calvert and Price, 1970), which therefore suggests that mineralogy is not the only factor determining nodule composition. However, these nodules, together with other manganese-rich varieties from similar environments, have probably formed rapidly (Ku and Glasby, 1972) by the precipitation of manganese which may have undergone diagenetic remobilisation. Under these conditions, the manganese has probably become separated from its commonly associated minor elements. However, assuming normal slow growth rates (Bender et al., 1966; Barnes and Dymond, 1967; Ku and Broecker, 1969) and an adequate supply of minor elements, the mineralogical composition of nodules is probably one of the principal factors determining nodule minor element chemistry.

The mineralogy of the iron oxide phases in nodules is known in much less detail than that of the manganese minerals. Buser (1959), Arrhenius (1963) and Andrushchenko and Skornyakova (1969) have reported the presence of goethite and hydrogoethite, the latter occurring as laminae intimately intergrown with manganese oxides and as irregular collomorphic structures. Johnson and Glasby (1969) have shown by Mössbauer spectroscopy that iron is principally present in the form of $Fe^{3+}$ compounds in nodule samples from the Indian Ocean, and have concluded that the spectra which they obtained were probably those of iron phases consisting of a mixture of goethite and lepidocrocite. By contrast, Hrynkiewicz et al. (1972) have found the Mössbauer spectra of nodules from the South Pacific to be very similar to those obtained from $Fe(OH)_3$ gel, suggesting that in these nodules the iron may be present in a colloidal state. Carpenter et al. (1972) have reported that the concentration of $Fe^{2+}$ in nodules from both marine and freshwater environments is $< 1\%$, but also conclude that unambiguous identification of iron minerals in nodules by Mössbauer spectroscopy is not possible.

In addition to their principal authigenic minerals, nodules also usually contain a number of minor and accessory phases. These include rutile, anatase, barite, celestobarite, calcite, clay minerals, detrital silicates, phosphates and zeolites. Two of the most important non-ferro-manganese oxide minerals in nodules are montmorillonite and phillipsite. These are commonly associated with the products of submarine vulcanism, such as palagonite and volcanic glass, and probably form as a result of the alteration of these materials.

## 28.5. GEOCHEMISTRY

### 28.5.1. COMPOSITION

Data on the composition of manganese nodules are summarized in Table 28.2 where the average compositions of nodules from the Pacific, Atlantic and Indian oceans are listed. However, these data are by no means comprehensive. For example, they do not include analyses of near-shore shallow marine and continental margin nodules (which are dealt with in Section 28.5.3.1.), nor do they include some elements for which very few determinations are available. The average compositions of nodules from each of the three major oceans and the Southern Ocean, are given in Table 28.3 together with the World Ocean average.

The data presented in Tables 28.2 and 28.3 smooth out the large compositional variations often found between, and within, individual nodules. The maximum and minimum contents of Mn, Fe, Ni, Co and Cu, five of the more abundant elements in nodules, vary considerably both within each ocean and from one ocean to another (Table 28.4). Furthermore, even in Pacific nodules alone, other elements such as K, Si, Ti, Ga, Zr, Mo and Pb vary by more than a factor of 10 (Mero, 1965). Compositional variations within single nodules are also sometimes quite considerable (Willis and Ahrens, 1962; Cronan and Tooms, 1967b; Ahrens et al., 1967; Ku and Broecker, 1969). These can, in part, be attributed to variations in the detrital content of the nodules (Cronan and Tooms, 1967b). However, many nodule to nodule differences reflect variations in the mineralogy and composition of the authigenic phases (Section 28.6). Ku and Broecker (1969) found that a tabular nodule had an eight-fold variation in $^{230}$Th and in $^{231}$Pa between its upper and lower surfaces, whereas in a spherical nodule the concentrations of these nuclides were almost uniform. It was suggested that these differences were a result of the spherical nodule being able to overturn more easily than the tabular one, thus enabling the elements to accumulate more uniformly over its surface. However, Cronan and Tooms (1967b) found differences in the chemical composition of the outer layers of near-spherical nodules from a single site on the Carlsberg Ridge. They concluded that a shape theoretically suitable for uniform accumulation of elements does not necessarily lead to a uniform composition. Nevertheless, in spite of the often considerable compositional variations within individual nodules, bulk chemical analyses of morphologically similar types from single stations usually show reasonable consistency among themselves (Skornyakova *et al.*, 1962; Mero, 1965; Lorber, 1966; Cronan and Tooms, 1967b; Glasby, 1970). Thus, it is reasonable to consider individual nodules as being more or less representative of the bulk compositions of the nodules at the sites from which they were collected.

28.5.2. ELEMENT ENRICHMENT

One of the most striking features of the chemistry of the deep-sea ferro-manganese nodules is their enrichment in many elements which are usually present in the earth's crust in much lower concentrations. The magnitude of these enrichments is evident from Table 28.3 in which the deep-ocean average nodule composition has been expressed as a ratio to the average crustal abundances of the elements concerned. On the basis of these data the elements can be divided into three groups (i) B, P, V, Mn, Fe, Co, Ni, Cu, Zn, Sr, Y, Zr, Mo, Ag, Cd, Ba, La, Yb, W, Ir, Hg, Tl, Pb, and Bi, which are enriched in nodules relative to their normal crustal abundances; (ii) Na, Mg, Ca, Sn, Ti, Ga, Pd, and Au, which are neither significantly enriched nor depleted; (iii) Al, Si, K, Sc and Cr, which are somewhat depleted.

The degree of enrichment among the enriched elements varies considerably. Thus: Mn, Co, Mo and Tl are concentrated more than one hundred-fold relative to their average crustal abundances; Ni, Ag, Ir and Pb from fifty to one hundred-fold; B, Cu, Zn, Cd, Yb, W and Bi from ten to fifty-fold; and P, V, Fe, Sr, Y, Zr, Ba, La and Hg less than ten-fold. No elements are very strongly depleted in nodules relative to their normal crustal abundances, probably because all nodules contain some detrital materials which contain most of the elements that are at low concentrations in the authigenic phases.

There are several factors which determine the composition of deep-sea nodules. These include: the availability and chemical behaviour of elements in the marine environment; the adsorptive and crystallochemical properties of the authigenic phases in the nodules; the rate of accretion of the nodules, and the physicochemical nature of their environment of deposition. Most of the element enrichments in nodules can probably be ascribed to various combinations of these factors.

TABLE 28.2

*Average compositions of various groups of ferro-manganese oxide deposits (in wt. %)*

| | 1 | 2 | 3 | 4 | 5 | 6 | 7 | 8 | 9 | 10 | 11 | 12 | 13 | 14 |
|---|---|---|---|---|---|---|---|---|---|---|---|---|---|---|
| B | — | 0·005 | — | — | — | 0·029 | — | — | — | — | — | — | — | — |
| Na | — | 1·92 | — | — | — | 2·06 | — | — | — | — | 1·88 | — | — | — |
| Mg | — | 0·265 | — | — | — | 1·76 | — | — | 2·78 | 1·90 | 1·87 | 3·60 | — | — |
| Al | 0·73 | 3·23 | — | — | — | 3·27 | — | 3·27 | — | — | — | — | — | — |
| Si | 0·54 | 10·37 | — | — | — | 8·27 | — | 10·61 | — | 7·25 | — | 11·40 | — | — |
| P | — | 0·126 | — | — | — | 0·17 | — | 0·098 | — | — | — | — | — | — |
| K | — | 1·24 | — | — | — | 0·74 | — | 0·493 | — | — | 0·571 | — | — | — |
| Ca | — | 1·35 | — | — | 1·86 | 1·98 | — | 3·32 | — | 3·82 | 2·90 | 1·44 | 4·54 | — |
| Sc | — | 0·0006 | — | — | — | 0·001 | — | — | — | — | — | — | — | — |
| Ti | 0·81 | 0·431 | 0·690 | — | — | 0·66 | — | 0·446 | 0·290 | 0·421 | — | 0·421 | — | 0·681 |
| V | — | 0·035 | 0·044 | 0·203 | — | 0·054 | — | — | 0·04 | 0·059 | — | — | — | 0·044 |
| Cr | — | 0·0005 | 0·0011 | 0·0099 | — | 0·001 | — | — | 0·007 | — | — | — | — | 0·0014 |
| Mn | 19·00 | 17·03 | 17·50 | 14·25 | 27·76 | 21·06 | 17·35 | 12·80 | 12·11 | 23·46 | 15·97 | 14·38 | 17·07 | 14·88 |
| Fe | 13·80 | 12·17 | 11·87 | 12·10 | 11·89 | 11·97 | 10·77 | 14·96 | 23·34 | 16·34 | 21·55 | 14·05 | 13·10 | 13·16 |
| Co | 0·28 | 0·10 | 0·408 | 0·504 | 0·553 | 0·31 | 0·30 | 0·360 | 0·087 | 0·312 | 0·318 | 0·19 | 0·324 | 0·238 |
| Ni | 0·46 | 0·28 | 0·682 | 0·045 | 0·286 | 0·67 | 0·40 | 0·344 | 0·516 | 0·528 | 0·313 | 0·631 | 0·720 | 0·432 |
| Cu | 0·55 | 0·31 | 0·379 | — | — | 0·43 | 0·21 | 0·126 | 0·060 | 0·130 | 0·115 | 0·536 | 0·150 | 0·235 |
| Zn | — | 0·038 | — | — | 0·059 | 0·071 | — | 0·055 | — | 0·090 | 0·087 | 0·057 | 0·064 | — |
| Ga | — | 0·0013 | — | — | — | 0·001 | — | — | — | — | — | — | — | — |
| Sr | — | 0·064 | — | — | — | 0·086 | — | 0·1103 | 0·002 | — | — | 0·086 | — | — |
| Y | — | 0·002 | — | — | — | 0·033 | — | — | — | — | — | — | — | — |
| Zr | 0·0064 | 0·022 | — | — | — | 0·063 | — | — | — | — | — | — | — | — |
| Mo | — | 0·022 | 0·039 | 0·059 | — | 0·052 | — | 0·058 | 0·011 | 0·046 | — | 0·022 | — | 0·030 |
| Pd | — | — | — | — | — | — | 0·602$^{-6}$ | — | — | 0·574$^{-6}$ | — | — | 0·391$^{-6}$ | — |
| Ag | — | 0·001 | — | — | — | 0·0003 | — | — | — | — | — | — | — | — |
| Cd | — | 0·0007 | — | — | 0·0007 | — | — | — | — | 0·0011 | — | — | — | — |
| Sn | — | — | — | — | — | — | 0·00027 | — | — | — | — | — | — | — |

| | 1 | 2 | 3 | 4 | 5 | 6 | 7 | 8 | 9 | 10 | 11 | 12 | 13 | 14 |
|---|---|---|---|---|---|---|---|---|---|---|---|---|---|---|
| Te | — | — | — | — | — | 0·005 | — | — | — | — | — | — | — | — |
| Ba | 0·330 | 0·215 | — | — | — | 0·320 | — | — | — | — | 0·498 | 0·370 | — | 0·151 |
| La | 0·02 | — | — | — | — | 0·016 | — | — | — | — | — | — | — | — |
| Yb | — | — | — | — | — | 0·0031 | — | — | — | — | — | — | — | — |
| W | 0·006 | — | — | — | — | — | — | — | — | — | — | — | — | — |
| Ir | — | — | — | — | — | 0·939$^{-6}$ | — | — | — | 0·932$^{-6}$ | — | — | 0·81$^{-7}$ | — |
| Au | — | — | — | — | — | 0·266$^{-6}$ | — | — | — | 0·302$^{-6}$ | — | — | 0·15$^{-6}$ | — |
| Hg | 0·0002 | — | — | — | — | 0·320$^{-4}$ | — | — | — | 0·165$^{-4}$ | — | — | 0·010 | — |
| Tl | 0·01 | — | — | 0·0194 | — | — | — | — | — | 0·0077 | — | — | — | — |
| Pb | 0·017 | 0·054 | 0·321 | 0·089 | — | 0·10 | — | 0·040 | — | 0·099 | 0·150 | 0·132 | 0·140 | — |
| Bi | 0·001 | — | — | 0·0005 | — | — | — | — | — | 0·0005 | — | — | 0·0014 | 0·045 |

1. Pacific Ocean data from Goldberg (1954) in Manheim (1965).
2. Pacific Ocean data from Riley and Sinhaseni (1958) in Manheim (1965).
3. Pacific Ocean data from Cronan and Tooms (1969).
4. Pacific Ocean data from Glasby (1970), from the southwestern Pacific.
5. Pacific Ocean data from Ahrens et al. (1967).
*6. Pacific Ocean data from Mero (1965) and Skornyakova and Andrushchenko (1970).
7. Pacific Ocean data from Okada and Shima (1970), Harriss et al. (1968), Lakin et al. (1963), Harriss (1968), and Smith and Burton (1972).
8. Atlantic Ocean data from Mero (1965).
9. Atlantic Ocean data from Bacon (1967).
10. Atlantic Ocean data from Willis and Ahrens (1962), Ahrens et al. (1967), Harriss (1968) and Harriss et al. (1968).
11. Atlantic Ocean data from Cronan (1972a) and unpublished analyses.
12. Indian Ocean data from Mero (1965).
13. Indian Ocean data from Ahrens et al. (1967), Harriss (1968), and Harriss et al. (1968).
14. Indian Ocean data from Cronan and Tooms (1969).

* The average values in column 6 are taken largely from Skornyakova and Andrushchenko (1970), because these authors did not publish the averages of their own data but combined them with those of Mero (1965).

Note: Superscript indicates powers of ten, e.g. $^{-6} = \times 10^{-6}$.

TABLE 28.3

Average abundances of elements in ferro-manganese oxide deposits from each of the major oceans (in wt%), together with the world average, crustal abundance (in wt%), and enrichment factor for each element

| | Pacific Ocean | Atlantic Ocean | Indian Ocean | Southern Ocean* | World Ocean Average | Crustal Abundance† | Enrichment Factor |
|---|---|---|---|---|---|---|---|
| B | 0·0277 | — | — | — | — | 0·0010 | 27·7 |
| Na | 2·054 | 1·88 | — | — | 1·9409 | 2·36 | 0·822 |
| Mg | 1·710 | 1·89 | — | — | 1·8234 | 2·33 | 0·782 |
| Al | 3·060 | 3·27 | 3·60 | — | 3·0981 | 8·23 | 0·376 |
| Si | 8·320 | 9·58 | 11·40 | — | 8·624 | 28·15 | 0·306 |
| P | 0·235 | 0·098 | — | — | 0·2244 | 0·105 | 2·13 |
| K | 0·753 | 0·567 | — | — | 0·6427 | 2·09 | 0·307 |
| Ca | 1·960 | 2·96 | 3·16 | — | 2·5348 | 4·15 | 0·610 |
| Sc | 0·00097 | — | — | — | — | 0·0022 | 0·441 |
| Ti | 0·674 | 0·421 | 0·629 | 0·640 | 0·6424 | 0·570 | 1·13 |
| V | 0·053 | 0·053 | 0·044 | 0·060 | 0·0558 | 0·0135 | 4·13 |
| Cr | 0·0013 | 0·007 | 0·0014 | — | 0·0014 | 0·01 | 0·14 |
| Mn | 19·78 | 15·78 | 15·12 | 11·69 | 16·174 | 0·095 | 170·25 |
| Fe | 11·96 | 20·78 | 13·30 | 15·78 | 15·608 | 5·63 | 2·77 |
| Co | 0·335 | 0·318 | 0·242 | 0·240 | 0·2987 | 0·0025 | 119·48 |
| Ni | 0·634 | 0·328 | 0·507 | 0·450 | 0·4888 | 0·0075 | 65·17 |
| Cu | 0·392 | 0·116 | 0·274 | 0·210 | 0·2561 | 0·0055 | 46·56 |
| Zn | 0·068 | 0·084 | 0·061 | 0·060 | 0·0710 | 0·007 | 10·14 |
| Ga | 0·001 | — | — | — | — | 0·0015 | 0·666 |
| Sr | 0·085 | 0·093 | 0·086 | 0·080 | 0·0825 | 0·0375 | 2·20 |
| Y | 0·031 | — | — | — | — | 0·0033 | 9·39 |
| Zr | 0·052 | — | — | 0·070 | 0·0648 | 0·0165 | 3·92 |
| Mo | 0·044 | 0·049 | 0·029 | 0·040 | 0·0412 | 0·00015 | 274·66 |
| Pd | $0·602^{-6}$ | $0·574^{-6}$ | $0·391^{-6}$ | — | $0·553^{-6}$ | $0·665^{-6}$ | 0·832 |

| | | | | | | | |
|---|---|---|---|---|---|---|---|
| Ag | 0·0006 | — | — | — | — | 0·000007 | 85·71 |
| Cd | 0·0007 | 0·0011 | — | — | 0·00079 | 0·00002 | 39·50 |
| Sn | 0·00027 | — | — | — | — | 0·00002 | 13·50 |
| Te | 0·0050 | — | 0·182 | 0·100 | — | — | — |
| Ba | 0·276 | 0·498 | — | — | 0·2012 | 0·0425 | 4·73 |
| La | 0·016 | — | — | — | — | 0·0030 | 5·33 |
| Yb | 0·0031 | — | — | — | — | 0·0003 | 10·33 |
| W | 0·006 | — | — | — | — | 0·00015 | 40·00 |
| Ir | $0·939^{-6}$ | $0·932^{-6}$ | $0·811^{-7}$ | — | $0·935^{-6}$ | $0·132^{-7}$ | 70·83 |
| Au | $0·266^{-6}$ | $0·302^{-6}$ | $0·15^{-6}$ | — | $0·248^{-6}$ | $0·400^{-6}$ | 0·62 |
| Hg | $0·82^{-4}$ | $0·16^{-4}$ | $0·15^{-6}$ | — | $0·50^{-4}$ | $0·80^{-5}$ | 6·25 |
| Tl | 0·017 | 0·0077 | 0·010 | — | 0·0129 | 0·000045 | 286·66 |
| Pb | 0·0846 | 0·127 | 0·070 | — | 0·0867 | 0·00125 | 69·36 |
| Bi | 0·0006 | 0·0005 | 0·0014 | — | 0·0008 | 0·000017 | 47·05 |

* Data from Goodell *et al.* (1970).

† Data from Taylor (1964).

*Note:* Superscript numbers denote powers of ten, e.g. $^{-6} = \times 10^{-6}$.

TABLE 28.4

*Variability of Mn, Fe, Ni, Co and Cu in ferro-manganese oxide concretions from the Atlantic, Pacific and Indian Oceans (in wt. %)*

| | Atlantic* | | | Pacific† | | | Indian‡ | | |
|---|---|---|---|---|---|---|---|---|---|
| | Maximum | Minimum | Ratio | Maximum | Minimum | Ratio | Maximum | Minimum | Ratio |
| Mn | 37·69 | 1·32 | 28·55 | 41·1 | 8·2 | 5·01 | 29·16 | 11·67 | 2·49 |
| Fe | 41·79 | 4·76 | 8·77 | 32·73 | 2·4 | 13·6 | 26·46 | 6·71 | 3·94 |
| Ni | 1·41 | 0·019 | 74·21 | 2·37 | 0·16 | 14·81 | 2·01 | 0·167 | 12·03 |
| Co | 1·01 | 0·017 | 59·41 | 2·58 | 0·014 | 184·28 | 1·04 | 0·068 | 15·29 |
| Cu | 0·884 | 0·022 | 40·18 | 1·97 | 0·028 | 70·35 | 1·38 | 0·029 | 47·58 |

* Data of Cronan (1972a).
† Data of Mero (1965) and Cronan and Tooms (1969).
‡ Data of Cronan and Tooms (1969).

Excluding oxygen, manganese is the principal constituent of nodules. It attains its high concentration as a result of its fractionation and separation from the principal rock forming elements during processes of weathering, transport, deposition and diagenesis. One of the more important characteristics of nodules is their greater enrichment with manganese than iron relative to their crustal abundances. According to Krauskopf (1957), the separation of these elements in secondary environments is largely a function of pH and Eh. Increase in either of these parameters leads to the precipitation of iron from solution before manganese. In general, redox potentials increase from lacustrine and near-shore areas to the deep sea, and iron is selectively removed in the former environments leaving manganese to be enriched in the latter. A possible example of this process on a limited scale is to be found in Lake Ontario (Cronan and Thomas, 1972), where manganese and iron-bearing waters traverse a redox gradient and deposit first iron and then manganese as the redox potential increases. However, manganese is not always enriched relative to iron in deep-sea nodules. Those forming in areas receiving a high input of Fe-bearing terrigenous detritus (Manheim, 1965), or near submarine volcanic sources of iron (Bonatti and Joensuu, 1966), often have Fe/Mn ratios greater than unity. The enrichment of iron relative to manganese in nodules from volcanic areas near the centres of the oceans, e.g. the Mid-Atlantic Ridge (Cronan, 1972a), can probably be best explained by local volcanic sources of iron (Section 28.5.3.2.).

The hydrous manganese and iron oxides in nodules are characterized by high specific surface areas, and are capable of strong interaction with cations in solution. According to Stumm and Morgan (1970), adsorption of metal ions onto these oxides can explain the high concentrations of Cu, Ni, Zn and Pb in ferro-manganese nodules, and possibly also those of other elements. Krauskopf (1956) found that Cu, Zn and Pb can be rapidly removed from sea water by adsorption onto iron and manganese oxides, and that although Ni and Co are less easily adsorbed, manganese dioxide adsorbs them the more efficiently, than do iron oxides.

Although it may be possible to account for the initial fixation of many minor metals in manganese nodules by adsorption, it is unlikely that this mechanism can account for the very high concentrations of some minor elements in nodules occasionally found in certain environments. For example, Ni and Cu are enriched, to a greater or lesser extent, in all deep-ocean nodules relative to their normal crustal abundances, but, they reach their highest concentrations in the deep-water varieties rich in todorokite. As mentioned above this may result from substitution of $Ni^{2+}$ and $Cu^{2+}$ in the todorokite lattice (Section 28.4). In this context, it is interesting to note that McKenzie (1971) has shown that synthetic preparations of todorokite can accept large concentrations of these elements. The strong enrichment of

I

cobalt and lead in some sea-mount nodules in the central areas of the ocean, which it is difficult to explain on the basis of adsorption alone, may also be attributed to substitution. The enrichments of these two elements may be due to their oxidation to higher oxidation states, and substitution within $\delta$-$MnO_2$ (Section 28.4).

Little is known about the causes of the enrichments of most of the less abundant minor elements in nodules. Molybdenum follows manganese very closely (Cronan, 1969), and, unlike elements such as nickel and copper which also follow manganese, it is enriched in the Mn-rich, Fe-depleted, continental margin nodules which are formed as a result of the diagenetic remobilization of manganese at depth in the sediments (Section 28.5.3.1.). Thus, molybdenum may undergo post-depositional remobilization and reprecipitation. In contrast, vanadium tends to follow iron, possibly because it is more strongly adsorbed by iron oxides than by manganese oxides (Krauskopf, 1956). Thallium in nodules has some geochemical features in common with cobalt (Willis and Ahrens, 1962), although no statistical correlation between these elements, or between Tl and either Fe or Mn, has been observed. Both have been found to be strongly depleted in Fe-poor continental margin nodules relative to their normal abundance in the concretions. Barium can be present in nodules either in the manganese minerals, or in the forms of barite and celestobarite (Arrhenius, 1963). Barium is sometimes enriched in nodules from seamounts (Cronan and Tooms, 1969), possibly as a result of biological mechanisms (Church, 1970), or from supply from volcanic sources (Arrhenius and Bonatti, 1965), or the formation of psilomelane (Skornyakova and Andrushchenko, 1970). The enrichments of Bi, Cd, W, Ag, and Hg in nodules can probably best be explained by their adsorption onto the ferro-manganese oxide phases (Krauskopf, 1956). Other processes have been suggested to account for trace element enrichment. For example, Harriss et al. (1968) have suggested that palladium and iridium may be derived from noble metal-enriched meteoritic material. A process which has received increasing attention in recent years is submarine hydrothermal activity which might account for the enrichment of some volatile elements in nodules. Ferro-manganoan sediments associated with the World Mid-Oceanic Ridge System are enriched in mercury, arsenic and other volatile elements (Boström et al., 1969; Boström and Fisher, 1969; Boström and Valdes, 1969; Horowitz, 1970; Cronan, 1972b), and nodules from these areas might also be expected to contain above average concentrations of these elements.

28.5.3. REGIONAL GEOCHEMISTRY

28.5.3.1. *Shallow marine and continental margin nodules*

Ferro-manganese oxide deposits in near-shore and continental margin environments are sufficiently different from deep-sea nodules to warrant separate consideration. The former are characterized by: (a) their Mn/Fe ratios which are very variable, although uniformly high ratios occur in nodules overlying sediments which have a low redox potential just below the sediment surface; (b) their low contents of many relatively minor elements; (c) their comparatively rapid growth rates, (d) their relatively high contents of organic matter; (e) their relatively low O:Mn ratios. Some of these characteristics are seen in the analyses of nodules from Loch Fyne and from the continental margins off British Columbia and Baja California, (see Table 28.5). These nodules can be divided into two groups. Those from Hodgkins

TABLE 28.5

*Mn, Fe, Ni, Co and Cu in selected near-shore ferro-manganese oxide deposits (in wt. %)*

|    | 1 | 2 | 3 | 4 | 5 | 6 | 7 | 8 | 9 |
|----|------|------|------|------|------|------|------|------|------|
| Mn | 19·56 | 14·91 | 41·22 | 42·47 | 28·12 | 30·02 | 33·90 | 33·92 | 34·12 |
| Fe | 23·15 | 28·61 | 4·90 | 4·44 | 5·39 | 2·24 | 1·69 | 1·99 | 1·18 |
| Ni | 0·179 | 0·259 | 0·050 | 0·024 | <0·005 | 0·01 | 0·111 | 0·110 | 0·069 |
| Co | 0·437 | 0·191 | 0·016 | 0·016 | 0·014 | 0·026 | 0·010 | 0·007 | 0·0055 |
| Cu | 0·037 | 0·04 | 0·008 | 0·003 | 0·0012 | 0·0022 | 0·057 | 0·086 | 0·052 |

1. Encrustation from Hodgkins Seamount (53° 31′ N, 136° 08′ W).
2. Encrustation from Dellwood Seamount (50° 37·6′ N, 130° 48·6′ W).
3. Nodule from Jervis Inlet, British Columbia.
4. Nodule from Georgia Strait, British Columbia.
5 and 6. Nodules from Loch Fyne, Scotland (data from Cronan, 1967).
7, 8 and 9. Nodules from the continental borderland off Baja California
   (data from Cronan and Tooms, 1969).

and Dellwood Seamounts contain more iron than manganese, whereas the remainder contain the converse. The latter group are also characterized by lower nickel and cobalt contents than the former.

The highly variable Mn/Fe ratios in shallow marine and continental margin nodules can probably be ascribed to a number of factors. According to Manheim (1965), iron generally becomes more predominant over manganese in nodules as continental influences increase. This is thought to be a result of the much greater amounts of iron relative to manganese in continental run-off. However, Glasby (1970) has noted that the high iron content of some nodules from near-shore environments appears to be related to the presence of phosphate ions and has suggested that iron-rich shallow water nodules might result from the precipitation of ferric phosphate in preference to ferric oxide or hydroxide. Many nodules from near-shore environments are, in fact, richer in phosphorus than are their deep-ocean counterparts (Manheim, 1965; Winterhalter, 1966), and electron probe studies have shown that this phosphorus is concentrated in the iron-rich layers of these nodules (Winterhalter and Siivola, 1967). However, these workers also noted that the concretions which they examined contained more iron than could be bound in ferric phosphate, and concluded that phosphorus in these concretions was more likely to be present as a result of the scavenging effect of ferric hydroxide on the $PO_4^{3-}$ anion than through its precipitation as ferric phosphate.

The very low manganese concentrations found occasionally in nearshore concretions might be a result of their formation under redox conditions too low for the precipitation of manganese dioxide. In such cases, the manganese present may simply exist as $Mn^{2+}$ adsorbed onto the ferric oxides (Stumm and Morgan, 1970). Collins and Buol (1970) have noted that when iron is precipitated before manganese, some of the divalent manganese in solution is removed by adsorption onto the hydrated iron oxide phases.

Nodules containing high concentrations of manganese relative to iron, such as those exemplified in Table 28.5, are probably formed as a result of the diagenetic remobilization of manganese under reducing conditions. Such conditions are encountered, for example, at shallow depths in sediments off Baja California (Lynn and Bonatti, 1965) as a consequence of the burial of organic matter. In such areas, manganese is reduced to the soluble divalent state and migrates upwards towards the sediment surface where it is reoxidized and reprecipitated. This process could effectively separate manganese from iron. The latter would require a lower redox potential to reduce it, and might remain behind in the sediments in the form of iron sulphide (Cheney and Vredenburgh, 1968).

The generally low concentrations of many minor elements in near-shore and continental margin nodules may be related to both their rates of accumulation and the effects of any organic influences upon them. According to Manheim (1965), the high growth rates of near-shore nodules, compared with those of the nodules of deep-sea areas can be accounted for in terms of the higher concentrations of manganese and iron in the waters of the former environment. Their low minor element contents are a result of their high growth rates coupled with lesser variations in the concentrations of the minor elements between the waters of the two environments. Similarly, Mero (1965) has ascribed the low Ni, Cu and Co contents of manganese-rich Pacific margin nodules to the rapid precipitation of manganese dioxide before it has time to scavenge significant concentrations of these minor elements from sea water. In general, the more rapid the growth rates of nodules are, the lower their content of certain minor elements seems to be. This conclusion can also be extended to some lake concretions in which some minor elements are inversely related to the rate of accumulation of the manganese phases (Cronan and Thomas, 1970). The second factor that may be of importance in contributing to the low minor element content of near-shore nodules is the possible complexation of minor elements by organic compounds. Price (1967) has suggested that the presence of water soluble organic compounds, such as amino and humic acids, in buried reducing sediments could cause the complexing of elements such as Cu, Zn and Pb and lead to their depletion in ferro-manganese oxide minerals.

The general association of near-shore and continental margin nodules with less oxidizing environments than deep-sea nodules is exemplified by their lower O:Mn ratios and higher concentrations of organic materials. Manheim (1965) found O:Mn ratios of 1·5–1·9 in shallow water nodules, in contrast to those of 1·8–2·0 in deep-sea varieties. These low O:Mn ratios reflect the presence of minerals of lower oxidation grade than those commonly abundant in deep-sea concretions.

### 28.5.3.2. Deep-sea nodules

#### (a) Inter-oceanic variations

Differences in the average compositions of nodules from the three major oceans may be summarized from data in Table 28.3 as follows. The Mn/Fe ratio is greater than unity in the Pacific and Indian Oceans, but less than unity in the Atlantic. Nickel and copper decrease in abundance in the nodules in the order Pacific > Indian Ocean > Atlantic varieties. Co, Mo, Pd, Ba, Au and Hg are lowest in the Indian Ocean, whereas P, Ti and Tl are lowest in the Atlantic.

The compositional ranges of Mn, Fe, Ni, Co and Cu in nodules are presented in Table 28.4, together with ratios of the maximum to minimum values for each element. Mn and Ni vary most in the Atlantic, whereas Fe, Co and Cu show their greatest variation in the Pacific. The apparently limited variability of these elements in Indian Ocean nodules probably reflects the limited sampling in this ocean.

Factors causing inter-oceanic differences in the compositions of nodules will, of course, largely be the same as those causing regional variations in nodule composition within each of the oceans, and will be discussed subsequently. However, one aspect deserves comment here. The lower Mn/Fe ratio and minor element content of concretions in the Atlantic relative to those from the Pacific may result from the higher sedimentation rates in the former. In general, high Mn concentrations in nodules, other than those from continental borderland and equatorial regions, appear to be correlated with low rates of sedimentation of terrigenous material (Skornyakova and Andrushchenko, 1970; Ewing et al., 1971; Horn et al., 1972a). In contrast, iron concentrations tend to increase in nodules as continental influence increases, except for those from volcanic areas.

(b) *Intra-oceanic variations*

(1) *Pacific Ocean.* The regional geochemistry of the manganese nodules of the Pacific has been the subject of several investigations. Mero (1962, 1965) and Skornyakova *et al.* (1962) first showed that the Pacific could be divided into zones according to the composition of their nodules (Fig. 28.6). Subsequent investigations by Skornyakova and Andrushchenko

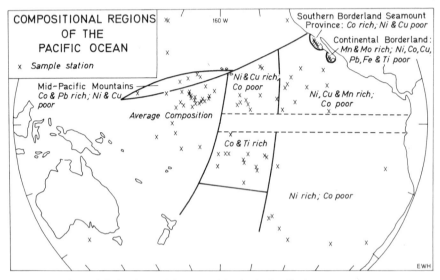

FIG. 28.6. Regions of distinctive nodule composition in the Pacific Ocean (average compositions are given in Table 28.6).

(1964, 1968, 1970), Barnes (1967), Cronan (1967), Cronan and Tooms (1969) and Goodell *et al.* (1970) have confirmed and amplified this conclusion. The average abundances of Mn, Fe, Ni, Co and Cu in nodules from the different areas of this ocean are given in Table 28.6.

Manganese is usually present in concentrations exceeding 20 % in nodules from the pelagic areas of the eastern Pacific having low sedimentation rates. Its concentration is variable in nodules of the equatorial zone, in which carbonate sedimentation is rapid, and generally low in the western Pacific. However, a manganese-related suite of elements has been reported by Goodell *et al.* (1970) in the south-western Pacific basin. The highest manganese concentrations in the nodules of the Pacific (> 30 %), occur in those from the American continental borderland, and are a result of the diagenetic remobilization of manganese (Section 28.5.3.1.). The behaviour of iron is the converse of that of manganese, as it is generally depleted in the nodules of the East Pacific and in particular in the continental borderland areas. It

TABLE 28.6

*Average abundances of Mn, Fe, Ni, Co and Cu in nodules from different regions of the Pacific and Indian Oceans From Cronan (1967) and Cronan and Tooms (1969) in wt. % on air dried weight*

|          | 1      | 2      | 3      | 4      | 5      | 6      | 7      | 8      | 9      | 10     | 11     |
|----------|--------|--------|--------|--------|--------|--------|--------|--------|--------|--------|--------|
| Mn       | 12·29  | 16·87  | 13·96  | 15·71  | 16·61  | 22·33  | 19·81  | 15·85  | 33·98  | 13·56  | 15·83  |
| Fe       | 12·00  | 13·30  | 13·10  | 9·06   | 13·92  | 9·44   | 10·20  | 12·22  | 1·62   | 15·75  | 11·31  |
| Ni       | 0·422  | 0·564  | 0·393  | 0·956  | 0·433  | 1·080  | 0·961  | 0·348  | 0·097  | 0·322  | 0·512  |
| Co       | 0·144  | 0·395  | 1·127  | 0·213  | 0·595  | 0·192  | 0·164  | 0·514  | 0·0075 | 0·358  | 0·153  |
| Cu       | 0·294  | 0·393  | 0·061  | 0·711  | 0·185  | 0·627  | 0·311  | 0·077  | 0·065  | 0·102  | 0·330  |
| Depth(m) | 4990   | 5001   | 1757   | 5049   | 3539   | 4537   | 4324   | 1146   | 3033   | 3793   | 5046   |

1. North Pacific
2. West Pacific
3. Mid-Pacific Mountains
4. Central Pacific
5. South Pacific
6. North-east Pacific

7. South-East Pacific.
8. Southern Borderland Seamount Province
9. Continental Borderland
10. Western Indian Ocean
11. Eastern Indian Ocean

increases in a southerly and westerly direction and reaches maximum values of 15–25% in the vicinity of some of the island groups of the South and West Pacific. Iron is also abundant in nodules found near the Phillipines Trench, on parts of the Macquarie Ridge and on the Pacific-Antarctic Ridge (Skornyakova and Andrushchenko, 1970; Summerhayes, 1967; Goodell et al., 1970).

In general, the minor elements in Pacific nodules are associated with iron or manganese. For example, Ni, Cu, Mo and Zn are enriched in Mn-rich nodules from the eastern Pacific,* and are depleted in the Fe-rich varieties from the west. Maximum Ni and Cu contents, exceeding 1% occur in todorokite-rich nodules from the deep water areas of siliceous ooze deposition in the northeast Pacific. Both elements generally decrease in a westerly direction. Excluding the continental borderlands, these elements reach their lowest concentrations in the elevated areas of the southern and western Pacific, Cu being especially low on seamounts. Co, Pb, Sn, Ti and V generally vary in a converse manner to Ni and Cu, and, with some exceptions, tend to follow iron (Skornyakova and Andrushchenko, 1970; Cronan and Tooms, 1969). Co and Pb are enriched in $\delta$-$MnO_2$-rich nodules from the elevated volcanic areas of the western and southern Pacific. Chromium differs from most elements in that its distribution in nodules is related to detrital silicate phases rather than to the authigenic minerals. It is depleted in nodules relative to its normal crustal abundance, probably because under the oxidizing conditions in which nodules form it would tend to be oxidized to soluble chromate ions. In the Pacific this element reaches relatively high concentrations, (i.e. > 20 ppm), in some nodules from volcanic areas which contain abundant, partially weathered, volcanic materials.

Factors determining regional variations in the composition of Pacific nodules have been discussed by several workers, including Mero (1965), Barnes (1967), Cronan and Tooms (1969), Price and Calvert (1970), Skornyakova and Andrushchenko (1970) and Cronan (1967, 1972c, 1974). Two important considerations are (i) the location of the nodule deposits relative to continental or submarine volcanic sources of elements and (ii) the redox conditions operative in their environment of deposition. The latter influences both the Mn/Fe ratio, and the mineraology of the nodules, and thus their minor element content (Section 28.4). For example, manganese concentrations and Mn/Fe ratios are high in some marginal and equatorial areas of the Pacific, because of the diagenetic remobilization of manganese under low redox conditions caused by the burial of organic matter as a result of rapid sedimentation (Lynn and Bonatti, 1965; Price and Calvert, 1970; see also Section 28.5.3.1.). Elsewhere, manganese appears to be more enriched in nodules from areas of relatively slow sedimentation than in regions in

* Other than those from the continental borderland.

which biogenic and terrigenous sedimentation rates are rapid (Opdyke and Foster, 1970; Skornyakova and Andrushchenko, 1970; Horn et al., 1972c; Cronan, 1972c). This suggests that diagenetic remobilization of manganese has little effect on nodule composition outside the marginal and equatorial areas of the ocean.

The influence of redox potentials and nodule mineralogy on the Ni, Co, Cu and Pb contents of nodules has already been discussed (Section 28.4). If continental margin areas are excluded, Ni and Cu are enriched in deep water nodules, whereas Co and Pb are most abundant in those from shallower depths, and particularly on seamounts. Regional variations in the depth of the Pacific affect redox conditions and hence nodule mineralogy, and so may indirectly be a factor of considerable importance in determining regional minor element variations in Pacific deep-sea nodules.

(2) *Indian Ocean.* Regional variations in the compositions of Indian Ocean nodules are known in less detail than those in the Pacific. However, the few samples from this ocean which have been analyzed indicate regional variations similar to those of Pacific nodules (Cronan and Tooms, 1969; Willis, 1970; Bezrukov and Andrushchenko, 1972). Manganese is richest in the east and generally low in the west, whereas Fe behaves conversely. The latter has intermediate to low concentrations over much of the eastern Indian Ocean, and also in some of the basins in the west, but its concentration is intermediate to high in the elevated volcanic areas of the west, such as the Carlsberg Ridge. Of the minor elements, Ni and Cu are generally most abundant in the east and in some of the basins in the west, but are intermediate to low in concentration in the elevated areas. Cobalt and lead behave conversely, being highest in the elevated areas in the west, intermediate in the intervening basins, and generally low over much of the pelagic eastern Indian Ocean. Average compositions of nodules from the eastern and western basins respectively are shown in Table 28.6.

Superimposed on these regional variations are local ones associated with the Carlsberg Ridge. Nodules from a small area on the flanks of the Ridge (Area 4C, $2° 48'$ N–$2° 45'$ N, $59° 52'$ E–$60° 03'$ E), were found by Cronan and Tooms (1967b) to show considerable compositional variations between morphologically different nodules from one site, and between morphologically similar nodules from two sites only a few miles apart. With regard to the latter, the nodules from one site were enriched in Ni and Cu (unlike most mid-ocean ridge nodules), whereas those from the other were relatively poor in these elements, but were richer in Fe, Co and Ti. Mineralogical differences were also observed; nodules from the first of these stations contained todorokite whereas $\delta$-$MnO_2$ was the principal phase at the other. These samples were collected from an area of extreme topographic contrasts and it is possible that local variations in the redox conditions of the environment of deposition caused both the mineralogical and chemical variations. Glasby (1970) analyzed additional material from the same area and also found distinct compositional variations between discrete morphological nodule populations. However, the variations were smaller than those reported by Cronan and Tooms (1967b) and, in part, the morphologies of the nodules were different. This sampling area (4C) is one of the most intensively sampled small areas of the ocean floor, and the variability of the composition of its nodules illustrates just how much variation there can be in nodule composition in an area of strong topographic contrasts.

(3) *Atlantic Ocean.* Although there are considerable variations in the compositions of Atlantic nodules (Table 28.4), regional variations in their compositions are less distinct than those in the other two oceans. Manganese is most abundant in nodules from the eastern and central South Atlantic, and least in those from the south-western Atlantic (Cronan, 1972a). Over much of the tropical and northern Atlantic, manganese concentrations range between 10 and 15%, but higher values occur in parts of the south-western and eastern North Atlantic. The highest iron concentrations occur in the nodules of the south-western Atlantic, a region where manganese is low. However, the level of iron is $< 20\%$ in the nodules of much of the South Atlantic. Iron varies irregularly in the North Atlantic nodules, but tends to be above average in those of the Mid-Atlantic Ridge (Cronan, 1972; Scott et al., 1972). The concentrations of Ni, Cu, Co and Zn tend to follow that of manganese in the nodules.

Superimposed on these regional trends there are local variations in the composition of concretions in at least two areas of dissected topography (Cronan, 1967, and unpubl. data). Encrustations from near the Kings Trough area, at approx. 42° 54′N, 20° 12′W, have variable contents of several elements. Mn, Ni and Cu form one suite, and Fe, Co and V another. In contrast, ferro-manganese oxide encrustations from the Mid-Atlantic Ridge near 45° N have less variable contents of Fe, Cu, Zn and Co than those from the Kings Trough area, but greater variations in Mn and Ni.

The existence of local variations in the compositions of nodules from small areas of elevated and dissected topography does not invalidate the conclusions drawn regarding oceanwide regional variations. Local variations on a similar scale have not been recorded in the literature to date for nodules from regions of uniform topography, and the available evidence indicates that these areas, which can be very extensive, contain nodules of fairly similar compositions relative to the overall variability in nodule composition (Mero, 1965; Cronan, 1967).

(c) *Sub-surface nodules.*

Sub-surface nodules are fairly abundant in Pacific sediments, and have also been found in those of the Atlantic and Indian Oceans. Little is known about these nodules, but available analyses indicate that those within the top few metres of the sediment differ little in composition from their surface counterparts (Cronan and Tooms, 1969, Goodell *et al.*, 1970). However, analyses of more deeply buried nodules from cores from the Deep Sea Drilling Project show greater variability. Most of those nodules are thought to have been buried as a result of the drilling process (McManus *et al.*, 1970), and are usually fairly similar in composition within any one core. However, two deeply buried nodules (72 and 131 m respectively) in DSDP 162 from the north-eastern tropical Pacific were significantly different in composition from those higher in the section. One of these nodules was rich in iron and impoverished in manganese and minor elements; on the basis of palaentological evidence it was considered to be *in situ*. Its total enclosure in a bleached mottle indicated that it may have formed as a result of the leaching of iron and manganese from the surrounding sediments.

(d) *Regional variations in nodule composition in relation to sea floor spreading*

It is evident from modern plate tectonic theory that ferro-manganese oxide concretions will be transported on the ocean floor as it moves away from spreading centres towards plate margins. Concretions initially precipitated at, or near, ridge crests will be moved to abyssal regions and eventually to subduction zones. Such movements will have an important effect on the composition of the concretions. The precipitates which are initially formed near the ridge crest would be expected to be relatively enriched in iron which has been derived from volcanic sources (Cronan, 1967; Corliss, 1971; Cronan, 1972c), and comparatively poor in manganese. However, as the deposits move away from the centres of spreading they will receive an increasing proportion of their constituents from "normal" sea water, and their Mn/Fe ratios should increase and so approach values typical of abyssal nodules in non-volcanic areas. In order to test this hypothesis fully, it would be necessary to analyse successive growth layers of nodules and encrustations collected along a traverse across the flank of an actively spreading ridge. However, some support for this theory can be obtained from bulk analyses of mid-ocean ridge nodules and encrustations from both the Atlantic and Indian Oceans. The encrustations of the ridge crest near $45°$ on the Mid-Atlantic Ridge, for example, have average Mn/Fe ratios much lower (0·62) than those from the King's Trough area (1·03) on its eastern flank (Cronan, unpublished data). Similarly, encrustations from near the crest of the Carlsberg Ridge have on average Mn/Fe ratio lower (0·71) than that in area 4C (1·23) on its western flank. It is evident, therefore, that there is a change in the composition of concretions away from active spreading centres which does support the hypothesis.

This hypothesis has important implications in the economic exploration for ferro-manganese oxide deposits. When attempting to locate deposits of a particular composition the characteristics of both the past and present environments of deposition should be considered. It also suggests a genetic relationship between nodules and ridge crest metalliferous sediments. These deposits show a fairly continuous range of Mn/Fe ratios, from very Fe-rich sediments on some ridge crests, to very Mn-rich nodules in some abyssal areas. Thus, the hypothesis implies that nodules and metalliferous sediments may be varieties of the same genetic class of deposit.

The fate of nodules and encrustations on reaching subduction zones at plate margins is not fully understood. Because of its instability under conditions of low redox potential $MnO_2$ would not be expected to survive very deep burial, and it should therefore dissolve, and $Mn^{2+}$ should migrate to the sediment surface where a portion of it might reoxidize and reprecipitate. Evidence for this is limited because few nodules or encrustations have been collected from the vicinity of subduction zones. However, nodules collected at three stations near the seaward side of the Peru trench were found to have an average Mn/Fe ratio of 20·2 (Cronan, unpublished data). However, the diagenetic remobilization of manganese required to produce this ratio might partly result from causes other than subduction (Section 28.5.3.1).

Should manganese nodules and encrustations not be subductable owing to their diagenetic instability, it might help to explain the "excess" manganese in pelagic sediments which has occupied marine geochemists since first reported by Murray and Renard (1891).

### 28.5.4. GROWTH RATES

Manganese nodule growth rates have already been mentioned briefly in Section 28.4; however, they merit fuller consideration. Since the time of their original discovery, manganese nodules have been generally thought to have a fairly slow rate of growth. Early radiometric dating techniques tended to support this conclusion, although inaccuracies in the methods used may often have led to considerable errors in the values obtained. Nevertheless, modern radiometric techniques, such as those employing ionium/thorium and K/Ar ratios and $Be^{10}$ decay, have confirmed that the average growth rate of deep-sea manganese nodules is of the order of a few mm per $10^6$ years (Tooms et al., 1969), However, these slow rates may be misleading, as growth is probably not continuous. Periods of accumulation may be separated by ones of little, or no, growth. Krishnaswamy and Lal (1972) have reviewed the evidence from radionuclide studies which suggest that nodules may be buried for as much as 90% of their life time as they may be alternately covered and exposed by bottom current induced sediment winnowing. If this is so, nodules may cease to grow for periods when buried, or may accumulate components from in the interstitial water environment at rates different from those operative at the sediment surface.

Near-shore and continental margin nodules have generally been thought to grow more rapidly than do the deep-sea varieties and growth rates of as much as several cm per thousand years have been suggested (Mero, 1965; Tooms *et al.*, 1969). In some instances this conclusion was based on the occurrence of sizeable nodules in areas whose data of submergence could be estimated (Manheim, 1965); in other cases the growth rate was estimated by radiometric dating techniques (Ku and Glasby, 1972). However, rapid nodule growth is not restricted to near-shore areas. Naval shells in the Pacific have been found to be thickly coated with ferro-manganese oxides (Goldberg and Arrhenius, 1958), and the precipitation of manganese oxides from hydrothermal solutions has actually been observed (Section 28.6.1).

Several factors are likely to affect the growth rates of manganese nodules. The most important of these is probably the rate of supply of their constituents to the sediment water interface. On this basis, the sometimes higher than normal growth rates of near shore nodules and those associated with submarine vulcanism can be explained. Such nodules are accumulating close to potential local sources of elements, and, in the case of many nearshore nodules, probably receive diagenetically remobilized manganese. By contrast, most deep sea nodules will derive their constituents from normal sea water, and thus would be expected to grow at slower rates. However, factors other than rate of supply may also be of importance in determining nodule growth rates. Evidence to be discussed in Section 28.6.3 suggests that physicochemical and biological factors may also be of importance in this regard. For example, the rapid growth of ferromanganese oxides around iron rich objects such as naval shells may be related to the ability of iron to catalyze the oxidation and precipitation of the ferromanganese oxides.

## 28.6. Origin of Ferro-manganese Oxide Deposits

A number of authors have pointed out that any attempt to explain the origin of manganese nodules and encrustations must involve answering three major questions (see for example, Menard, 1964; Riley and Chester, 1971). These are: (i) the source of the manganese and other elements in the concretions; (ii) the mechanism(s) by which these elements are transported to the sites of deposition; (iii) the precipitation and growth mechanisms involved in the actual accumulation of the deposits. The first of these problems has been a subject of controversy ever since Murray (Murray and Renard, 1891) concluded that the manganese might have a mainly volcanic origin. In contrast Renard (*loc. cit.*) considered that the manganese was principally derived from the continents. Although the second and third questions have not received so much attention they nevertheless, still present some difficulties. Each of these important questions will be considered separately.

### 28.6.1. SOURCES OF METALS

Dissolved and particulate manganese and manganese-bearing material is being supplied to the oceans by river and wind transport and coastal erosion. If it is assumed that there are no other contributions, the amount of manganese entering the oceans should balance that which is removed into nodules and marine sediments. Such mass balance calculations have been undertaken by Tooms et al. (1969) and by Varentsov (1971), and they indicate that at least a significant proportion of the manganese in nodules and encrustations is derived from continental sources. However, it should be emphasized that it is difficult to make accurate mass balance calculations for manganese and other elements. This is so because of uncertainties regarding, (i) the amount of manganese entering the world's oceans from the rivers, (ii) the amount of manganese contained in phases other than nodules, and (iii) the total amount of manganese present in the nodules on the floors of the oceans.

Evidence that submarine vulcanism is a factor in nodule formation has been discussed in detail by Bonatti and Nayudu (1965). In this context Bonatti and Joensuu (1966) and Varentsov (1971) have shown that several of the elements which are enriched in nodules are leached from submarine basalts during alteration on the sea floor, and Wilkniss et al. (1971) have observed that the leaching of fresh Kilauea pumice by sea water resulted in enrichment of the aqueous phase with manganese and iron. A proportion of the elements found in nodules, particularly those from volcanic areas, could therefore be provided by such alteration and leaching processes. Several elements could be liberated during this process. Some would reprecipitate almost immediately either in the form of segregations within the weathered rock (Nayudu, 1964; Cronan and Tooms, 1968), or as encrustations, whereas the remainder would be transported from the site of weathering, either to precipitate elsewhere or to remain in solution.

The influence of volcanic activity in the deep ocean is exemplified by the occurrence of extensive ferruginous deposits rich in manganese which immediately overlie the basement in many areas, (von der Borch and Rex, 1970; von der Borch et al., 1971; Cook 1971; Cronan et al. 1972). Niino (1959) has found manganese deposits associated with Mn-rich submarine hydrothermal springs off the coast of Japan. Similarly Zelenov (1964) has observed jets of hot iron and manganese-rich solutions debouching onto the sea floor through active fumaroles associated with the submarine Banu Wuhu volcano, Indonesia. These elements have precipitated on the rocks surrounding the vents; the deposits containing all the minor elements normally found in deep-sea manganese nodules. Furthermore, Elderfield (1972) has found manganese of volcanic origin in waters from Deception Island, Antarctica.

According to Ellis and Mahon (1964), hydrothermal waters are principally meteoric in origin, the dissolved matter which they contain being derived from the rocks through which the waters have passed. Accordingly, elements introduced into the sea by hydrothermal activity are most probably not derived from primary magmatic sources but from the rocks of the ocean floor itself. This conclusion has recently received support through studies of rocks from the Mid-Atlantic Ridge. Corliss (1971) found that the holo-crystalline interiors of submarine basalt flows were depleted in iron, and several other elements, relative to the rapidly cooled flow margins. He attributed these differences to the loss of these constituents of the magma from the interior of the flows. The elements concerned were thought to have been concentrated in residual solutions formed during crystallization so that they were accessible to mobilization by dissolution in the sea water which entered along contraction joints formed in the late stages of cooling. Much of the Mn-rich ferruginous sediment immediately overlying the basement in the oceans may form from such solutions.

The importance of ferruginous sediments overlying the basement in supplying elements to nodules at, or near, the sediment surface has yet to be established. If they were subjected to metamorphic or volcanic activity in the way suggested by Boström (1967), iron and manganese might be partially mobilized at depth and migrate upwards· Furthermore, the deposits might also undergo remobilization, when, as a result of sea floor spreading, they reach subduction zones at plate margins.

In summary, therefore, both continental and volcanic sources contribute manganese and associated elements for nodule formation. Indeed, evidence has been accumulating for some time that manganese from any source is a potential constituent of nodules and encrustations (Cronan, 1967; Cronan and Tooms, 1969; Glasby, 1970).

28.6.2. MECHANISMS TRANSPORTING THE METALS TO THE SEDIMENT–WATER INTERFACE

There are several mechanisms by which manganese and associated elements could be transported to the sediment-water interface. The most obvious of these is simply their transport in solution to the reaction site by normal processes of oceanic circulation and mixing. This is the most likely mechanism in areas which are swept clear of accumulating bottom sediment by currents, such as the tops of some seamounts and sites where bottom current velocities are high. An alternative mechanism for their transport to bottom waters involves organic agencies. Correns (1941) has suggested that manganese could be extracted from seawater by foraminifera, subsequently transported to deeper waters on the death of these organisms, and finally liberated on dissolution of their tests. In addition, Wangersky and Gordon (1965) have pointed out that the incorporation of $Mn^{2+}$ into particulate organic aggregates and its release at depth, after biological utilization of the organic phases, could be an important mechanism in enriching this element at the sediment-water interface. Manheim (cited in Ehrlich, 1968) has suggested that occlusion of detrital particles by nodules may account for the incorporation of some elements. This concept has been further developed by Ehrlich (1968) who concluded that rare earth elements may be incorporated into the nodules partly by surface to surface transfer from the particle to the nodule, and partly by occlusion of the entire particle. Surface to sediment transport was considered to be the principal process for slowly accumulating deep-ocean concretions (for which time is available for such a transfer to take place,) whereas, occlusion was thought to be more significant in their relatively rapidly accumulating near-shore counterparts.

The mechanisms discussed above only involve the transfer of elements from sea water to the sediment-water interface. Upward diffusion of manganese through the interstitial waters to the sediment surface is probably equally, if not more, important in some areas. This manganese could have a number of origins, which include buried manganese dioxide which had been subsequently reduced (Section 28.5.3.1), groundwater flowing into marginal marine areas, hydrothermal fluids injected from depth and, buried terrigenous and volcanic material leached within the sediment column.

28.6.3. PRECIPITATION AND GROWTH MECHANISMS

Stumm and Morgan (1970) have described the chemistry involved in the oxidation and precipitation of iron and manganese in natural environments. The oxidation of $Mn^{2+}$ is catalysed by a reaction surface, the $MnO_2$ so produced adsorbs additional $Fe^{2+}$ or $Mn^{2+}$ which in turn become oxidized. These authors have described experiments in which a calcite crystal was suspended in an aerated solution of bicarbonate which contained traces of $Fe^{2+}$ and $Mn^{2+}$. Ferric oxide (hydrated) coated the crystal surface and then became an adsorbent for $Mn^{2+}$, the latter was then oxidized and in turn reacted with $Fe^{2+}$ or additional $Mn^{2+}$.

The fact that initiation of manganese oxidation and precipitation requires a surface has also been recognized by several other workers. Goldberg and Arrhenius (1958) suggested that ferric oxide could provide such an initial active surface on which manganese could be oxidized in the formation of nodules, according to the equation:

$$2OH^- + Mn^{2+} + \tfrac{1}{2}O_2 = MnO_2 + H_2O$$

Support for this suggestion has been provided by electron probe studies. Burns and Brown (1972) found that iron oxides had been deposited around the nuclei of nodules prior to the deposition of the manganese oxides. Morgenstein (1972) found that ferro-manganese oxide accretion in parts of the Hawaiian archipelago was controlled by the distribution of sediments rich in iron oxide, which presumably provided the surface on which manganese deposition was initiated. Morgenstein and Felsher (1971) have suggested that in many nodules from volcanic areas the iron for the initial reaction surface is provided by the decomposition of the palagonite of the nodule core. The derivation of iron in this way by the alteration of volcanic debris would help to explain the fact that the majority of nodules accrete around volcanic nuclei. Furthermore, the rate of alteration of the debris, and consequently the rate of supply of iron, might be one factor determining nodule growth rates. This may provide an explanation for the occurrence of highly weathered volcanics having thick ferro-manganese oxide encrustations being sometimes found in association with less weathered volcanic debris that is less thickly encrusted with the deposits. However, the age of the volcanics, and hence the time available for ferro-manganese oxide deposition, is also of importance in this context (Section 28.2). Once manganese oxides start to accumulate they can adsorb more iron which, in turn, could provide the surface for additional manganese accumulation. In this way, the nodules and encrustations could continue to grow as long as there is an adequate supply of iron and manganese to the reaction surface.

NOTE ADDED IN PRESS

Since this chapter was written, a programme of research has developed out of the I.D.O.E. manganese nodule symposium held at Arden House, and published by the Lamont-Doherty Geological Observatory (Horn *et al.*, 1972b), emphasising the genesis of economic nodule deposits in the northeastern equatorial Pacific. Preliminary results have stressed the importance of biological factors and nodule mineralogy respectively in contributing to the genesis and compositional variability of the deposits.

REFERENCES

Agassiz, A. (1906). *Mem. Mus. Comp. Zool. Harvard*, **33**.
Ahrens, L. H., Willis, J. P. and Oosthuizen, C. O. (1967). *Geochim. Cosmochim. Acta*, **31**, 2169

Andrushchenko, P. F. and Skornyakova, N. S. (1969). *Oceanology*, **9**, 229.
Arrhenius, G. (1963). *In* "The Sea" (M. N. Hill, ed.), Vol. 3, p. 655. Interscience Publishers. New York.
Arrhenius, G. and Bonatti, E. (1965). *In* "Progress in Oceanography" (M. Sears, ed.), Vol. 3, p. 7. Pergamon Press, London.
Aumento, F. (1969). *Can. J. Earth Sci.* **6**, 1431.
Baas Becking, L. G. M., Kaplan, I. R. and Moore, D. (1960). *J. Geol.* **68**, 243.
Bacon, J. (1967). *U.S.A.E.C. Report C00-1540-6.* (Unpublished).
Barnes, S. S. (1967). *Science*, **157**, 63.
Barnes, S. S. and Dymond, J. R. (1967). *Nature, Lond.* **213**, 1218.
Bender, M. L., Ku, T. L. and Broecker, W. S. (1966). *Science*, **151**, 325.
Bezrukov, P. L. (1962). *Okeanologiya*, **2**, 1014.
Bezrukov, P. L. (1963). *Okeanologiya*, **3**, 540.
Bezrukov, P. L. and Andrushchenko, P. F. (1972). *Izv. Akad. Nauk. SSSR, Ser. Geol.* **7**, 3.
Bonatti, E. and Joensuu, O. (1966). *Science*. **154**, 643.
Bonatti, E. and Nayudu, Y. R. (1965). *Amer. J. Sci.* **263**, 17.
Boström, K. (1967). *In* "Researches in Geochemistry" (P. H. Ableson, ed.), Vol. 2, p. 421. John Wiley, New York.
Boström, K. and Fisher, D. E. (1969). *Geochim. Cosmochim. Acta*, **33**, 743.
Boström, K. and Valdes, S. (1969). *Lithos*, **2**, 351.
Boström, K. Peterson, M. N. A., Joensuu, O. and Fisher, D. E. (1969). *J. Geophys. Res.* **74**, 3261.
Bricker, O. (1965). *Amer. Mineral*, **50**, 1296.
Brown, B. A. (1972). *Amer. Mineral*, **57**, 284.
Burns, R. G. (1965). *Nature, Lond.* **205**, 999.
Burns, R. G. and Brown, B. A. (1972). *In* "Ferro-manganese Deposits on the Ocean Floor" (D. R. Horn, ed.). Lamont–Doherty Geol. Obs.
Burns, R. G. and Fuerstenau, D. W. (1966). *Amer. Mineral*, **51**, 895.
Buser, W. (1959). Preprints Int. Oceanogr. Congr. (M. Sears, ed.) *Amer. Assoc. Adv. Sci.* Washington, 962.
Buser, W. (1959). Preprints Int. Oceanogr. Congr. 962.
Buser, W. and Grütter, A. (1956). *Schweiz. Mineral Petrogr. Mitt.* **36**, 49.
Buser, W., Graf, P. and Feitknecht, W. (1954). *Helv. Chim. Acta*, **37**, 2322.
Calvert, S. E. and Price, N. B. (1970). *Contr. Mineral Petrol.* **29**, 215.
Carpenter, R., Johnson, H. P. and Twiss, E. S. (1972). *J. Geophys. Res.* **77**, 7163.
Cheney, E. S. and Vredenburgh, L. D. (1968). *J. Sediment Petrol.* **38**, 1363.
Church, T. M. (1970). Ph. D. Thesis, University of California.
Collins, J. F. and Buol, S. W. (1970). *Soil Sci.* **110**, 157.
Cook, H. E. (1971). *Geol. Soc. Amer. Abstracts with Programs*, **3**, No. 7, 530–531.
Corliss, J. B. (1971). *J. Geophys. Res.* **76**, 8128.
Correns, C. W. (1941). *Nachr. Akad. Wiss. Göttingen Math.-Phys. Kl*, **5**, 219.
Cronan, D. S. (1967). Ph.D. Thesis, University of London.
Cronan, D. S. (1969). *Chem. Geol.* **5**, 99.
Cronan, D. S. (1972a). *Nature, Lond.* **235**, 171.
Cronan, D. S. (1972b). *Can. J. Earth Sci.* **9**, 319.
Cronan, D. S. (1974). *In* "The Sea" (E. D. Goldberg, ed.), Vol. 5. Interscience Publishers, New York.
Cronan, D. S. (1974). *In* "The Sea" (E. D. Goldberg, ed.), Vol. 5. Interscience Publishers, New York.

Cronan, D. S. and Thomas, R. L. (1970). *Can. J. Earth. Sci.* **7**, 1346.
Cronan, D. S. and Thomas, R. L. (1972). *Bull. Geol. Soc. Amer.* **83**, 1493.
Cronan, D. S. and Tooms, J. S. (1967a). *Deep-Sea Res.* **14**, 117.
Cronan, D. S. and Tooms, J. S. (1967b). *Deep-Sea Res.* **14**, 239.
Cronan, D. S. and Tooms, J. S. (1968). *Deep-Sea Res.* **15**, 215.
Cronan, D. S. and Tooms, J. S. (1969). *Deep-Sea Res.* **16**, 335.
Cronan, D. S., van Andel, T. H., Heath, G. R., Dinkelman, M. G., Bennett, R. H., Bukry, D., Charleston, S., Kaneps, A., Rodolfo, K. S. and Yeats, R. S. (1972). *Science*, **175**, 61.
Ehrlich, A. M. (1968), Ph.D. Thesis (M.I.T.).
Elderfield, H. (1972). *Mar. Geol.* **13**, M1.
Ellis, A. J. and Mahon, W. A. J. (1964). *Geochim. Cosmochim. Acta*, **28**, 1323.
Ewing, M., Horn, D., Sullivan, L., Aitken, T. and Thorndike, E. (1971). *Oceanology Int.* December.
Frondel, C., Marvin, U. and Ito, J. (1960). *Amer. Mineral*, **45**, 1167.
Giovanoli, R., Feitknecht, W. and Fischer, F. (1971). Unpublished Ms.
Glasby, G. P. (1970). Ph.D. Thesis, University of London.
Glasby, G. P. (1972). *Mar. Geol.* **13**, 57.
Goldberg, E. D. (1954). *J. Geol.* **62**, 249.
Goldberg, E. D. (1961). *In* "Oceanography" (M. Sears, ed.), p. 583. Amer. Assoc. Adv. Sci., Washington.
Goldberg, E. D. (1963). *In* "The Sea" (M. N. Hill, ed.), Vol. 2, p. 3. Interscience Publishers, New York.
Goldberg, E. D. and Arrhenius, G. O. S. (1958). *Geochim. Cosmochim. Acta*, **13**, 153.
Goodell, H. G., Meylan, M. A. and Grant, B. (1970). *In* "Antarctic Oceanology I" (J. L. Ried ed.), p. 27, Amer. Geophys. Union, Baltimore.
Grant, J. B. (1967). *Geol. Soc. Amer. Spec. Paper*, **115**, 80.
Grill, E. V., Murray, J. W. and MacDonald, R. D. (1968). *Nature, Lond*, **219**, 358.
Harriss, R. C. (1968). *Nature, Lond.* **219**, 54.
Harriss, R. C., Crocket, J. H. and Stainton, M. (1968). *Geochim. Cosmochim. Acta*, **32**, 1049.
Hays, J. D., Saito, T., Opdyke, N. D. and Burckle, L. R. (1969). *Bull. Geol. Soc. Amer.* **80**, 1481.
Horn, D. R., Ewing, M., Horn, B. M. and Delach, M. N. (1972a). *Ocean Industry*, **7**, 26.
Horn, D. R., Ewing, M., Horn, B. M. and Delach, M. N. (1972b). *In* "Ferro-manganese Deposits on the Ocean Floor" (D. R. Horn, ed.). Lamont–Doherty Geol. Obs.
Horowitz, A. (1970). *Mar. Geol.* **9**, 241.
Hrynkiewicz, A. Z., Pustowk, A. A. J., Sawicka, B. D. and Sawicki, J. A. (1972). *Phys. Status Solidi, A.* **9**, 159.
Johnson, C. E. and Glasby, G. P. (1969). *Nature, Lond.* **222**, 376.
Jones, L. H. P. and Milne, A. A. (1956). *Mineral Mag.* **31**, 283.
Kaufman, R. and Siapno, W. D. (1972). *In* "Ferro-manganese Deposits on the Ocean Floor" (D. R. Horn, ed.). Lamont–Doherty Geol. Obs.
Krauskopf, K. B. (1956). *Geochim. Cosmochim. Acta*, **9**, 1.
Krauskopf, K. B. (1957). *Geochim. Cosmochim. Acta*, **12**, 61.
Krishnaswamy, S. and Lal, D. (1972). *In* "The Changing Chemistry of the Oceans", (D. Dyrssen and D. Jagner, eds.), 365 pp. Almqvist and Wiksell, Stockholm.
Ku, T. L. and Broecker, W. S. (1969). *Deep-Sea Res.* **16**, 625.
Ku, T. L. and Glasby, G. P. (1972). *Geochim. Cosmochim. Acta*, **36**, 699.

Lakin, H. W., Thompson, C. E. and Davidson, D. F. (1963). *Science*, **142**, 1568.

Laughton, A. S. (1967). *In* "Deep Sea Photography" (J. B. Hersey, ed.), p. 196. John Hopkins Press, Baltimore.

Levinson, A. A. (1960). *Amer. Mineral*, **45**, 802.

Levinson, A. A. (1962). *Amer. Mineral*, **47**, 790.

Lorber, H. R. (1966). M.Sc. Thesis, University of California.

Lynn, D. C. and Bonatti, E. (1965). *Mar. Geol.* **3**, 457.

Manheim, F. T. (1965). *In* "Symposium on Marine Geochemistry" (D. R. Schink and J. T. Corless, eds.). Marine Lab. Occasional Publ. 3, p. 217. University of Rhode Island.

McKenzie, R. M. (1971). *Mineral. Mag.* **38**, 493.

McManus, D. A., Burns, R. E., Weser, O., Vallier, T., von der Borch, C. C., Olsson, R. K., Goll, R. M. and Milow, E. D. (1970). "Initial Report of the Deep Sea Drilling Project", Vol. 5, 827 pp. U.S. Govt. Printing Office, Washington.

Menard, H. W. (1964). "The Marine Geology of the Pacific." 271 pp. McGraw-Hill, New York.

Menard, H. W. and Shipek, C. J. (1958). *Nature, Lond.* **182**, 1156.

Menard, H. W., Goldberg, E. D. and Hawkes, H. E. (1964). Scripps Institute of Oceanography. Unpublished Ms.

Mero, J. L. (1962). *Econ. Geol.* **57**, 747.

Mero, J. L. (1965). "The Mineral Resources of the Sea", 312 pp. Elsevier, Amsterdam.

Moore, T. C. Jr. and Heath, G. R. (1966). *Nature, Lond.* **212**, 983.

Morgenstein, M. (1972). *In* "Ferro-manganese Deposits on the Ocean Floor" (D. R. Horn, ed.). Lamont–Doherty Geol. Obs.

Morgenstein, M. and Felsher, M. (1971). *Pacific Sci.* **25**, 301.

Murray, J. (1876). *Proc. R. Soc. Lond.* **24**, 471.

Murray, J. and Renard, A. F. (1891). "Report on the Scientific Results of the Exploring Voyage of H.M.S. *Challenger*", Vol. 3, 525 pp. H.M.S.O., London.

Nayudu, Y. R. (1964). *Bull. Volcan.* **27**, 391.

Niino, H. (1959). *Preprints Int. Oceanogr. Congr.* **1**, 646.

Okada, A. and Shima, M. (1970). *J. Oceanogr. Soc. Japan*, **26**, 151.

Opdyke, N. D. and Foster, J. H. (1970). *Mem. Geol. Soc. Amer.* **126**, 83.

Price, N. B. (1967). *Mar. Geol.* **5**, 511.

Price, N. B. and Calvert, S. E. (1970). *Mar. Geol.* **9**, 145.

Riley, J. P. and Chester, R. (1971). "Introduction to Marine Chemistry", 465 pp. Academic Press, London and New York.

Riley, J. P. and Sinhaseni, P. (1958). *J. Mar. Res.* **17**, 466.

Sano, M. and Matsubara, H. (1970). *Suiyokai-shi*, **17**, 111.

Scott, R. R., Rona, P. A., Butler, L. W., Nalwalk, A. J. and Scott, M. R. (1972). *Nature Lond.* **239**, 77.

Sillén, L. G. (1961). *In* "Oceanography" (M. Sears, ed.). p. 549. Amer. Assoc. Adv. Sci., Washington.

Skornyakova, N. S. and Andrushchenko, P. F. (1964). *Lithol. Miner. Resourc.* **5**, 21.

Skornyakova, N. S. and Andrushchenko, P. F. (1968). *Okeanologiya*, **8**, 865

Skornyakova, N. S. and Andrushchenko, P. F. (1970). *In* "The Pacific Ocean" (P. L. Bezrukov, ed.), Vol. 6, p. 202. Nauka, Moscow.

Skornyakova, N. S., Andrushchenko, P. F. and Fomina, L. S. (1962). *Okeanologiya*, **2**, 264.

Smith, J. D. and Burton, J. D. (1972). *Geochim. Cosmochim. Acta*, **36**, 621.

Sorem, R. K. (1967). *Econ. Geol.* **62**, 141.

Sorem, R. K. (1972). Paper presented in "Working Group for Manganese". Symposium at 24th Int. Geol. Congr., Montreal.

Sorem, R. K. and Gunn, D. W. (1967). *Econ. Geol.* **62**, 22.

Straczek, J. A., Horen, A., Ross, M. and Warshaw, C. M. (1960). *Amer. Mineral.* **45**, 1174.

Strunz, H. (1970), "Mineralogische Tabellen" Akad. Verlag, Leipzig.

Stumm, W. and Morgan, J. J. (1970). "Aquatic Chemistry", 583 pp. Wiley–Interscience, New York.

Summerhayes, C. P. (1967). *N.Z. J. Geol. Geophys.* **10**, 1372.

Taylor, S. R. (1964). *Geochim. Cosmochim. Acta*, **28**, 1273.

Tooms, J. S. (1972). *Endeavour*, **31**, 113.

Tooms, J. S., Summerhayes, C. P. and Cornan, D. S. (1969). *Oceanogr. Mar. Biol.* **7**, 49.

Varentsov, I. M. (1971). *Soc. Min. Geol. Japan. Spec. Issue*, **3**, 466.

von der Borch, C. C. and Rex, R. W. (1970). *In* McManus, D. A. *et al.*, "Initial Report of the Deep Sea Drilling Project", Vol. 5, p. 541. U.S. Govt. Printing Office, Washington.

von der Borch, C. C., Nesteroff, W. D. and Galehouse, J. S. (1971). *In* Tracey Jr., J. I. *et al.*, "Initial Report of the Deep Sea Drilling Project, Vol. 8, p. 829. U.S. Govt. Printing Office, Washington.

Wangersky, P. J. and Gordon Jr., D. C. (1965). *Limnol. Oceanogr.* **10**, 544.

Wilkniss, P. E., Warner, T. B. and Carr, R. A. (1971). *Mar. Geol.* **11**, M39.

Willis, J. P. (1970). M.Sc. Thesis, University of Cape Town.

Willis, J. P. and Ahrens, L. H. (1962). *Geochim. Cosmochim. Acta*, **26**, 751.

Winterhalter, B. (1966). *Geotek. Julk*, **69**, 1.

Winterhalter, B. and Siivola, J. (1967). *C.R. Soc. Geol. Finlande*, **39**, 161.

Wiseman, J. D. H. (1937). *Sci. Rep. John Murray Exped.* **3**, 1.

Yoshimura, T. (1934). *J. Fac. Sci. Hokkaido Univ.* **4(2)**, 289.

Zelenov, K. K. (1964). *Dokl. Acad. Sci. USSR, Earth Sci.* **155**, 91.

Chapter 29

# Biogenous Deep Sea Sediments: Production, Preservation and Interpretation

## WOLFGANG H. BERGER

*Scripps Institution of Oceanography, University of California, San Diego, La Jolla, California 92037, U.S.A.*

## 29.1. Introduction

### 29.1.1. historical background

One hundred years ago John Murray, the naturalist on board H.M.S. *Challenger,* collected samples of shell-bearing plankton and biogenous sediments which he was to study in detail during the following decades. As a result of this expedition, and later less extensive cruises, he was able to outline the main features of the distribution of shell-secreting organisms and their remains on the sea floor. His principal conclusions were: (1) there is a general latitudinal zonation of species associated with tropical, subtropical, temperate, subpolar and polar waters; (2) diversity of species decreases away from the tropics, and abundance of organisms varies as a function of the nutrient supply; (3) the dissolution of siliceous tests occurs in many oceanic areas; (4) there is a foraminiferal and a pteropod compensation depth on the ocean floor, above which dissolution of many species of forams and pteropods takes place enriching the underlying sediment assemblages with resistant forms (Murray and Renard, 1891; Murray, 1897). These major topics—biogeography of shelled plankton, distribution of carbonate sediments, and the preservation patterns of the organisms within them, have since been studied in much more detail and some aspects of them will be discussed below.

Little is known about the life cycles of shell-bearing organisms, the manner in which empty tests are derived from living populations, the way in which they settle to the sea floor, and how long they remain there. However, some progress has been made towards solving these problems, and it is now apparent that the process transforming freshly produced shells into sedimentary deposits must largely be inferred from inventories of life, death and sediment assemblages (Fig. 29.1). Much information has been obtained from the study of Pleistocene deep-sea stratigraphy, including the determina-

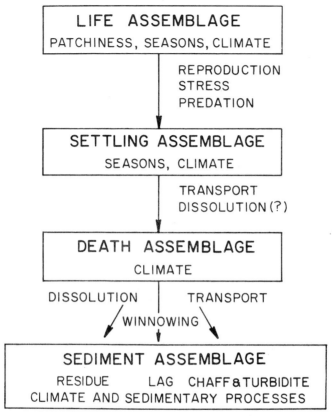

FIG. 29.1. Path from the living planktonic assemblage to the sediment assemblage found on the ocean floor. Factors influencing aspects of assemblages are indicated within the boxes. The last box contains the four main kinds of sediment assemblages as produced by partial dissolution (residue-), by winnowing (lag-), by differential transport and resedimentation of fine material (chaff-) and by gravity-induced turbid flow (turbidite assemblage). The greek-derived terms for life-, death- and sediment- (or burial-) assemblages are bio-, thanato-, and tapho-coenosis. "Thanatocoenosis" is frequently used to denote sediment assemblages in general, although this is etymologically incorrect. To avoid misunderstanding, the common terms are used here. For the distinction between palaeoecologic and taphonomic concepts see Lawrence (1968).

tion of the rates of sedimentation—see for example, Schott (1935), Bramlette and Bradley (1942), Arrhenius (1952), Phleger *et al.* (1953), and Turekian (1971). The results of the drilling program of the JOIDES group have further extended our knowledge to include Tertiary and Cretaceous deep-sea sedimentation.

The present chapter discusses the production of shells, the conditions governing shell burial, and the interpretation of Quaternary and pre-

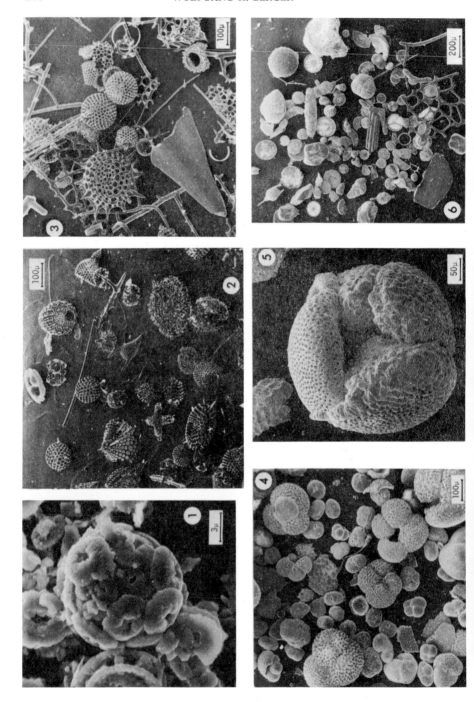

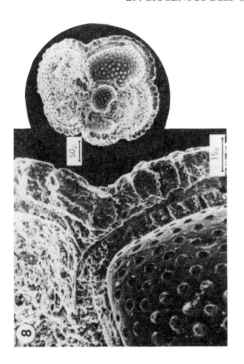

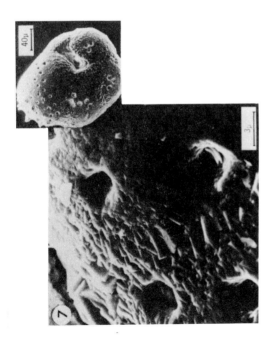

PLATE 29.1. Associations of common biogenous particles in deep-sea sediments. (SEM Geol. Inst. Kiel, photo W.H.B./C. Samtleben. Carbon coating, 10 kV; prepared from Leg 14 material, Central Atlantic; see Berger and von Rad, 1972). Fig. 1. Coccosphere of placoliths (Miocene). Note etching, overgrowth. Fig. 2. Excellently preserved siliceous assemblage with diatoms (note index fossil *Craspedodiscus*) and delicate radiolarians (Oligocene). Fig. 3. Well preserved fish tooth in radiolarian residue assemblage (Eocene). Fig. 4. Foram assemblage (Quarternary). Note state of preservation (FS ≃ 4; lysocline assemblage). Fig. 5. Highly resistant foram (*Globoquadrina venezuelana*) with thick cortex, partially decortified, from a residual Miocene assemblage (FS ≃ 7). Fig. 6. Mixed well preserved (diatoms) and poorly preserved (orosphaerid fragments) siliceous assemblage from hemipelagic mud (Miocene). Fig. 7. Typical small foram in moderately dissolved Quaternary assemblage. Note cortex formation and thickening at sutures. Fig. 8. Foram fragment from a Miocene assemblage. Cortex encloses entire test and appears unrelated to the construction of any particular chamber wall.

Quaternary biogenous sediments. Attention is mainly confined to the calcareous and the siliceous plankton of the open-ocean (i.e. coccolithophores, diatoms, silicoflagellates, foraminifera and radiolarians; see Plate 29.1), but pteropods will also be considered. Symposia have been held recently on these and other groups (see, for example, Brönnimann and Renz, 1969; Funnell and Riedel, 1971; Farinacci, 1971). See also Berger and Roth (1975) for a survey of recent literature.

### 29.1.2. THE GENERAL FERTILITY-DEPTH PATTERN

Consideration of the factors, such as production and preservation of shell material, which control the distributions of the major facies leads to the suggestion that there is a fertility-depth pattern in the context of which the oceanic record may be considered (Olausson et al., 1971). In such a pattern, well preserved calcareous tests in the sedimentary record are indicative of deposition in shallow depths from water of low fertility; conversely, well preserved siliceous remains are suggestive of high fertility in the overlying water (see Fig. 29.2). The aragonite compensation depth (ACD) (i.e. the depth contour which separates aragonitic from aragonite-free sediments) is between 1 and 2 km in the western tropical South Pacific, and between 2 and 3 km in the western tropical North Atlantic. Data for other regions are scanty. In the tropical North Pacific the ACD is only a few hundred metres deep. The lysocline separates well preserved from poorly preserved calcareous

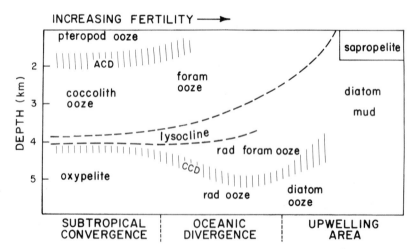

FIG. 29.2. Model of biogenous sediment distribution within a general fertility-depth frame. ACD, Aragonite Compensation Depth; CCD, Calcite Compensation Depth. Depths given are typical for the southern and equatorial Pacific and they do not necessarily apply elsewhere.

assemblages. The calcite compensation depth (CCD; the depth at which the supply of calcite to the ocean floor is balanced by the rate of dissolution) is the facies boundary between calcareous ooze and pelagic clay. These depths vary within, and between, oceans (Sections 29.4.1 to 29.4.5). Other terms which are found in the literature include "oxypelite" which is an iron oxide-rich red, brown or yellow clay, essentially barren of fossils, and "sapropelite" which is rich in organic matter, iron sulfides, and fossils.

## 29.2. SYSTEMATICS, LIFE HISTORIES, SHELL FORMATION

### 29.2.1. NUMBER OF SPECIES

The taxonomy of shelled plankton, alive and fossil, is based on the morphology of their skeletons. Although this is convenient for geologists it arouses scepticism in biologists, particularly when skeletal parts which are thought to have originated from different species, or even genera, can be shown to have been produced by the same organism (e.g. coccolithophores, Parke and Adams, 1960; diatoms, Holmes and Reimann, 1966; forams, Myers, 1943; silicoflagellates, van Falkenberg and Norris, 1970). A knowledge of evolutionary lineages (forams: Parker, 1967; radiolarians: Riedel, 1971b) should greatly improve our ability to assign fossils to their correct classification. It should be pointed out that laboratory studies are necessary to delineate the variability occurring in living populations.

Tappan and Loeblich (1971b) give the numbers of living species of shelled plankton as follows (numbers of species actually identified in recent large-scale sedimentary distributional studies are in brackets*); calcareous nannoplankton, 382 (12); Centrales (diatoms), 1500 (60); silicoflagellates, 58 (10); Spumellaria (radiolarians), ~250 (50); Nasselaria (radiolarians), ~650 (50); planktonic Foraminifera, 72 (35). In addition, there are about 50 species of shell-bearing pelagic gastropods. The marked difference between the maximum numbers given by Tappan and Loeblich (1971b) and the numbers of those actually found in deep-sea sediments may be partly due to taxonomic artificialities (especially for forams and silicoflagellates), but major causes are preservation effects and the scarcity of records (particularly for coccolithophores, diatoms and radiolarians). See Tappan and Loeblich (1973) for a review of plankton evolution.

### 29.2.2. COCCOLITHOPHORES

Coccoliths, which are the remains of the skeletal parts of calcareous nanno-

* Data for coccolithophores (McIntyre and Bé, 1967); diatoms, mostly centric, (Kanaya and Koizumi, 1966); silicoflagellates (Gemeinhardt, 1934); polycystine radiolarians (Nigrini, 1967); forams (Parker and Berger, 1971).

plankton or (coccolithophores), are major constituents of calcareous pelagic oozes. They are calcitic particles, mainly of silt size (1–50 μm, commonly less than 10 μm), which show a great variety in their morphology (see Deflandre and Fert, 1954; Halldall and Markali, 1955). They are derived from calcareous planktonic algae which are enveloped by coccoliths forming a "coccosphere" (Murray and Renard, 1891; Lewin, 1962b). The number of coccoliths per sphere ranges from less than 10 to more than 30 (Plate 29.1, photo. 1). The oldest coccoliths have been found in early Jurassic sediments; reported Permian occurrences are not generally accepted.

Coccolithophores are included in the Haptophyceae of the Chrysophyta (golden-brown algae) because at least some members of the order Coccolithophorales ("Coccolithinae", "Coccolithophoridae") possess the flagellar apparatus and the organic surface scales characteristic of the Haptophyceae (see Paasche, 1968a). Perch-Nielsen (1971) has given a useful survey of Tertiary species and genera, with sketches of some 400 forms. The range of variation is illustrated in Section 29.5.8. Valid taxa and literature are listed in the "Annotated index and bibliography of the calcareous nannoplankton" by Loeblich and Tappan (1966, 1968, 1969, 1970a, 1970b, 1971c), and a catalogue of calcareous nannofossils is being published by Farinacci (1969, and following years).

Both planktonic and benthonic coccolithophores exist in the sea. There are also a few fresh water species. Despite the great abundance of oceanic coccolithophores little is known about their life histories. The function of coccoliths is unknown. Species cultured in the laboratory, not surprisingly, are hardy forms typical of inshore areas (*Cricosphaera* group), or species with wide tolerance limits (*Emiliania huxleyi, Coccolithus pelagicus*). They are not necessarily representative of the other more restricted oceanic species ("oceanic" is here used in contradistinction to "coastal"). *Cricosphaera* species apparently have both planktonic and benthonic phases. Both motile and non-motile phases exist in the widespread pelagic species *C. pelagicus* (Parke and Adams, 1960). Each phase has its own characteristic coccoliths and can reproduce itself by simple fission. The conditions triggering the phase change are not known, but both stages are frequently found together suggesting that there is a ready transition from one to the other.

In the sediments the coccoliths from the non-motile phase ("heterococcoliths") are virtually the only ones preserved. Coccoliths from the weakly calcified motile phase ("holococcoliths") disintegrate when the organic matrix decomposes (Gartner and Bukry, 1969). Thus, conditions leading to a transition from a non-motile to a motile phase, if such conditions exist, would be recorded as a disappearance of the species in the sediment. Similarly, transition to a naked stage (lacking coccoliths) would also be recorded as a disappearance. These possibilities are of some interest since Hasle

(1960), McIntyre and Bé (1967) and McIntyre *et al.* (1970) did not find *Coccolithus pelagicus* in the Southern Ocean, although it is abundant in the cold waters of the North Atlantic. However, McIntyre and Bé (1967) did find *C. pelagicus* in several Southern Ocean cores, and they suggested that this species disappeared during post-glacial times. Naked and poorly calcified stages are not restricted to *C. pelagicus*, but also appear in, for example, *E. huxleyi* (Klaveness and Paasche, 1971).

Maximum growth rates recorded for *Cricosphaera elongata* and *Emiliania huxleyi* are two divisions per day which are comparable to those of other fast-growing planktonic algae (Fig. 29.3). Coccoliths are formed within the

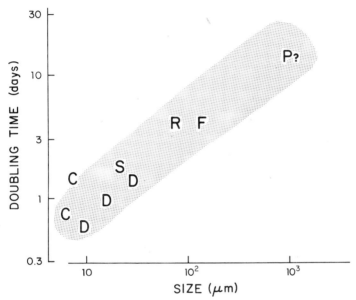

FIG. 29.3. Typical near-optimum doubling times for various shelled plankton. C, D, S, laboratory experiments. R, F, estimated from shell output. C, coccolithophores. D, diatoms; S, silicoflagellates; R, radiolarians; F, foraminifera; P, pteropods (actually pertains to *Clione*, a predator of *Limacina*, see Lalli, 1970; Conover and Lalli, 1972). Sources cited in text. Shaded region generalized from Sheldon *et al.* (1972).

cell (Parke and Adams, 1960; Paasche, 1962, 1969; Wilbur and Watabe, 1963; Manton and Leedale, 1969), and may be lost during growth (Wilbur and Watabe, 1963; see also Paasche, 1968b, p. 179). This scaling-off process may be important in the formation of sediments. The coccoliths remain within an organic membrane even after extrusion by the cell. Such a covering, if retained, would undoubtedly enhance preservation of the coccolith during transportation to the sea floor and on the floor itself.

Paasche (1968b), using a clone in which all cells formed coccoliths, found that the amount of coccolith carbonate produced per unit cell volume was almost independent of temperature and light conditions, although light appears to be a prerequisite for coccolith formation. However, Blankley (1971) has shown that light is not necessary for those coccoliths which grow heterotrophically (e.g. *E. huxleyi*; *Cricosphaera carterae*).

A possible effect of nutrient availability on coccolith formation is implied in the observations by Wilbur and Watabe (1963, p. 108) that certain strains of *E. huxleyi* only produced coccoliths in a nitrogen-deficient medium and not in a "normal" medium. According to Paasche (1968a, p. 80), experiments on *E. hyxleyi* suggest that coccolith formation proceeds inside the cell even when external coccoliths dissolve in a $CaCO_3$-undersaturated medium. This finding disagrees with Murray's (1897) often quoted hypothesis, that the degree of saturation of ocean water with $CaCO_3$ affects shell formation. *C. pelagicus* may deposit several times more carbon in its (hetero) coccoliths than it assimilates photosynthetically (Paasche, 1969).

Considerable morphological variation in the coccoliths of *E. huxleyi* has been observed both in the laboratory (Watabe and Wilbur, 1966), and in nature (Hasle, 1960; McIntyre and Bé, 1967). The variations appear to be temperature-dependent. The warm water type is delicate and open-structured, whereas the cold water type is solid and dense both for *E. huxleyi* and *Umbilicosphaera mirabilis* (McIntyre and Bé, 1967, pp. 569 and 572).

## 29.2.3. DIATOMS

Diatoms exist in virtually all environments where there is water and light, e.g. in soils, lakes, lagoons and the ocean. Their frustules, which are especially abundant in those marine sediments which underlie water of high fertility, consist of opal—a hydrated form of silica—and range in size from a few micrometres to about 2 mm. In the littoral zone and on the continental shelf benthic forms are common. Usually, these are much more silicified than planktonic ones, and have a better chance of preservation. The oldest known diatoms are probably of Jurassic age and the earliest well preserved assemblages are found in Late Cretaceous sedimentary rocks.

Diatoms belong to the Class Bacillariophyceae of the Chrysophyta. Their frustules are constructed on a pill box principle; the lid is the *epitheca* and the bottom is the *hypotheca*. Both these structures are *valves* and the connective band is called the *girdle*. Diatoms are classified into *centric* and *pennate* forms. The former have a structure which radiates from a central point, or which is concentric around it; the latter have bilateral symmetry about a median line. Most pennate diatoms with a benthic habitat bear raphes which are clefts along the apical axis (see Hustedt, 1961; Hendey, 1964; Cupp, 1943).

Most planktonic marine diatoms are centric (Castracane, 1886). The following genera of pennate diatoms occur abundantly in oceanic plankton; *Thalassionema, Thalassiothrix, Asterionella, Nitzschia* and *Navicula*. Of these, members of the *Navicula* and *Nitzschia* have raphes, the others do not, but usually have a pseudoraphe (axial area without cleft). In deep-sea sediments raphe-less centric forms are the most important and include discoid diatoms (circular; e.g. *Coscinodiscus*); gonioid diatoms (bipolar, triangular, polygonal; e.g. *Eucampia, Hemidiscus, Biddulphia*); and solenoid diatoms (greatly elongated by separation of valves; e.g. *Rhizosolenia*). For the identification of diatoms in deep-sea sediments the reader is referred to the references listed in Kanaya and Koizumi (1966) (see also Initial Reports DSDP, Vols 18, 19 and 24); for taxonomy the monographs by Hustedt (1930, 1959) should be consulted. Some stratigraphically useful forms are illustrated in Section 29.5. 8.

By comparison with the other plankton groups discussed in this chapter diatom life cycles have been well studied (see von Stosch, 1950, 1951, 1958; Lewin and Guillard, 1963; von Stosch and Drebes, 1964; Holmes and Reimann, 1966; Holmes, 1967; Werner, 1971a, 1971b; and Drebes, 1972). The forms studied, like the coccolithophores, are the hardier representatives of the group, and may not be typical of diatoms as a whole.

Diatoms reproduce by simple cell division and by auxospore formation. Typically, after mitotic division of the diploid diatom each daughter cell retains one valve which it uses as an epitheca (outer valve) and a hypotheca is newly generated. One daughter cell is the same size as the mother cell, but the other is somewhat smaller. On average, therefore, the cell size of the progeny decreases. Only when a cell has reached a certain size does auxospore formation follow (see Fig. 29.4). If vegetative reproduction goes beyond this stage, the cells may become too small for auxospore formation. For example, the initial size of *Coscinodiscus asteromphalus* is 200 µm and this diatom reproduces by cell division until it attains a size of 80–90 µm, at which stage auxospore formation sets in. If mitotic division continues, it will cease and the cells will die when the cells become 55–60 µm in size. Each mitotic division decreases the valve diameter by 1·5 µm (Werner, 1971a).

The fully grown auxospore is generally about three times longer (or wider) than the mother cell. Two mitoses follow, each producing an abortive nucleus which disappears. Although the cytoplasm does not divide during these mitoses a new shell is formed after each one. The entire process of auxospore formation therefore results in a number of empty shells. These are both the shells of the parent individuals which formed the gametes, and the valves formed during auxospore growth, which are shed. The resulting large cell is then ready to start the cycle of vegetative reproduction.

K

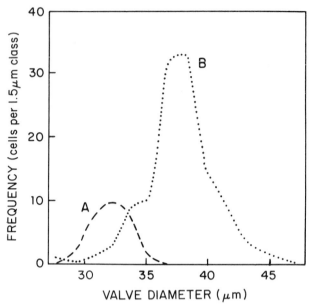

Fig. 29.4. Frequency distribution of cell sizes in a culture of *Coscinodiscus concinnus*. A, cells containing, or producing, gametes. B, vegetative cells. Adapted from Holmes (1967).

There is considerable variation between species with regard to reproductive mechanisms within the general auxospore-vegetative fission cycle. The cycle has important implications for the size distribution of diatom frustules in sediments (Arrhenius, 1952). Arrhenius found more or less regular and periodic variations in the size distribution of *Coscinodiscus nodulifer* frustules in deep-sea cores from the equatorial Pacific, and he suggested that variations in fertility are a causal factor. It is not clear, however, to what extent these size distributions reflect differential transport and differential preservation of the frustules since the evidence for fertility variations is not convincing (see Sections 29.3 and 29.4).

The formation of resting spores also has implications for sedimentation. They are cysts of diatoms with little outward manifestation of life, and arise from spontaneous transformation of a diploid diatom cell. This process involves shrinking and the secretion of a new silica wall within the frustule. The structure of the resting spore is entirely different from that of the normal frustule which is discarded when the spore leaves. Pelagic diatoms, especially centric ones inhabiting shelf regions, are commonly able to form such spores. The siliceous tests of the spores are thick-walled and robust; qualities which assist the spores to settle to the sea floor to await a better opportunity for growth. A spore has a better chance of becoming part of the sediment than does a normal frustule.

Silicification of diatoms has been the subject of much study (see Lewin, 1962a; Reimann *et al.*, 1966; Busby and Lewin, 1967). On deposition, silica shells are entirely enclosed by organic material within a membrane-bounded vesicle. The degree of silicification varies greatly as shown by the following values expressed as $SiO_2$ (% dry weight) derived from diatoms grown in culture medium: *Navicula pelliculosa,* Si-deficient culture (4·3); *Navicula pelliculosa,* not Si-limited (21·6); *Skeletonema costatum* (30·6); *Coscinodiscus sp.* (47·3); *Phaeodactylum tricornutum* (0·4–0·6), (Lewin and Guillard, 1963).

Doubling times for diatoms are of the order of one day for actively growing populations (Fig. 29.3). According to Lewin (1962a) rapidly dividing cells usually deposit thinner shells than do slowly dividing ones. At least in some species, silica uptake appears to be a function of time rather than growth. A similar phenomenon may explain shell thickening in planktonic foraminifera (Berger, 1969b) and, by analogy, in other shelled plankton. However, rapid division of diatoms in the ocean is associated with increased silica supply, which, in turn, favours a higher degree of silicification.

Silicification proceeds very rapidly; new cell walls may be deposited and completely silicified within 10 to 20 minutes after division of the protoplast (Lewin, 1962a, citing Reimann). The degree of hydration of the amorphous silica (opal) making up the tests varies from one species to another and even within the same species (Lewin, 1962a). This may be important in the differential preservation of diatoms under conditions of slow dissolution (Kamatani, 1971; Huang and Vogler, 1972). Another important factor affecting the preservation is the specific surface area of the frustules. Lewin (1961) found values of 123 and 89 $m^2 g^{-1}$ for *Navicula pelliculosa* (a fresh water diatom) and *Coscinodiscus asteromphalus* respectively.

### 29.2.4. SILICOFLAGELLATES

Silicoflagellates are a secondary component of diatom and radiolarian oozes. They were first described by Ehrenberg (1838). Their apparent geological time span is somewhat less than that of diatoms, extending from the mid-Cretaceous. Silicoflagellates are planktonic marine organisms, but apparently tolerate brackish water since they have been reported from the Black Sea (Glezer, 1966) and the Baltic, albeit as stunted forms (Gemeinhardt, 1934). Ebridians and archaemonads (fossil only) are siliceous organisms which are found with diatoms and silicoflagellates; their taxonomic position is unknown (see Tappan and Loeblich, 1971a).

Silicoflagellates, like coccolithophores, are a group within the golden brown algae. They comprise the flagellate-bearing unicellular phytoplankton with internal siliceous skeletons. The skeletons are made of hollow rodlets and are between 20 and 50 μm in diameter. Only two genera of silicoflagellates are important in modern sediments, *Dictyocha* and *Distephanus*, which are typical of warm water and cold water masses respectively (Gemeinhardt, 1934; Mandra, 1969; Martini, 1971b; see also Poelchau (1974) for a taxonomic and ecologic summary).

A comprehensive account of modern and fossil silicoflagellates is given in Glezer's monograph (1966). For a complete index (up to 1967) to the taxonomy of both silicoflagellates and ebridians see Loeblich *et al.* (1968). Lipps (1970) has summarized the ecology and evolution of silicoflagellates.

*Dictyocha fibula* has recently been cultured in the laboratory by van Valkenburg and Norris (1970) who found an average generation time of 49 hours, which is in agreement with expectations for phytoplankton of this size range (Fig. 29.3). Swimming cells both with, and without, skeletons and non-swimming resting stages were observed in clone culture. Both cell diameter and spine length were found to increase under optimal growth conditions. At least some cells were able to shed their skeletons under one set of conditions and grow new ones under a different set. Great variability of skeletons was noted supporting earlier suggestions, based on field observations, that such variations do occur among silicoflagellate species (see e.g. Gemeinhardt, 1934).

The degree of silicification of silicoflagellates, like that of diatoms, can vary considerably. Robust forms are mainly typical of coastal areas (Gemeinhardt, 1934), and in turn the skeletons within the sediment tend to be more robust than those in the water (ibid. p. 300 ff.), introducing the same bias into silicoflagellate fossil assemblages that has been amply demonstrated for those of other groups.

### 29.2.5. PLANKTONIC FORAMINIFERA

Planktonic forams are the most conspicuous contributors to biogenous deep-sea sediments. Murray (1897) was the first to recognize the fundamental difference between pelagic and benthic forams using material collected during the Challenger Expedition. Systematic descriptions of the Challenger material were made by Brady (1884; see also Barker, 1960). Planktonic forams give a sandy texture to modern lime ooze since their size ranges from about 30 μm to 1 mm. Because of their usefulness in stratigraphy, foraminifera have been extensively studied for the last half century.

Benthonic foraminifera may be calcitic, tectinous, agglutinated or aragonitic. Planktonic forams are calcitic. They are essentially restricted to normal marine waters, whereas benthic forms also occur in both hypersaline and brackish waters. The known geological range of the foraminifera spans the Phanerozoic; the oldest pelagic forams are reported from the Jurassic. The rapid development of planktonic forams during Cretaceous and subsequent periods has been said to be responsible for a shift of lime deposition from the shelves to the deep-sea (Kuenen, 1950; p. 393). This idea, though interesting and frequently quoted, appears to be untenable because it neglects the great importance of coccolithophores in pre-Recent calcareous oozes (Bramlette, 1958).

Foraminifera are phagotrophic protists within the class Rhizopodea which contains, among other organisms, the amoebas. The pelagic foraminifera are classified into a superfamily, the Globigerinacea, a grouping which implies that they all derived from one benthic ancestor, or at least from a few closely related ones (but see Hofker, 1968). The transition from benthic to planktonic habitat presumably involved an intermediate stage as it does with the living species "*Tretomphalus bulloides*" which is planktonic, but which also has a benthic phase (*Rosalina*).

The classification of planktonic forams is based on the morphology of their shells, and this gives rise to difficulties similar to those noted above for the other shelled plankton. Very closely related forms have been identified as separate genera (e.g. *Globigerinoides sacculifer* and *Sphaeroidinella dehiscens*; Bé, 1965). Conversely, the same species assignment has been given to rather distantly related forms (e.g. the several species designated as *Orbulina universa*; Rhumbler, 1911; Parker, 1962; Bandy, 1966). A comprehensive introduction to foraminiferal classification has been given by Loeblich and Tappan (1964), and a voluminous catalogue of foraminifera is available for systematic work (Ellis and Messina, 1940 and subsequent supplements). The most widely used work of reference for identification and classification of Recent planktonic foraminifera is the richly illustrated paper by Parker (1962) which is based on the study of sediments. Her classification is also being used in plankton studies (see the identification key of Bé, 1967). Parker divides the foraminifera essentially into two groups; those having spines in life, and those lacking them. This two-fold grouping also roughly splits the forams into "shallow" water (spined) and "deep" water (non-spined) forams, and into those susceptible to dissolution (spined forams) and those which are resistant to it (non-spined forms). Subsequent research on wall structure (Lipps, 1966), and on the amino acid composition of organic shell material (King and Hare, 1972), has confirmed Parker's classification.

Little is known about the life cycle of planktonic foraminifera from direct observations. This cycle is inferred both from indirect evidence of the distribution of the organisms in the water column and on the sea floor, and from analogy with benthic forams. Benthic life histories have been reviewed by Jepps (1956), Loeblich and Tappan (1964), Boltovskoy (1965) and Grell (1967). Recent laboratory observations on benthic forams have been made by Bradshaw (1961), Sliter (1965, 1970), Lee *et al.* (1966), Lee and Zucker (1969), Lipps and Erskian (1969), Röttger (1972) and Röttger and Berger (1972). Reproduction in benthic foraminifera is characterized by an alternation of asexually and sexually produced individuals. These are called gamonts and schizonts respectively, since the former produce gametes and the latter reproduce by multiple fission. The reproductive process leaves an empty shell in either mode of reproduction. Lee and his associates (Freudenthal *et al.*, 1964; Lee *et al.*, 1965) have made cytological studies of several planktonic foraminifera giving special consideration to the zooxanthellae in *Globigerinoides ruber*. This foram is typical of subtropical areas and may largely rely on symbiotic algae for food in low fertility central gyres.

A large number of holoplanktonic foraminifera, both fossil and living, appear to possess analogous structures to that of the reproductive floating chamber of *Tretomphalus*, mentioned above (see Hofker, 1959). The prime example of an apparent reproductive chamber is the terminal all-enclosing chamber of *Orbulina,* the early stage of which is a globigerine foram (Fig. 29.5). Gametes have been observed within the orbuline stage (LeCalvez,

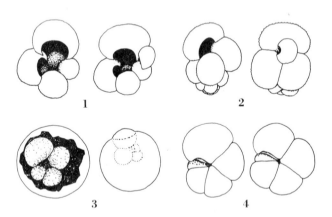

FIG. 29.5. Foraminifera from the California Current. Left specimen of each pair is the normal form; right specimen is the kummerform variant, i.e. those having small terminal chambers. 1. *Globigerina bulloides*, 2. *Globigerinoides ruber*, 3. *Orbulina universa*, 4. *Globorotalia hirsuta.* From Berger (1970c).

1936). In order to account for the lack of individuals actively reproducing in shallow water plankton samples Rhumbler (1911) suggested that forams reproduce below the photic zone, although there seems to be no evidence to support this idea. Thick-walled large forms at depth could conceivably fulfil the role of resting cysts ready to repopulate the water with their offspring when they are returned to a favourable environment by mixing or upwelling (cf. Ericson and Wollin, 1962). Possibly, encrusted individuals simulate a schizont benthic stage. A benthic schizont planktonic gamont cycle has been proposed for *Globigerinoides ruber* (Christiansen, 1965), and may exist for other planktonic forams as well. Emiliani (1954, 1971) has reported the interesting fact that small forms of certain species taken from deep-sea sediments tend to have $^{18}O/^{16}O$ ratios different from those of large ones. Unfortunately, these data cannot be used to elucidate normal life processes because it cannot be assumed that the sediment assemblage is a sample of a growing population (see Fig. 29.6).

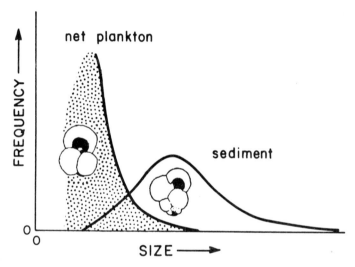

FIG. 29.6. Size and morphology distribution of a planktonic foram (*Globigerinoides ruber*) in the water and in the sediment. The net plankton is characterized by normal forms and the sediment by shells with small terminal chambers ("kummerforms"). Size and morphology differences indicate that the sediment assemblage is not an unbiased sample of a growing population. Modified from Berger (1971b).

The rate of reproduction of planktonic foraminifera has been studied in the field by comparing concentrations of living populations with their empty shell production ("shell output") (Berger, 1969a, 1970c, 1971a, Berger and Soutar, 1967, 1970). The relationship between apparent shell output rates and foraminiferal reproductive rates requires assumptions to be made about the life span, the number of offspring, the predation pressures in the various age groups, the probability of empty shell production at various life stages, the size-age correlation, and the settling rates (Berger, 1971b).

The best estimate for the maximum residence time of planktonic forams > 150 μm in the surface waters is one week (Berger, 1971b). The doubling time for a foraminiferal population must be shorter than this, i.e. a few days (Fig. 29.3). Because of higher predation pressure in the smaller sizes, small species probably have faster turnover rates than do large ones, other factors being equal.

Shell formation takes place within the organism by calcification on an organic membrane (Towe and Cifelli, 1967). In most planktonic foraminifera construction proceeds according to a relatively simple plan (Rhumbler, 1911 ; Berger, 1969b). The proloculus is surrounded by a few initial chambers and each subsequent chamber is added to the previous one at a certain angle and with a certain overlap during most of the growth period. In addition, each chamber diameter tends to have a fixed ratio to that of the previous chamber. This ratio is called the "Q-ratio" and its variations during growth, or between populations of the same species, have ecological implications (Berger, 1969b, 1971a). The main parameters of shell growth, or ones derived from them, are amenable to statistical manipulation and can be used to advantage in ecological, evolutionary and palaeoenvironmental studies (Scott, 1966, 1967; Cifelli and Smith, 1970; Olsson, 1971). A final chamber tends to be added in many species when growth approaches termination; it does not conform to the parameters governing the morphology of the rest of the shell (Bé, 1965). The development of specimens with the last chamber smaller than the previous one ("kummerform"; see Fig. 29.5) appears to be related to the slowing of growth as a result of maturation and reproduction (Berger, 1970c, 1971a; Olsson, 1973) and also to environmental stress (Berger, 1971b; Hecht and Savin, 1972).

Wall structure in planktonic foraminifera has been the subject of extensive research (see Loeblich and Tappan, 1964) which has been greatly enhanced by the advent of the scanning electron microscope (Reiss and Luz, 1970; Towe, 1971). Bé and his associates have given much attention to the two-phase calcification process in many planktonic forams (Bé and Ericson, 1963; Bé and Lott, 1964; Bé and Hemleben, 1970). The first phase is characterized by the lamellar formation of an entire shell to a certain terminal size during which the walls of earlier chambers gradually increase in thickness. In the second phase a "calcite crust" is secreted which consists of a thick calcite layer that covers the entire exterior of the test in late ontogeny. This cortex is by no means common; it is usually found in individuals living deeper in the water, at depths at which foraminiferal concentrations are low. The cortified individuals, however, have an excellent chance of becoming part of the sediment; their proportions are greatly enriched through differential dissolution removing tests without a cortex (Plate 29.1–Fig. 5). Thus, dissolution enhances the deep-water aspect of a species assemblage, and this has important implications both for the oxygen isotope record (Berger, 1971b) as well as for the interpretation of other palaeoecological data. Cortified forams are typical of the Neogene (Plate 29.1–Figs. 5, 7, 8). It is probable that environmental changes which occurred during the Tertiary favoured development of heavy shells in some foraminifera. These changes may have included increased mixing, shallowing of the thermocline, and cooling of deep waters (see Berger and von Rad, 1972). Other morphological features of the shell, in addition to constructional plan and wall structure, are of considerable interest in sedimentology. Thus, coiling directions have proved of use in phylogenetic and palaeoclimatic studies (Bolli, 1971). Statistically significant differences in the mode of coiling of various size fractions of the same species may be helpful in delineating subpopulations which attain different sizes (Thiede, 1971a). Shell porosity differences have been used to delineate climatic variations (Wiles, 1967; Bé, 1968; Pflaumann, 1972); relatively porous tests indicating warm water, and vice versa (see also Zobel, 1968). The less porous cold water forms are better preserved on the sea floor.

### 29.2.6. RADIOLARIANS

The word "radiolarians" commonly designates (Haeckel, 1887) marine protozoans bearing skeletons made of strontium sulfate (acantharians), of opaline silica (polycystins: spumellarians and nassellarians), or of a siliceous-organic mixture (phaeodarians or tripyleans). Acantharians are not preserved in sediments, and phaeodarians are only rarely so. Modern classifications (see for example, Honigberg et al., 1964) do not include the "Acantharia" with the "Radiolaria". In the present account "radiolarians" (Plate 29.1–Fig. 2) always excludes the acantharians and, unless otherwise stated, the phaeodarians.

Radiolarians are ubiquitous members of the marine plankton at all depths. They form an important component of certain deep-sea sediments, especially the siliceous oozes which, in low latitude regions, consist mainly of their skeletons. Common sizes of these skeletons range from a few tens to several hundreds of micrometres. The oldest known radiolarian occurrences are of Cambrian age. For a brief review of the history of the investigation of Cenozoic radiolarians the reader is referred to Riedel and Sanfilippo (1973).

Radiolaria are a subclass of the protozoan class Actinopodea (Honigberg et al., 1964). In Riedel's (1971b) classification, Ehrenberg's (1856) group "polycystins" retains systematic standing (order Polycystina), and is still a convenient way of indicating those radiolarians that make skeletons of hydrated silica which is unadmixed with organic material. To cope with the wealth of radiolarian forms Haeckel (1887) introduced an admittedly arti-

ficial system which is followed in the "Treatise on Invertebrate Palaeonto-
logy" (Campbell, 1954). This is only now being revised (Riedel, 1971b;
Petrushevskaya, 1971d) to provide a more natural system based on phylo-
geny. The compilation of a comprehensive catalogue is in progress (Foreman and Riedel, 1972).

Information on the radiolarian life cycle is far from complete. The occurrence of binary fission, sexual reproduction involving flagellate gametes
and colony formation in collosphaerids by repeated division, were known
to Haeckel (1887). The latter colonies (Haeckel, 1887; Plate 6 ff.) are large
enough to be seen with the unaided eye and contain dense aggregations of
individuals in a thick jelly-like envelope. Brandt (1885), and more recently
Strelkov and Reshetnyak (1971), have studied these colonies which appear
to be the result of asexual reproduction. Specimens with strongly differing
morphology can occur in the same colony. Naked forms also exist.

An extensive account of cytological studies on spumellarians and phaeodarians, as well as some observations on their reproduction, are given by
Hollande and Enjumet (1960), Cachon and Cachon (1971) and Kling,
(1971a). The process of skeleton formation in radiolarians is poorly understood. The skeleton is formed within the protoplasm presumably with the
aid of a plasmatic matrix. In spherical radiolarians, the oldest part of the
skeleton is generally at the centre; in nassellarians the cephalis and thorax
form first, then subsequent segments. Throughout growth, existing skeletal
parts thicken, but semi-silicified spines are not observed (see Hollande and
Enjumet, 1960; Petrushevskaya, 1971c). Species of the genus do not secrete
skeletons.

Several investigators have discussed the differences in distribution between
delicate and massive radiolarian skeletons. Haeckel (1887) found that, in
general, the radiolarians in surface waters are distinguishable from the
abyssal ones by the more delicate and slender structures of their skeletons.
He also pointed out that there are numerous surface water radiolarians
which have either an incomplete skeleton or none at all (ibid. p. CLIII).
Haeckel drew attention to the fact that the abyssal radiolarians with "their
small size and their heavy massive skeletons" are strikingly reminiscent of
the fossil radiolarians of Barbados and the Nicobar Islands (ibid, p. CLV).
Similar observations on the "extraordinary solid nature of the skeleton" of
polycystins caught at great depth were made by Haecker (1907) who studied
samples from closing net tows. He also noted a reduction in the number of
spines with depth. Apparently in response to Haeckel's remark on the
subject, he wrote (p. 116), "In any case, the solid and massive nature of the
skeleton favours the preservation of the empty shells in the deep sea sediments, and it is not unexpected that it is just these thick-walled deep-living,
forms which are found even in older sediments in a well preserved condition."

The rate of reproduction of radiolarians is not known. In colonial collo-sphaerides the presence of one thousand individuals presupposes eleven doublings which have to be completed before the colony perishes or gets transported outside of its normal habitat. Doubling rates are likely to be measured in days (or possibly weeks), therefore, rather than months. A minimum replacement value can be obtained by comparing radiolarian standing stock with radiolarian shell output, by analogy with the foramini-feral turnover calculations. Off southern California a concentration of 250 radiolarians ($>61$ μm) per m$^3$ in the upper waters (Casey, 1971a) compares with about 4000 individuals per 20 ml of Santa Barbara sediment, correspond-ing to a rate of supply of between 200 000 to 400 000 specimens m$^{-2}$ yr$^{-1}$ (Berger and Soutar, 1970). Assuming that the bulk of the productive popu-lation resides in the upper 100 to 200 m of the water column, by analogy with the foraminifera in this region (Berger and Soutar, 1967), the corresponding standing stock from which to draw the radiolarian shell supply is 25 000 to 50 000 radiolarians per m$^2$ of water column. Thus, the maximum residence time of radiolarians $>61$ μm in the upper waters is between 3 weeks and 3 months (stock divided by supply). On the basis of similar reasoning, Casey *et al.* (1971) arrived at the somewhat shorter replacement time of 2 weeks, which agrees with that proposed for foraminifera in the same area (Berger and Soutar, 1970). From this general agreement it is suggested that the life spans of these two major planktonic protist groups are rather similar, at least with regard to the forms inhabiting the upper water layers (Fig. 29.3).

## 29.3. PRODUCTIVITY AND SHELL OUTPUT

### 29.3.1. GENERAL FERTILITY PATTERNS

Productivity and its controlling factors have been reviewed by Harvey (1957), Raymont (1963), Ryther (1963), Steemann Nielsen (1963), and Strickland (1965) (see also Chapter 14). A number of useful compilations have also been produced by Bogorov (1967) and Wooster (1970). There is a high correlation between the overall distributions of nutrients in the euphotic zone, primary production, and the standing stock of phytoplankton and zooplankton (Reid, 1962). The simplified primary production map shown in Fig. 29.7 gives an approximate indication of all these distributions.

Primary production depends mainly on the availability of sunlight and nutrients. In about half of the ocean, nutrients have been virtually exhausted from the euphotic zone by incorporation into phytoplankton via photo-synthesis. The phytoplankton are consumed by marine animals, the excreta and dead bodies of which tend to move downwards, transporting nutrient

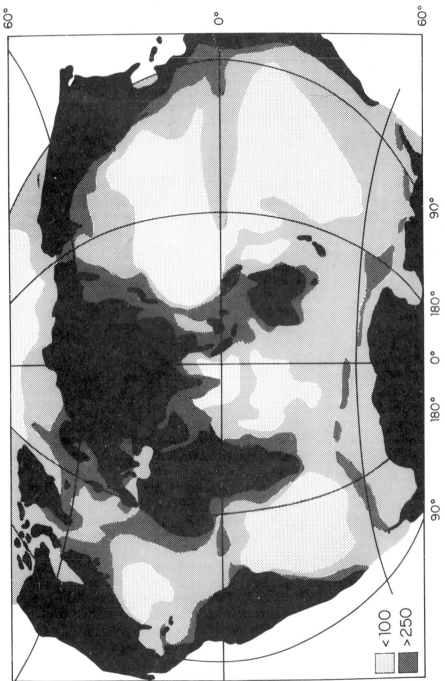

FIG. 29.7. Primary production in the World Ocean (in mg C m$^{-2}$ day$^{-1}$). Simplified from Ko-

elements below the mixed layer. Nutrients are bacterially regenerated at depth from this particulate matter (Redfield *et al.*, 1963). A deep, permanent thermocline in the low fertility areas forms a barrier between the nutrient-rich waters at depth and the illuminated surface waters, thus preventing the replenishment of the euphotic zone with nutrients. Consequently, productivity is believed to be less than 100 mg C m$^{-2}$ day$^{-1}$ over 35% of the World Ocean (Fig. 29.8). Venrick *et al.* (1973) have recently stressed the importance of the

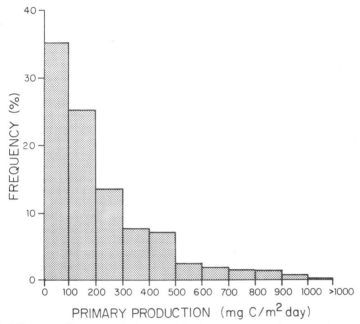

FIG. 29. 8. Frequency histogram of primary production rates, averaged out over the year for the World Ocean. Adapted from Koblentz-Mishke *et al.* (1970). Low values may be over-represented (see text).

contribution to production of a chlorophyll-rich layer below the "euphotic zone" which extends down to the 1% light level. Thus, the value of 100 mg C cm$^{-2}$ day$^{-1}$ may be low (see also Ivanenkov *et al.*, 1972). Areas in which the thermocline is shallow or indistinct, at least during part of the year, tend to be rich in nutrients. These are the regions of divergence or intense mixing (e.g. equatorial areas, west-wind drift and polar front regions), and the boundary regions between land and sea (e.g. neritic areas, coastal upwelling zones, and the waters immediately surrounding oceanic islands). Thus, high fertility zones form a ring around an ocean basin (more intensely developed and wider in extent in eastern than in western boundary areas) with transverse latitudinal strips along the oceanic divergences. It is now appreciated that the limiting role of illumination becomes very important in high latitudes.

### 29.3.2. CLIMATIC REGIMES

On land, climatic regimes are largely defined by temperature and rainfall. In general, there is a latitudinal temperature zonation with superimposed average patterns of rising and falling air masses controlling precipitation. Where light and temperature favour growth, rising air brings rain producing fertile regions. Similarly, in the ocean there is a latitudinal temperature zonation on which are superimposed the patterns of ascending and descending water masses which control the nutrient supply to the lighted surface waters. Light and temperature permitting, rising water results in high fertility. Thus, important factors controlling productivity and biogenic sediment output are the temperature (including the effects of irradiation and seasonality) and the concentrations of nutrients (which are controlled by vertical motions, mixing, closeness to land, boundary conditions etc.). Other environmental factors sometimes emphasized in the geological literature, such as humidity, salinity and water depth, are more or less incidental to planktonic shell output in normal marine waters. Their efficiency as descriptive parameters derives mainly from correlations with the primary factors of temperature, irradiation and nutrient concentration.

The contributions of the various climatic regimes to total production vary from ocean to ocean (Table 1 in Koblentz-Mischke *et al.*, 1970) and in all oceans except the Pacific the coastal waters play the dominant role. The great importance of upwelling to the chemistry and fertility of the oceans is well known (Wooster and Reid, 1963; Ryther, 1969; Smith, 1969; Costlow, ed., 1971). Effects other than upwelling, especially mixing, also bring nutrient-rich waters to the euphotic zone in coastal areas, whence they disperse towards the open ocean.

### 29.3.3. MAJOR PLANKTONIC TRENDS

The fundamental differences in the planktonic populations between the extreme end members of the oceanic fertility spectrum (i.e. the coastal high-fertility regime within a few hundred miles of the shore, and the oceanic low fertility regime with a permanent deep thermocline) are summarized in Table 29.1. The coastal regime is characterized by diatoms and the oceanic one by coccolithophores (Hentschel, 1936; see Fig. 29.9). This dichotomy has been repeatedly stressed by Hulburt (1970) and others (see, for example, Murray and Irvine, 1891). In general, phytoplankton adapted to upwelling regimes apparently grow much better in nutrient-rich waters than do truly

TABLE 29.1

*Major plankton trends, coastal vs. oceanic*

| | Coastal | Oceanic |
|---|---|---|
| Production | High productivity<br>High standing stock | Low productivity<br>Low standing stock |
| Types of organisms | Adapted to sudden increase in fertility, rapid growth response | Adapted to low nutrient concentrations |
| Food chain | Short food chain<br>Metazoans eat much of the algal production | Long food chain<br>Protista eat much of the algal production |
| Phytoplankton* in general | Immobile, large forms abundant | Mobile, small forms dominant |
| Shelled phytoplankton | Diatoms dominant (siliceous sediment output) | Coccolithophores important (calcareous sediment output) |
| Diatoms | Heavy frustules abundant | Weakly silicified frustules dominant |
| Coccoliths | *E. huxleyi* strongly dominant | Diverse assemblages |
| Forams | *G. quinqueloba*<br>*G. bulloides*<br>*G. dutertrei*<br>strongly dominant | Diverse assemblages |

* Nannoplankton ($< 20 \mu m$) are apparently the most important producers everywhere over a wide range of fertility (see Pomeroy and Johannes, 1966, 1968; Malone, 1971; Watt, 1971), but microphytoplankton ($> 20 \mu m$) make a proportionally greater contribution to standing stock and production in nutrient-rich regions (Tundisi, 1971; and refs. therein).

oceanic forms, although the latter in turn reach equivalent growth potentials at much lower nutrient concentrations. An example of the different ways in which the growth of plankton responds to nutrient availability has been given by Eppley (1970) (see Fig. 29.10). It is possible that biological conditioning by dissolved organic substances—marine and terrigenous—also plays some role in producing the contrast between coastal and oceanic phytoplankton (Prakash, 1971).

The distribution of *Emiliania huxleyi* and the positive response of this species to nutrient enrichment are in contrast to those of other coccolithophores (Hulburt and Corwin, 1969), and may indicate that it behaves in many ways more like a diatom than a typical oceanic coccolithophore, such as *Gephyrocapsa* or *Syracosphaera*. Thus, the difference between coastal diatoms and oceanic coccolithophores should be generally greater than that

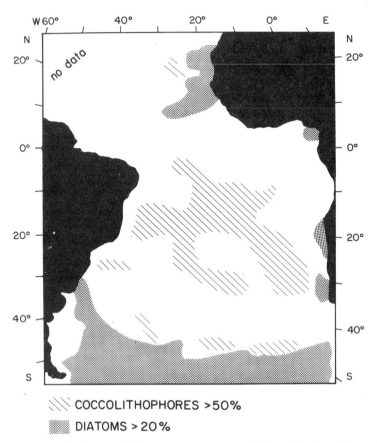

FIG. 29.9. Coccolithophores and diatoms as a percentage of the total microplankton. Where coccolithophores dominate, diatoms usually constitute less than 5%. There are essentially no coccolithophores south of the Antarctic Convergence. Data from Hentschel (1936).

suggested by Fig. 29.10. The effects of competition can be excluded when diatom vs. coccolithophore abundance patterns are considered (Hulbert, 1970). The relatively slow coccolithophore response to fertility increase also appears to be reflected to some degree in the ratio of forams to coccolithophores. Foram abundances appear to increase slightly faster than do those of coccolithophore, in going from barren to fecund areas (Fig. 29.11). The implications of this apparent trend for biogenous sedimentation were first stressed by Olausson (1961a).

The effects of the major biogeographical dichotomy "coastal-oceanic", or more correctly high fertility-low fertility, are superimposed on those produced by temperature gradients. Unicellular algae, like other organisms,

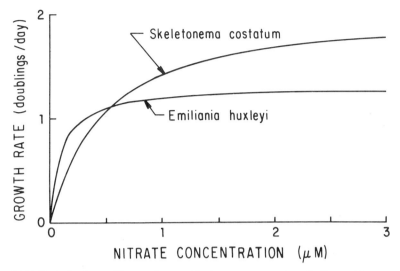

FIG. 29.10. Variation in specific growth rate with nitrate concentration of a diatom (*S. costatum*) and a coccolithophore (*E. huxleyi*) in culture, exemplifying different responses of algae to nitrate enrichment. Adapted from Eppley (1970).

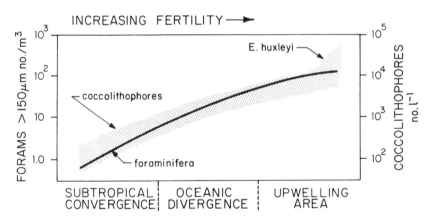

FIG. 29.11. Abundance ranges of forams and coccolithophores in the euphotic zone. Abundance of forams given arbitrarily as a single line rather than as a zone. Note that *Emiliania huxleyi* does not conform to the general trend of the coccolithophores. Data from the various authors cited in text.

have various temperature optima for growth; that is they are adapted to live in either warm, temperate or cold environments. This fact, which forms the basis for paleotemperature determinations, also has implications for the maximum potential growth rate. Cold water species reproduce more slowly than do warm water species at their respective optimum temperatures (Fig. 29.12). If other conditions were optimal for growth, therefore, the sediment output would be expected to be greater with low latitude populations than with those of high latitudes. It is not clear, whether this temperature effect is important at the present time in shaping the patterns of biogenous sedimentation because of the overriding effects of nutrient and light availability. When assessing the potential fertility of ancient oceans through

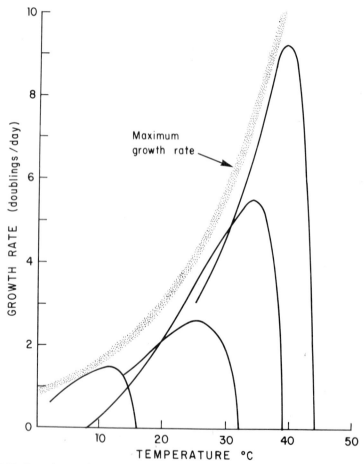

FIG. 29.12. Growth rates of some species of phytoplankton as a function of temperature. At optimum conditions for growth, warm-adapted plankton have a greater rate of production than do cold-adapted ones. Curves based on responses of certain diatoms and chlorophytes (left to right *Detonula confervacea, Ditylum brightwellii, Dunaliella tertiolecta, Chlorella pyrenoidosa*). Adapted from Eppley (1972).

geological time, however, the temperature factor may have to be taken into account.

## 29.3.4. CONCENTRATIONS

The abundance values of phytoplankton (usually expressed as cells per litre) depend greatly on the method of measurement employed. Membrane filter samples give considerably higher results (Kozlova and Mukhina, 1967) than do net samples, but give figures which are very much smaller than those obtained with water samples which have been carefully settled by the Utermöhl technique, as is proved by Venrick's (1969) diatom data. Fortunately, these discrepancies are of little concern here because most pelagic diatoms are weakly silicified and dissolve rapidly in the upper water layers after death (see Antia et al., 1963). For this reason they virtually never reach the sediment, or even the base of the thermocline. The phytoplankton abundance values given in Table 29.2 are typical of those found by membrane filtration.

TABLE 29.2

*Typical concentrations of shelled protists in the upper 50 to 100 m of the water column*

| | Oceanic | | Upwelling | |
| | Convergence | Divergence | Circumpolar | Coastal |
|---|---|---|---|---|
| *Phytoplankton* | | | | |
| Cells per litre (microfilter) | | | | |
| Diatoms | $1-10^2$ | $10^2-10^4$ | $10^3-10^6$ | $10^4-10^7$ |
| Percent of phytomass | $\sim 50$ | $\sim 80$ | $\sim 90$ | $\sim 99$ |
| Silicoflagellates | $0.3-10$ | $2-50$ | $5-10^3$ | $10-10^4$ |
| Percent of siliceous | $10-90$ | $\sim 1$ | $\sim 0.3$ | $\sim 0.1$ |
| Coccolithophores | $50-\sim 10^4$ | $10^2-\sim 10^5$ | $0-\sim 10^6$ | $10^3-\sim 10^7$ |
| Percent of phytomass | $\sim 15$ | $\sim 15$ | $<0.05$ | $\sim 0$ |
| *Zooplankton* | | | | |
| Cells per m³ (microfilter and net) | | | | |
| Forams (total) | $<100$ | $\sim 10^4$ | ? | $10^4-10^5$ |
| Forams $> 150\,\mu m$ | $0.1-10$ | $10-100$ | $0.1-100$ | $10-100$ |
| Radiolarians (total) | $<100$ | $10^3-10^5$ | $<100$ (?) | $10^4-10^5$ |
| Radiolarians $> 150\,\mu m$ | $0.1-10$ | $30-300$ | n.d. | $30-300$ |

Phytoplankton data from Blasco (1971); Bogorov (1967); Gemeinhardt (1931); Hasle (1959, 1960, 1969); Hentschel (1936); Hulburt (1962, 1963, 1964, 1966, 1967); Hulburt and Mackenzie (1971); Kozlova and Mukhina (1967); Nival (1965); Okada (1970); Okada and Honjo (1970); Semina (1963). Zooplankton data from Bé (1969); Beers and Stewart (1967, 1971); Berger (1969a); Bradshaw (1959); Cifelli and Sachs (1966); Hentschel (1936); Petrushevskaya (1966, 1971a, b).

The trends in abundances of shelled phytoplankton parallel those of the total biomass. Diatoms are much more abundant than either coccolithophores (Hentschel, 1936) or silicoflagellates" (Table 29.2) because their abundance is affected by nutrient concentrations to a much greater extent than are those of either coccolithophores or silicoflagellates.

The relatively high proportions of coccolithophores in the rather sparse phytoplankton population of regions of "oceanic convergence" are similar to those in waters of subtropical gyre centres (e.g. the Sargasso Sea), and also of the Mediterranean which is a sea with an anti-estuarine circulation and hence a low nutrient content. The phytoplankton assemblage of the Mediterranean is completely dominated by one coccolithophore, *Cyclococcolithina fragilis,* which constitutes at least 65% of the phytomass at the surface, and 90% of it in the aphotic zone (Bernard, 1963). In the same waters, diatoms are only on average ca. 10% of the phytomass (Bernard, 1969).

The maximum phytoplankton abundances are found in coastal waters (e.g. fjords, estuaries). The Southern Ocean has among the highest open-ocean concentrations, but the abundance of coccolithophores suddenly approaches zero south of the Antarctic Convergence, a fact first noted by Murray (see Hasle, 1960). The boundary between subarctic and arctic waters also delimits the extent of coccolithophore distribution, and has been used to map Pleistocene migrations of the polar front (McIntyre et al., 1972). The variations in abundance shown within the categories of Table 29.2 are largely the result of patchiness and seasonal fluctuations as well as of differences in sampling method. By far the greatest concentrations are usually found in the upper 100 m or, in upwelling areas, in the upper 50 m (or less) of the water column (Fig. 29.13).

Over wide areas of the ocean the maximum production rate exceeds the annual average by a factor of four (see Bogorov, 1967) because of variations in both nutrient supply (at low latitudes) and incident solar radiation (at high latitudes). Thus, in two peak months about as much production can be achieved as during the rest of the year. The sediment output will be correspondingly skewed toward species, sizes and morphologies produced during peak production. Variations of phytoplankton concentrations may be considerable especially in subantarctic, Arctic and Antarctic waters (Hasle, 1959; Walsh, 1969).

The concentrations reported by various investigators for foraminifera range from less than 1 specimen per 1000 m$^3$ to 100 per litre, a spread of $> 10^8$, largely because of the use of different measuring techniques (e.g. centrifugation and the employment of 500 μm mesh nets). Within the common mesh size range (20 to 300 μm) a change of mesh by a factor of $a$ results in the catch changing by a factor of approximately $a^3$ (Berger, 1969a). Incredibly, concentrations are sometimes reported, and often quoted, with-

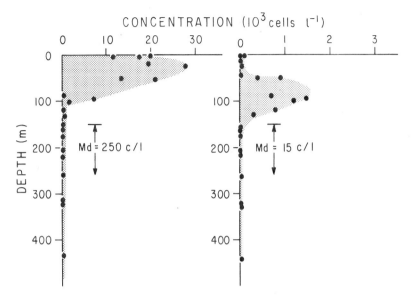

FIG. 29.13. Depth distribution of coccolithophores in the equatorial Pacific. Left: Small cocco-lithophores. Right: *C. fragilis*. Md = median. Data from Hasle (1959).

out specifying the mesh size used to filter the water; such numbers are essentially worthless.

Radiolarian abundances appear comparable to those of forams both in the coastal region off Southern California (Beers and Stewart, 1967), and in the fertile eastern tropical Pacific (see Fig. 29.14). Abundances of the two groups also seem to be similar in the relatively low fertility areas of the western central Atlantic (Cifelli and Sachs, 1966).

Foraminifera and radiolarians together constitute a sizeable proportion, but not a majority, of the total microzooplankton (Fig. 29.14). Concentration fluctuations range roughly over three orders of magnitude (Table 29.2), which is comparable to the range of oceanic phytoplankton concentrations, but not to that of coastal phytoplankton. It may be speculated that the utilization of zooxanthellae by certain foraminifera and radiolarians decreases their concentration ranges in comparison with those of other microzooplankton. It is thought that this symbiosis allows forams, such as *Globigerinoides ruber*, and radiolarians, such as certain colonial spumellarians, to flourish in low fertility waters where other heterotrophs fail. Khmeleva (1967) estimated that the photosynthetic activity of zooxanthellae associated with the radiolarian population in the Gulf of Aden considerably exceeded that of the free-living phytoplankton.

Patchiness (Hardy, 1955; Wiebe, 1970) and seasonality greatly influence

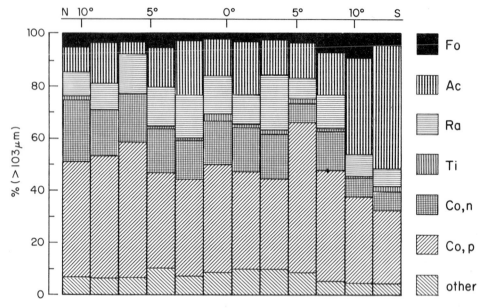

FIG. 29.14. Percentages of micro-zooplankton > 103 μm in plankton sample from the euphotic zone in a north–south section of the eastern equatorial Pacific near 105° W. Legend: Fo = Foraminifera; Ac = Acantharia; Ra = Radiolaria; Ti = Tintinnida; Co,n = Copepods (naupliar); Co,p = Copepods (post-naupliar); Other = other metazoans. Adapted from Beers and Stewart (1971).

the abundance and composition of zooplankton samples. In the Santa Barbara Basin, for example, blooms which result from upwelling and wind stirring favour the fast-growing cold-adapted foram species *G. quinqueloba* which is able to utilize quickly the increased food supply, and this results in a larger standing crop and greater shell output. As a consequence, the sediment is strongly skewed towards this "opportunistic" species, giving the sediment a cold-water aspect (Berger, 1971b). Extensive evidence for seasonal fluctuations in the plankton crop in the western North Atlantic has been presented by Tolderlund and Bé (1971), and Casey (1971a; 1973) has studied the seasonal distribution of radiolarians off Southern California. In addition to seasonal variations there are considerable year-to-year (Tolderlund and Bé, 1971) and decade-to-decade fluctuations (Berger, 1971b) in the abundance patterns of planktonic foraminifera and, by inference, in those of radiolarians. These variations have to be considered when the distribution of species in the plankton crop is compared with that in the sediment.

Seasonal variations and depth distributions of plankton are intimately linked. Submerged "waiting" populations are widespread in oceanic diatom populations (see Venrick, 1969, 1971). In contrast to the resting stages of the neritic meroplanktonic forms these diatoms are weakly silicified. Venrick (1969) has suggested that heavy shells would be a disadvantage to holoplanktonic oceanic forms because they would cause them to sink out of reach of the winter turbulence which could return them to active growth. For analogous seasonal fluctuations of silicoflagellates in the Mediterranean see Nival (1965).

The vertical distributions of planktonic foraminifera and radiolarians are rather similar (Fig. 29.15) and show maxima in, or just below, the euphotic zone in which the free phytoplankton and the zooxanthellae grow. At great depths, radiolarians are virtually the only indigenous shelled plankton, although their concentration is low.

### 29.3.5. SHELL OUTPUT

It cannot be assumed that each living shelled plankton individual has an equal chance of supplying an empty shell to the underlying sediment. Before this chance can be assessed it is necessary to understand those processes which produce empty shells (e.g. predation, reproduction, starvation), the rates at which such shells are produced and the mechanisms by which they are delivered to the sea floor.

An estimate of the quantity of shells formed might be obtained by multiplying the amount of organic production by the average ratio of shell matter to organic matter. For example, it can be argued that most of the suspended opaline silica in the open ocean is a product of precipitation by diatoms, which are also a chief primary producer. Thus, the residence times of organic matter and of siliceous matter in surface waters are presumably of comparable magnitude. Using this approach a semi-quantitative map can be drawn showing the annual production of silica in $g\,m^{-2}\,yr^{-1}$ (see Lisitzin, 1972, Fig. 142). The ratio "of amorphous silica to organic carbon" used for this figure was between 1·85 (Goldberg et al., 1971) and 2·3 (Lisitzin, 1972, p. 152),* and this suggests a typical fixation rate of about $200\,g\,SiO_2\,m^{-2}\,yr^{-1}$, with a range from $<100$ (central gyres) to $>500\,g\,m^{-2}\,yr^{-1}$ (Antarctic). The main problems in making such a map as that produced by Lisitzin include: (1) differences in the cycling rates of organic matter and silica in the upper waters; (2) seasonal variations in primary productivity, especially in high latitudes; (3) any seasonal or geographical variations in the silica/carbon ratios in the organisms; (4) the role of silica fixation by radiolarians, including phaeodarians. Fixation does not, of course, directly reflect delivery to the ocean floor since the proportion of recycled to delivered shells

---

* These ratios may be compared with the value of 1 :1 found for *Coscinodiscus* sp. by Lewin and Guillard (1963) (see Section 29.2.3).

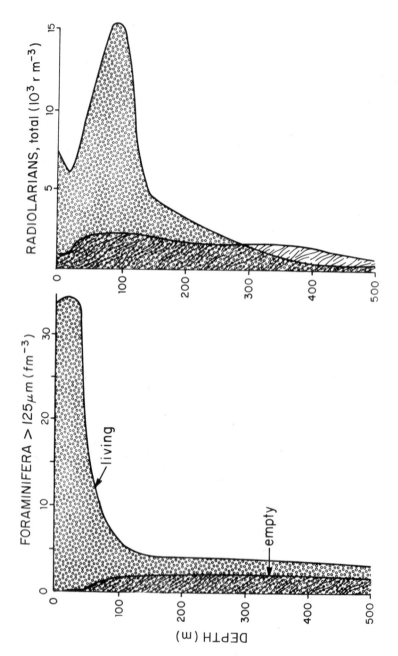

FIG. 29.15. Typical depth distributions of forams and radiolarians in upper waters. Forams: Santa Barbara Basin, August 1966 (Berger, 1971b). Radiolarians: Equatorial Pacific near 154° W (Petrushevskaya, 1971a). Note that virtually all the empty shells are produced in the uppermost part of the water column as the concentration no longer increases below 100 m.

is in itself correlated with fertility variations. In essence then, this approach amounts to redrawing a primary production map on the basis of the assumption that silica fixation is closely correlated with production. Of the annual silica fixation of $\sim 200 \text{ g m}^{-2} \text{ yr}^{-1}$ only about $10 \text{ g m}^{-2} \text{ yr}^{-1}$ reaches the sea floor (see Section 29.4.8), and only $1 \text{ g m}^{-2} \text{ yr}^{-1}$ can be incorporated into the sediments to balance the river input suggested by Livingstone (1963).

The ratio method is not applicable to carbonate sedimentation because of the importance of foraminifera and pteropods, which are not primary producers. Of course, it may still be possible to find some empirical factor to relate carbonate fixation to primary production, although comparisons between concentrations of suspended carbonate and organic matter cannot be used. Such a factor would be expected to vary considerably from one climatic regime to another.

The connection between standing stock, turnover rates and sedimentation rates was first suggested early in this century by Lohmann (cited in Bramlette and Bradley, 1942) who estimated the average concentration of coccolithophores to be 500 million $\text{m}^{-2}$ of the euphotic zone, doubling about every 3 days and producing coccoliths for sedimentation at a rate of about $1 \text{ mm}/1000 \text{ yrs}$. Bramlette (1961) applied similar logic to the foraminifera. By comparing sedimentation rates of calcite with standing stock patterns of forams in areas where loss by dissolution is negligible Berger (1971b) found that calcareous sedimentation is more or less proportional to the standing stock. For every 10 forams $>125 \,\mu\text{m}$ per $\text{m}^3$ in the upper waters (which is about average) there is a supply of total calcite (including coccoliths) of approximately $2 \text{ g cm}^{-2}/1000 \text{ yrs}$. Only about one-sixth of this amount must remain in the sediments in order to balance the calcium input by rivers; for this reason the remaining 5/6 of the supply must redissolve (see Section 29.4.11). To calculate the actual production rate of empty shells from living foram populations the productive population in the upper waters ($10 \text{ forams m}^{-3} \times 200 \text{ m} = 2000 \text{ forams m}^{-2}$) is related to the number of forams per g of sediment ($\sim 10\,000$) and to the sedimentation rate ($2 \text{ g cm}^{-2}/1000 \text{ yrs}$). This corresponds to an output of 200 000 shells $\text{m}^{-2} \text{y}^{-1}$. Thus, the standing stock has to be replaced 100 times a year, or about every four days. This estimate is roughly in accord with those derived for Santa Barbara Basin sediments (Berger and Soutar, 1970) and for settling foram assemblages (see Berger, 1971b). Radiolarian output can be similarly estimated (see Section 29.2.6).

The relationship between standing stock and sedimentation rates for diatoms is less meaningful than is that for forams and radiolarians because only a small fraction of the diatoms found in the water are actually incorporated into the underlying sediments. Thus, one of the dominant forms in the water column, *Chaetoceros*, is only present as spores and fragments on the sea floor (Calvert, 1966a). If the calculation is applied to single species which are abundant in sediments the replacements times which result are reasonable (e.g. *Coscinodiscus centralis pacificus* (cell size 35 μm) has a 10 day replacement time (A. Soutar, personal communication)). Such replacement times should be considerably greater than doubling times because of the destruction of tests through predation and dissolution. An estimate of these losses can be made for the Antarctic if the diatom standing crop and its sedimentation rate are assumed to be those given in Lisitzin's summary (1972, Table 21). A sedimentation rate of up to $1 \, g \, cm^{-2}/1000 \, yrs$ has been reported for biogenic siliceous material (ibid., p. 163), of which $1 \, g$ contains $10^8$ frustules (ibid., Table 21). Plankton cell concentrations are about $10^7 \, m^{-3}$ ("neritic to oceanic complex") for a productive standing stock of roughly $10^8$ diatoms $m^{-2}$. This stock delivers a flux of $10^8$ frustules per $cm^2/1000 \, yrs$ or $10^9 \, m^{-2} \, yr^{-1}$, i.e. it has to be replaced about ten times, presumably during the Antarctic growth season. Thus, minimum replacement times are of the order of 10 days, and the sediment output must be a sizeable fraction of the gross production, that is 10 percent, or more, for doubling times of a day or longer.

Enhancement of diatom accumulation rates by redeposition from elsewhere is apparently indicated for certain varved sediments in the Gulf of California. Calvert (1966a, b) has calculated that the rates of silica accumulation are considerably in excess of $1000 \, g \, m^{-2} \, yr^{-1}$. This value appears acceptable as a silica fixation rate in the euphotic zone of a fertile area (see Lisitzin, 1972), but not as a normal rate of deposition unless extremely high fertility is combined with a high transfer efficiency in the areas studied. Bramlett (1946) has suggested that a systematic redeposition of diatoms from burrowed areas in anaerobic areas is important for the deposition of Miocene diatomites.

### 29.3.6. TRANSFER OF SHELLS TO THE SEA FLOOR

There is quite a close correspondence between the geographical distribution of forams, radiolarians, coccoliths and diatoms found in sediment assemblages and those living in the productive waters above the sediment (Murray, 1897; Riedel, 1963; McIntyre, 1967). For coccoliths and diatoms (as well as clay minerals) this correspondence is puzzling because the slow sinking rates of these minute particles should result in them being widely distributed on the ocean floor. The same problems are also encountered in explaining the distribution of the clay minerals (see Chapter 26). Bramlette (1961) has suggested that accelerated sinking by faecal transport may be one explanation of this phenomenon (for a detailed review of this subject the reader is referred to Smayda, 1971). No such problem is posed by coarser particles which settle at rates of the order of 1 cm s$^{-1}$, and which reach the deep ocean floor within a week or so (Table 29.3). In addition, almost all horizontal transport leading to expatriation is likely to take place in the upper few hundred metres parallel to faunal boundaries (Hamilton, 1957). Another

TABLE 29.3

*Typical settling rates of empty shells*

Ranked from slow to fast. All figures are approximations. Within each group rates vary within a factor of at least 2 or 3, depending on the thickness of the shell and the morphology.*

| | Size | m day$^{-1}$ |
|---|---|---|
| Coccolithophores | | |
|    Solitary | ~10 μm | 0·3–13 |
|    Aggregate | up to 1 mm | 10–6000 |
| Diatoms | | |
|    *Skeletonema* | ~20 μm | ~1 (max. 7) |
|    *Coscinodiscus* | 70 μm | 15 |
|    *Ditylum* | 60 μm | 7 |
|    *Ethmodiscus* | 1 mm | 500 |
| Radiolarians | | |
|    Various Forms | 30–60 μm | 50 |
| | 60–120 μm | 100–200 |
| | 240 μm | 500 |
| Foraminifera | | |
|    Various Forms | 62–125 μm | 250 |
| | 125–177 μm | 500 |
| | 177–250 μm | 1000 |
| | >250 μm | 2000 |
| Pteropods | mm range | 1000–2000 |
| Faecal Pellets | | |
|    Euphausid | n.d. | 100–1000 |
|    Unspecified | 120 × 50 to 200 × 100 μm | 100–300 |
|    Copepod | 100 × 45 to 200 × 45 μm | 100–200 |

* Sources: Berger and Piper, 1972; Bernard, 1963; Fowler and Small, 1972; Smayda, 1969, 1970, 1971; Smayda and Boleyn, 1966; Zeitzschel, 1965. Radiolarian values are preliminary estimates from a few observations by Berger and Piper (1972) which are incidental to foram experiments.

important factor influencing the correspondence between sediment patterns and plankton distribution is the partial preservation of the trace of shell output on the ocean floor which will considerably reduce the track width of the production zone on the sea floor (Fig. 29.16).

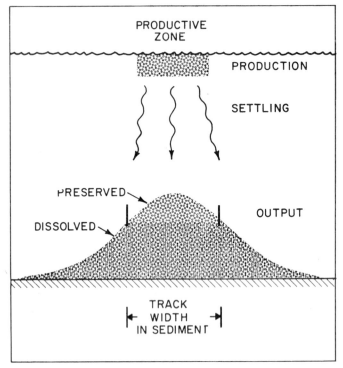

FIG. 29.16. Conceptual model of correspondence between production zone and sediment assemblage. Shells are dispersed during transfer to the sea floor (output) which produces a Gaussian distribution of shells under the shell producing area. Dissolution narrows the Gaussian profile by removing the tail ends.

The occurrence of varved diatomaceous sediments (see Smayda, 1971) can be used as a strong argument for the accelerated sinking of diatoms especially if it is assumed, by analogy with the laminated sediments of Lake Zurich (Nipkow, 1927), that the diatoms are deposited seasonally. However, Calvert (1966a) has ascribed lamination in the sediments of the Gulf of California to the effects of a seasonal influx of terrigenous material superimposed on a uniform biogenous deposition.

An efficient transfer of diatoms appears to be necessary to prevent them being extensively dissolved while they fall through the silica corrosion zone (Fig. 29.17). Accelerated sinking also appears necessary to enable the otherwise very slowly sinking coccoliths to traverse the calcite corrosion zone (Fig. 29.17), although enclosure in a membrane, or any coating with organic matter (Chave, 1965; Chave and Suess, 1967; Wangersky, 1969; Suess, 1970), would retard dissolution. The contribution to sediments made by those

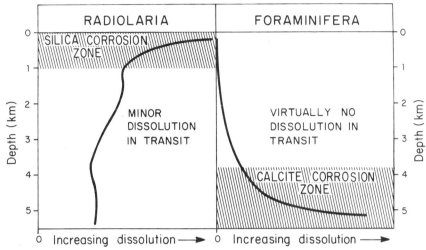

FIG. 29.17. Corrosion zones for siliceous and calcareous particles, based on field experiments in the North Pacific near 19° N, 169° W (Berger, 1967, 1968b). Boundaries have been drawn at depth of maximum rate change, they are not the same as facies boundaries in the ocean.

plankton which live at great depths would appear to be insignificant (coccoliths: Hentschel, 1936; Bernard, 1963, 1969 (see however, Fournier, 1968); radiolarians: Haeckel, 1887, p. CLV; forams: Phleger, 1951b; Berger, 1970c, 1971a).

Vertical density currents, which result when the density of a micro-layer of water and its contained organisms exceeds that of the underlying layer have been studied experimentally by Bradley (1965). In view of the long depth range for which excess density would have to be maintained through the water column in the ocean, this mechanism seems an unlikely one to account for deep-sea sedimentation.

Aggregate and colony formation is a common phenomenon for certain radiolarians (collosphaerids), diatoms (e.g. *Chaetoceros* colonies; and gallertic masses in *Thalassiosira*, see Schrader, 1972b), and coccolithophores (*C. fragilis*, palmelloid phase). However, accelerated sinking has only been reported in the last instance (Bernard, 1963).

Downwelling occurs where surficial water masses converge, or where densities are increased through cooling or evaporation, or both. Cascading of surface waters in the Mediterranean has been variously cited as a possible cause for downward transport of phytoplankton. Recently, David and Stommel (1970) have shown that large scale sinking of cooled surface waters to depths considerably below the photic zone could occur. In an oceanic convergence off Baja, California, downward displacement of living foraminifera by several hundred metres has been noted (Berger, 1971a). However, the subsequent fate of such displaced plankton is not clear.

There is no doubt about the importance of faecal pellet transfer in diatom deposition. Copepods produce pellets at a rapid rate, especially when phytoplankton are dense and "luxury feeding" occurs (see summary by Smayda, 1971). However, weakly silicified diatoms, such as *Chaetoceros*, are likely to be destroyed by ingestion to an extent which will depend on the feeding habits of the particular predator, which may vary considerably (see Mullin, 1966, for copepods). Under these circumstances they will usually make no contribution to the faecal pellets and hence to the sediments. For example, ingestion of plankton by the meropelagic crab *Pleuroncodes planipes* results in the complete mechanical destruction of foraminifera, radiolarians and diatoms. At times this crab is extremely abundant off Baja, California, (Longhurst *et al.*, 1967) and can noticeably clear the water of plankton (Berger, 1971a). Thus, the crab effectively destroys the potential record of certain seasonal blooms.

Fortunately for the geological record, not all predation destroys shells. Certain copepods meticulously pack undamaged frustules into membrane bags, thus allowing them to descend in a form well protected from corrosive waters (Schrader, 1971a, 1972a). Schrader found considerable concentrations ($\sim 2$ per m$^3$) of such pellets packed with diatoms down to a depth of 3500 m off Portugal. Thus, it is clear that those species which are eaten and packed into pellets by "recording" rather than by "destructive" predators are favoured in sedimentation. The pellet transport process described by Schrader (1971a) also has important implications for chemical oceanography, since it offers a mechanism for the large scale fractionation of phosphate from silicate in the water column especially in upwelling regions (see Section 29.4.10).

The effect of predation on the sedimentation of coccoliths and small radiolarians and foraminifera has not been studied (see, however, Honjo in Sliter *et al.*, 1975 and Roth *et al.*, 1975). Occasionally, one or the other of these plankton is mentioned as occurring in the guts of salps, medusae, siphonophores, pteropods, heteropods and copepods, and Haeckel (1887) has used such gut contents as a rich source of radiolaria. The sedimentary record should therefore reflect (among other factors) such predation effects. It cannot be assumed that "recording predation" and "destructive predation" will be equally distributed with respect to major fertility and temperature regimes, or with respect to seasonal variations within these. It is likely that changes in fertility regimes through geological time will have had an important effect on sedimentation through changes in the shell transfer pattern. In this connection, the puzzling change in coccolith/foram ratios at the end of the Tertiary is one phenomenon which needs investigation (see Section 29.5.4). For the sand-sized foraminifera it can be stated quite categorically that predatorial transfer has little or no importance because of the characteristic difference in size distributions in water and sediment, and the difference in morphologies of living and sedimented forams (Fig. 29.6).

## 29.4. DISSOLUTION AND SHELL PRESERVATION

### 29.4.1. MAJOR FACIES PATTERNS, OCEANIC FRACTIONATION

The areas of the deep-sea floor which are covered by pelagic sediments have been estimated by Sverdrup et al. (1942) (see Table 29.4), and revised estimates

TABLE 29.4.

*Relative areas of World Ocean covered by pelagic sediments (in percent)**

|  | Atlantic | Pacific | Indian | Total extent |
|---|---|---|---|---|
| Foram ooze | 65·1 | 36·2 | 54·3 | 47·1 |
| Pteropod ooze | 2·4 | 0·14 | — | 0·6 |
| Diatom ooze | 6·7 | 10·1 | 19·9 | 11·6 |
| Radiolarian ooze | — | 4·6 | 0·5 | 2·6 |
| Oxypelite ("red clay") | 25·8 | 49·1 | 25·3 | 38·1 |
| Relative size of ocean (%) | 23·0 | 53·4 | 23·6 | 100·0 |

* Data from Sverdrup et al. (1942). Pacific pteropod ooze area from Bezrukov (1970). Area of deep sea floor = 268·1 × 10⁶ km².

have been made by Ronov and Vernadskiy (1968, total sea floor) and by Bezrukov (1970, Pacific). The facies proportions depend on the classification adopted. In this chapter the more or less informal classification by Sverdrup et al. (1942), which goes back to the Challenger work, will be followed. In this, calcareous oozes contain more than about 30% shell carbonate and siliceous oozes contain "a large percentage" of opaline silica. This crude subdivision will suffice for a discussion of the major patterns of preservation and dissolution. For recent discussions of bio-sediment classification see Olausson (1960a), Bezrukov (1970) and Chapter 24. Sedimentation rates of such sediments have been compiled by Lisitzin (1972).

About half of the deep-sea floor is covered by calcareous oozes, one seventh by siliceous oozes, and the rest by yellow, brown and reddish deep-sea clay ("oxypelite"). The pelagic environment embraces a little over half of the Earth's surface (268 of 510 $\times$ $10^6$ km$^2$).* This means that calcareous ooze is the most widespread crustal covering material ($\sim 25\%$) of the planet's surface. There is a distinct tendency for carbonates to accumulate in the Atlantic, and for siliceous oozes to accumulate in the Pacific and Indian Oceans. This major difference in the distribution patterns of the biogenous sediments can be ascribed to fractionation of both silica and lime between ocean basins as a result of deep-sea circulation patterns (see Fig. 29.18). The North Atlantic is dominated by an anti-estuarine circulation and the North Pacific by an estuarine one (Redfield et al., 1963). This has important consequences for the concentration profiles of biologically concentrated elements. All such elements tend to become concentrated at depth by extraction from solution into solids and then may be transferred bodily downwards. Any basin such as the Mediterranean, which exchanges deep water for shallow water, will therefore be depleted in nutrient elements. Typically, such circulation stems from excess evaporation, so that the waters have a high salinity. Bottom waters in such basins are "young" and oxygenated, and tend to be close to saturation with respect to calcite. This is also true of the North Atlantic, although to a lesser extent than the Mediterranean. Conversely, any basin which exchanges shallow water for deep water will have waters that (1) are: greatly enriched in nutrients, (2) are "old" and enriched in $CO_2$, and (3) tend to be undersaturated with calcite at relatively shallow depths. Both production and preservation favour silica deposition in such basins. This is the general situation in the North Pacific. Analogous observations have been used by Seibold (1970) to explain the distribution of biogenous sediments in marginal seas.

Anaerobic deposits are typical of estuarine basins. The preservation of the planktonic skeletons within them varies greatly because $CO_2$ production at the sediment-water interface leads to dissolution of carbonate. However, sulphate reduction increases alkalinity (Berner et al., 1970) and this can lead to the calcareous fossils being excellently preserved (Berger and Soutar, 1970; Soutar, 1971). The calcareous record strongly depends, therefore, on the presence of a thin oxidizing layer on top of the sulphate reduction zone, and on the time which a shell spends in this corrosive top layer. The effect on siliceous skeletons is not clear; for example, there are no significant differences between the diatom and radiolarian assemblages of the oxygenated and anoxic uppermost sediments in the Santa Barbara Basin (Berger and Soutar, 1970). However, von Stackelberg (1972) has reported that diatoms and radiolarians are virtually absent from the anaerobic laminated sediments of the Indian continental margin in the Arabian Sea. He attributes this to dissolution of opaline skeletons in the alkaline environment.

* See Berger (1974) for a World Ocean sediment map.

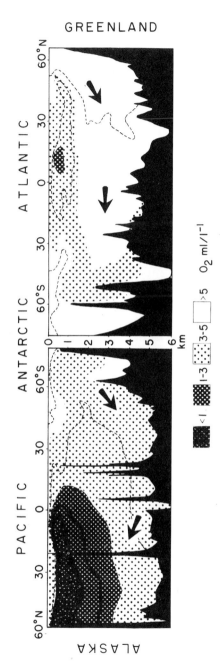

Fig. 29.18. Oxygen concentration along longitudinal profiles in the central Pacific and western Atlantic, illustrating the estuarine circulation in the North Pacific and the anti-estuarine circulation in the North Atlantic. Phosphate, nitrate, silicate and carbon dioxide distributions show analogous patterns (low values in the North Atlantic, high ones in the North Pacific). From Berger (1970b).

L

29.4.2. CALCITE COMPENSATION DEPTH

A flux of alkalinity equivalent to $15 \, g \, CaCO_3 \, m^{-2} \, yr^{-1}$ to the deep ocean waters can be deduced from alkalinity profiles and apparent deep water residence times (Berger, 1970b; Broecker, 1971a; see also Wattenberg, 1935; Pytkowicz, 1968; Li et al., 1969). By far the greater part of this abyssal flux is thought to derive from dissolution of shells on the sea floor, although a contribution from sinking calcareous particles cannot be ruled out entirely (see Section 29.4.6). The dissolution takes place both on the continental slope and in the abyssal depths of pelagic areas. Since dissolution rates increase with depth calcareous shells are typically preserved on the elevated areas of the deep-sea floor, such as the flanks of mid-ocean ridges, mountain ranges and volcanoes. The boundary zone between calcareous sediments and "non-calcareous" ones containing only a few percent of carbonate is more or less well defined, and tends to follow the depth contours in any one area, although it cuts across such contours on a regional basis. This boundary which was discovered by Murray and Renard (1891) is called the "calcium carbonate compensation depth" (Bramlette, 1961), or the "calcite compensation depth (CCD)", because at this level the rate of supply of calcareous shells equals the rate of dissolution. Thus, the CCD lies within the calcite corrosion zone (Fig. 29.17). Where the shell supply is low the CCD is in the upper part of this zone, but it migrates downward as the supply increases. (See papers in Hay (1974), Hsü and Jenkyns (1974) and Sliter et al. (1975) for detailed discussions of the CCD).

Both the percentage of carbonate in the sediments and the number of foraminifera per gram decrease as the CCD is approached. This is clearly shown in large scale distribution maps of calcite (Turekian, 1965 (Atlantic); Bezrukov, 1969, 1970 (Pacific); Lisitzin, 1971b (World Ocean)), and of planktonic foraminiferal numbers (Belyaeva, 1964 (Indian); 1970 (Atlantic). 1969 (Pacific)) in deep-sea sediments. The apparently good correspondence between sediment pattern and bathymetry is due to the averaging and extrapolation of existing data. The usefulness of such maps is greatly increased, therefore if bottom contours and stations are given.

The carbonate concentration in the sediment is a poor indicator of dissolution for several reasons: (1). The loss of carbonate $L$ from a deposit (expressed as a percentage of the total sediment) is given by

$$L = 100 \, (1 - R_0/R), \qquad (29.1)$$

where $R_0$ and $R$ are the initial and final percentages of insoluble material. Thus, for small values of $R_0$ the percent of non-carbonate $R$ is still small after considerable loss has occurred (see also Heath and Culberson, 1970), and the profile of carbonate percentages with depth is rather insensitive to the solution rate model adopted (Berger, 1971b). (2). The carbonate percentages depend on dilution by non-carbonate material which is important in coastal areas, abyssal plains and the fertile belts of high silica supply. (3). Redeposition effects introduce considerable scatter into non-averaged carbonate vs. depth plots (see Turekian, 1965; Fig. 2). For these reasons, as well as the regional variation in depth of the compensation level, correlations between percent carbonate and depth are rather poor (Smith *et al.*, 1968; see also earlier workers cited by Revelle, 1944). The striking bimodality of carbonate distributions (Fig. 29.19) is to be expected from equation (29.1),

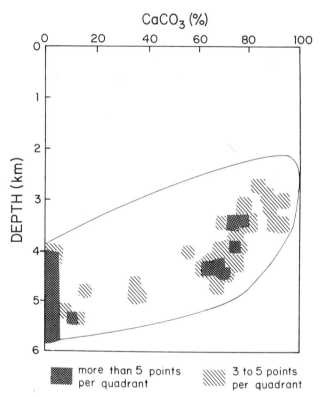

FIG. 29.19. Per cent carbonate concentrations in pelagic sediments of the World Ocean. They show a striking bimodality, confirming the reality of the CCD facies boundary. Each quadrant corresponds to a 250 m depth interval and to a 5% CaCO₃ interval. Data from Smith *et al.* (1968). The line encloses 95% of all observations. Original points have been omitted.

and supports the choice of the 30% carbonate limit in defining a carbonate ooze.

The depth distribution of the CCD is as follows (see Fig. 29.20): *western*

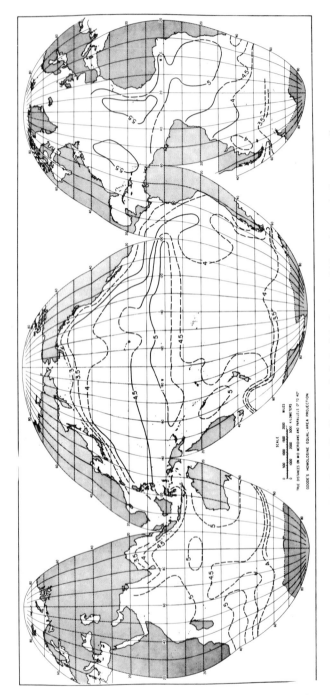

Fig. 29.20. Depth distributions (in km) of the level at which carbonate in deep-sea sediments decreases to a few per cent. Above the CCD given, the probability of the sediment being rich in carbonate is high, and conversely, for samples below the regional CCD the probability is high that carbonate will be rare or absent from the sediment. Compiled from all available sources (see text) including World Sediment Data Bank at SIO. Jane Frazer assisted in retrieving information from the Data Bank. Contours based on averaging by 10° squares. Solid line: CCD based on more than 20 control points per 10° square. (From Berger and Winterer in Hsü and Jenkyns, 1974).

*Atlantic trough*, 5·5 km at 30° N, about 5 km at the equator and at 30° S, 4·6 km at 40° S, rising again southwards of this ; *eastern Atlantic trough*, 5·6 km at 30° N, 5·3 km at the equator, 5·5 km in the Angola Basin, ~5 km south of the Walvis Ridge, rising toward the polar front (3·5 to 4·0 km). For its distribution in the Pacific see Section 29.5.9. The tropical Indian Ocean has CCD levels usually lying very close to 5 km. The seafloor of the Arctic Ocean is accumulating calcareous sediments at the present time (Hunkins *et al.*, 1971); the CCD is not defined. Sources of the above information are Murray and Renard (1891), Revelle (1944), Olausson (1960,b,d), Ericson *et al.* (1961), studies cited by Smith *et al.* (1968), and carbonate maps by Russian workers (see for example: Emelyanov and associates for the Atlantic; Lisitzin and Petelin for the Pacific; see also Lisitzin, 1972).

### 29.4.3. CALCITE DISSOLUTION FACIES*

Carbonate dissolution patterns contain information about the chemistry and fertility of the oceans, past and present. To extract this information it is necessary to understand modern sediments in terms of the present oceanic system. The carbonate system of the ocean is being intensely investigated by both laboratory methods and field studies (see Ben-Yaakov and Kaplan, 1969, 1971; Chave and Suess, 1970; Edmond and Gieskes, 1970; Gieskes, 1970; Hawley and Pytkowicz, 1969; Li *et al.*, 1969; Lyakhin, 1968; Millero and Berner, 1972; Morse and Berner, 1972; Pytkowicz, 1968; Suess, 1970; see also Chapter 9). At present we do not fully understand the fundamental processes involved in this system (Wangersky, 1972), although we are able to map the distributions of such properties as total $CO_2$, specific alkalinity, apparent pH etc., and are able to measure the equilibrium constants involved.

In contrast, the mapping of ocean floor dissolution patterns has received little attention. Murray (in Murray and Renard, 1891) has discussed evidence for pteropod and foram dissolution on the sea floor, and found that the respective compensation levels vary in depth, within and between, ocean basins, and that considerable dissolution occurred above these levels. He realized that an organic cover on shells is likely to retard dissolution (ibid., p. 267), but that the development of carbonic acid through the oxidation of such organic matter would accelerate dissolution (see also Murray, 1897). Schott (1935), Arrhenius (1952), Phleger *et al.* (1953), Olausson (1965) and Ruddiman and Heezen (1967) all recognized the importance of differential dissolution to the composition of foram assemblages. Arrhenius (1952) established the degree of fragmentation as an important criterion in recognizing preservation patterns (Fig. 29.21), and it has since proved useful in

* This topic has recently been discussed in Sliter *et al.* (1975).

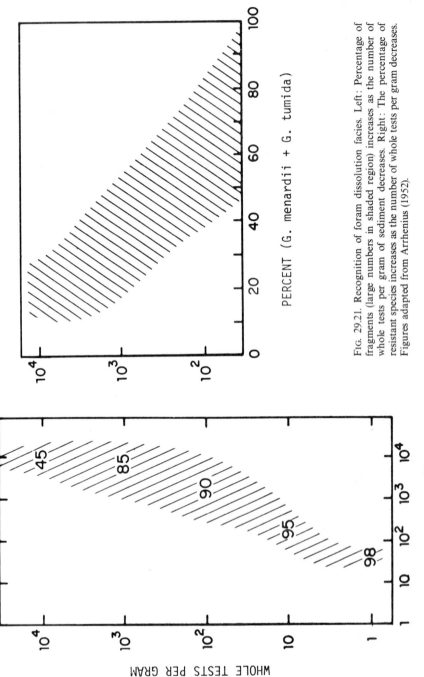

FIG. 29.21. Recognition of foram dissolution facies. Left: Percentage of fragments (large numbers in shaded region) increases as the number of whole tests per gram of sediment decreases. Right: The percentage of resistant species increases as the number of whole tests per gram decreases. Figures adapted from Arrhenius (1952).

delineating such patterns (Cita, 1971; Thiede, 1971b; Berger and von Rad, 1972).

Dissolution enriches the foram assemblages in the resistant forms, as is true for the other groups—coccoliths, radiolarians, and diatoms. The degree of enrichment with resistant forms is a measure of the amount of dissolution that has taken place. The minimum loss (in per cent) can be calculated using equation (29.1); $R_0$ being the initial proportion of solution-resistant forms and $R$ the final one. Resistant forms are partly, or all, from the Neogene Quaternary planktonic genera *Globorotalia*, *Globoquadrina*, *Pulleniatina*, *Sphaeroidinella* and *Turborotalita*. The proportion of benthonic foraminifera can also be used either singly, or in addition to the resistants. $R_0$ must be known reasonably accurately for this procedure to be useful. If $R_0$ is not known, the species are ranked by resistance into the following rank numbers: (1) very susceptible, e.g. *Hastigerina* (not including *Globigerinella*), *Globigerina rubescens*, *Globigerinoides ruber*; (2) susceptible, i.e. virtually all other spiny species; (3) intermediate, e.g. *Globigerinita*; and (4) resistant— see above. A solution index ($S$) can be calculated from the expression:

$$S = \sum_i (p_i \times r_i) \qquad (29.2)$$

where $p_i$ is the proportion of species $i$, and $r_i$ is the rank of this species with respect to dissolution. Several species can have the same rank. The index can be normalized to a median value of 1·0 by dividing $S$ by $r_{max}/2$. This indexing method is very general and can be employed for other parameters related to fauna such as temperature and depth provided that the species have been ranked with respect to these parameters. If desired, the indices can be calibrated against minimum solution losses calculated from equation (29.1). Similarly, temperature indices can be calibrated against actual surface water temperatures by correlating one with the other.

If a detailed faunal analysis is not available, dissolution facies can be recorded by foram solution codes which are numbers assigned to preservation states, one through nine from least to most dissolved, and which are preceded by the prefix FS (Table 29.5). FSX (for dissolution stage 10) denotes complete disappearance of calcitic remains. In some cases for a more general assessment a coarser scale than the FS scale of Table 29.5 is desirable. For this purpose the following terms are used: "P-facies" (*P*lenty of delicate forms; *P*reserved assemblage, FS1 to FS3); "L-facies" (*L*owered abundance of delicate forms, *L*ysocline zone, maximum equitability; FS4); "R-facies" (*R*are delicate forms, *R*esistant forms greatly enriched; FS5 to FS6); "N-facies" (*N*o delicate forms, *N*annofossils relatively enriched; FS7 to FS9) (see Plate 29.1 and Figs. 29.4 and 29.5). In the preservation terminology of Hsü and Andrews (1970), FS1 and FSX are "alytic" and "hololytic" facies, respectively. Details of dissolution assessment are given elsewhere

TABLE 29.5

*Solution facies based on foraminifera (minimum requirements)**

| FS criteria | Approx. loss % |
|---|---|
| 1. Foraminifera undissolved (aragonitic pteropods present) | 0 |
| 2. *Hastigerina* present. Many globigerinids bear spines. $R < 5\%$ | ~10 |
| 3. More spined than non-spined species. $R = 5\%-25\%$ | <50 |
| 4. Maximum equitability of assemblages.' $R = 25\%-50\%$ | 50–80 |
| 5. Solution obvious. Whole tests > fragments. $R > 50\%$ | 80–90 |
| 6. Fragments > whole tests. Plankt. whole tests > calc. benth. forams. | >90 |
| 7. Calcareous benthonics > planktonics. (Plankt. $\neq 0$) | >95 |
| 8. No whole planktonics. Calcareous benthonics present | >98 |
| 9. Fragments of calcareous foraminifera only | >99 |
| X No calcareous fragments | 100 |

\* FS values (foram solution codes) 7 and 8 are applicable to the deep-sea only, since near the continents the supply of benthonics is very large in some areas. *Hastigerina* does not include *Globigerinella*. The resistant forms (R) are species belonging to the late Cenozoic genera *Turborotalita, Sphaeroidinella, Globoquadrina, Pulleniatina, Globorotalia,* sensu Parker (1967; see text). The percent ranges of R given are based on modern subtropical to tropical assemblages. In temperate assemblages the initial R is greater, due to abundant *Globorotalia truncatulinoides* and *G. inflata*. Equitability is a measure of diversity and may be defined as $E = 1 - \Sigma p_i^2$ (see text).

(foraminifera: Berger, 1971b, 1972; Berger and von Rad, 1972; coccoliths: Roth in Berger and von Rad, 1972; Bukry, 1973).

29.4.4. FORAMINIFERAL LYSOCLINE

The rank-index method (equation 29.2) was employed to determine the relative degree of dissolution of foram assemblages in surface samples of the central Atlantic taken by the Meteor Expedition and counted by Schott (1935). A depth level was found which separates well preserved from poorly preserved assemblages, and this dissolution facies boundary was termed the "lysocline" (Berger, 1968a; see Fig. 29.22). The lysocline, in this area, coincides with the abyssal thermocline marking the boundary between North Atlantic Deep Water and Antarctic Bottom Water. A lysocline surface, mapped on the basis of forams in sediments, is also present in the South Pacific (Berger, 1970a; Parker and Berger, 1971). Any relationship to bottom water flow is much more obscure here than it is in the Atlantic because of the greater uniformity of Pacific Deep Water. Detailed studies of the deep water structure (Edmond et al., 1971; Craig et al., 1972), together with laboratory experiments (Morse and Berner, 1972), should throw light

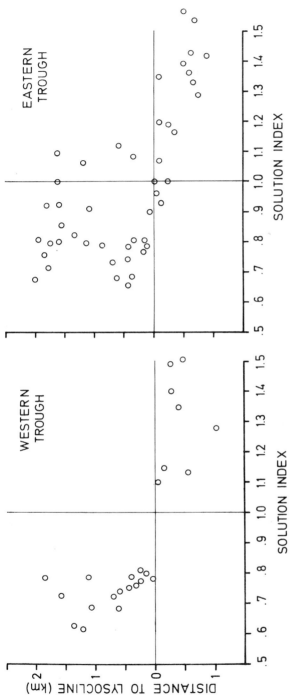

FIG. 29.22. Diagram showing the lysocline as the depth level separating well preserved from poorly preserved assemblages in the central Atlantic. The lysocline apparently coincides with the interface between the North Atlantic Deep Water and Antarctic Bottom Water. Note that considerable dissolution can take place at shallow depths in fertile coastal areas (eastern trough, off Africa). Adapted from Berger (1968a).

on this question. It is interesting to note that the abyssal discontinuity layer mapped by Craig *et al.* (1972) is in some places close to the surface at which dissolution of foraminifera is first obvious.

A complete agreement between the hydrographic lysocline found by Peterson (1966) (which is controlled by water chemistry and dynamics) and the sedimentary lysocline mapped on the ocean floor is not necessarily to be expected. It is the chemical processes occurring at, and near, the sediment-water interface which affect the calcareous assemblages. The differences in chemical composition between bottom waters and interstitial waters in contact with sediments are largely a function of the supply of organic matter and, as a consequence, should therefore be smallest in regions of low fertility. In these regions, the hydrographic and the sedimentary lysoclines should thus virtually coincide. Because of the low supply of carbonate to such areas the CCD will also be close to this level and R- and N-facies will almost disappear. Conversely, considerable separation of the sedimentary lysocline and the CCD occurs below fertile areas (Fig. 29.23a).

The lysocline surfaces are generally at shallower depths in the Pacific than they are in the Atlantic (Fig. 29.23b), and this (as well as an overall depth difference) is the fundamental reason for the difference in carbonate cover between these basins (Table 29.4). Within the basins, the lysocline surfaces

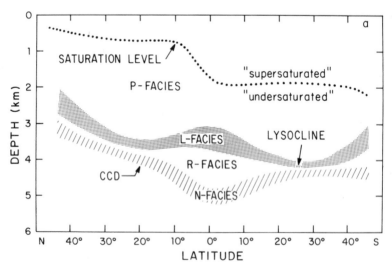

FIG. 29.23a. Model of foram dissolution facies in a N–S profile through the central Pacific. The position of the saturation level (Lyakhin, 1968) is independent of those of the lysocline and the CCD. Note the separation of the lysocline from the CCD in the fertile equatorial region. The lysocline position north of 10° S has not been well established. Modified from Berger (1971b).

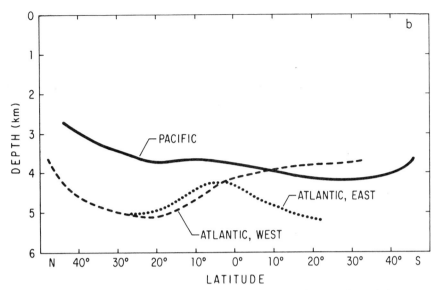

FIG. 29.23b. Approximate position of the lysocline in longitudinal profiles of the central Pacific and the two Atlantic troughs. Distance to the CCD varies. It is usually several hundred metres below the lysocline depth shown. From Berger (1970b).

slope rather steeply upwards as they approach the coastal regions and become diffuse and indistinct. In general, a sedimentary lysocline can be expected in areas where there is a well developed permanent thermocline. Where the thermocline is absent, weak or shallow, there is no well defined lysocline. This is because of the relatively large supply or organic matter to the sediments, and the concomitant $CO_2$ development near the sediment-water interface which leads to enhanced dissolution of carbonate (see Fig. 29.21, eastern trough).

### 29.4.5. COCCOLITH LYSOCLINE

So far, the sedimentary lysocline has been defined by foram data; see, however, Schneiderman in Smith and Hardenbol (1973), and Roth and Berger in Sliter *et al.* (1975). Entirely analogous arguments apply to coccoliths as well, with the exception that these very small calcareous particles are more susceptible to redeposition than are the larger foraminiferal tests. McIntyre and McIntyre (1971) have identified *Emiliania huxleyi, Cyclococcolithina leptopora*, "*Gephyrocapsa oceanica*", including both *G. oceanica* and *G. caribbeanica* and *Umbilicosphaera mirabilis* as resistant coccoliths which are found at those depths at which dissolution destroys other more fragile forms (see also Bukry, 1971; Ramsay, 1972). McIntyre and McIntyre deduced that, in general, the coccoliths are the most resistant of the carbonate secreting invertebrates (ibid., p. 260). It is true that coccoliths are relatively more abundant than are forams close to the CCD (Hsü and Andrews, 1970; Hay, 1970), and this may indicate that the former are on average more resistant. However, redeposition effects may also produce a similar pattern (see Section 29.5.4).

Analysis of the data by McIntyre and Bé (1967) for the placoliths (and Cyrtoliths), which constitute the most common group on the deep-sea floor, enables a dissolution ranking to be made for the Atlantic. These data were analyzed in the following way: for each core (excluding those close to the coast) the nearest neighbour was found. Pairings in which cores were more than 3° latitude apart and where the deeper core was from a depth of less than 3500 m were not considered. Thus, biogeographical variations were kept to a minimum, and it is possible to ascribe most of the difference in species found in such pairs to dissolution effects. For each species it was noted whether its abundance increased ("gain"), or decreased ("loss"), when going from shallow to deep water, or whether there was little effect ("draw"). By ranking the species with respect to the average score per pairing (given in parentheses) a rank order was obtained which was interpreted as a solubility ranking (Berger, 1973b): 1. *C. annulus* (0·17) 2. *C. fragilis* (0·18) 3. *U. tenuis* (0·24) 4. *C. tubifera* (0·25) 5. *E. huxleyi* (0·29) 6. *U. irregularis* (0·29) 7. *U. mirabilis* (0·32) 8. *R. stylifera* (0·33) 9. *H. carteri* (0·33) 10. *C. leptopora* (0·48) 11. "*G. oceanica*" (0·48) 12. *C. pelagicus* (0·62).

The three most resistant species, *C. leptopora*, "*G. oceanica*", and *C. pelagicus*, are clearly set apart in this ranking. In general, when these species are dominant in a sample, the species list is short, reflecting the fact that the others have been dissolved. The sum of these species is taken as *R*. The distribution of *R* shows remarkable agreement with the foraminiferal lysocline (Fig. 29.24), thus supporting the notion that the foram lysocline is also the coccolith lysocline. In Tertiary sediments the percentage of discoasters, which are on the whole more resistant than the rest of the coccoliths, should prove useful in mapping the palaeolysocline.

### 29.4.6. CALCITE DISSOLUTION DURING DESCENT

Kuenen (1950) and Bramlette (1961) have suggested that because exposure times on the sea floor are so much greater than those during settling most, if not all, dissolution occurs on the bottom. However, on the basis of his calcite dissolution experiments Peterson (1966) has suggested that dissolution during transit may be important, at least for the smaller forms. Berger and Piper (1972) have calculated foram dissolution values in the water column from dissolution rates measured in the field (Berger, 1967) and also from settling rates measured in the laboratory which were adjusted for the effect

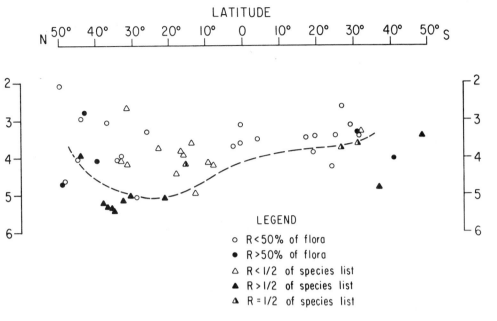

FIG. 29.24. Distribution of resistant coccoliths (R = C. *leptopora* + *Gephyrocapsa* sp. + C. *pelagicus*) with respect to the western Atlantic foraminiferal lysocline (dashed line). Species list is the number of species present based on analysis of data from McIntyre and Bé (1967). From Berger (1973b).

of temperature on viscosity. They found a maximum weight loss for small foraminifera of 6% in the North Pacific; a loss sustained mainly below 4000 m.

Observations on the gut contents of holothurians, asteroids, echinoids and ophiuroids in the Kurile-Kamchatka Trench (Saidova, 1968) bear on this problem. Thus, Saidova found abundant calcareous foraminifera, both planktonic and benthic, within the intestines of these mud-eating animals from depths below 4000 m. She emphasized that foraminifera were well preserved at both the front and hind parts of the gut, and that their abundance within the intestine contrasts markedly with their absence from the sediment. She stated that if the forams are not destroyed in the gut it can be assumed that they will be excreted in good condition. Her data may be interpreted as meaning that on the deep-sea floor, even to depths in excess of 6000 m, there is a very thin layer of calcareous foram tests which have recently been deposited; this hypothesis has recently been tested by Adelseck and Berger (1975). Apparently, this layer, although worked over by surface-feeding mud-eaters, is lost during coring.

### 29.4.7. SILICA BELTS AND SILICA RINGS

Areas of high fertility, characterized by a relatively shallow or- indistinct thermocline, typically have sediments rich in diatoms and/or radiolarians (Pratje, 1951; Riedel, 1959; Calvert, 1968). In such areas, the shell output is high, and the transfer of shells to the sea floor is enhanced by abundant faecal pellet production. Although not all, or even most, of the transferred material stays on the bottom, dissolution of the more fragile shells on the sea floor should act as a buffer and favour the preservation of the more robust shells. Rates of dissolution of opal, which increase with increasing temperature, also depend on the degree of undersaturation of the water (Krauskopf, 1959, Figs. 1, 2; Lewin, 1961, Figs. 1-9 and 22-25; Kato and Kitano, 1968; Hurd, 1972), as a result the silica corrosion zone which the shells must traverse (Fig. 29.17) will be much less pronounced where the thermocline is shallow. The increased sedimentation rates below fertile areas are also thought to favour preservation (Murray and Renard, 1891; Arrhenius, 1952; Riedel, 1959). Within the siliceous oozes, diatoms predominate in high latitudes and around continents and radiolarians dominate in tropical regions. Distributional maps of silica concentrations and silica deposition rates are given by Lisitzin (1967, 1971a, 1972). These maps are in accord with the general fertility pattern, with silica rings rimming the ocean basins and with silica belts along fertile latitudinal zones (Fig. 29.25).

### 29.4.8. SILICA DISSOLUTION FACIES

Dissolution patterns of siliceous sediments are less obvious than those of calcareous ones, because they have no clear-cut relationship to depth of deposition, and also because siliceous sediments are generally not very abundant. Maps showing both abundance and preservation patterns (Goll and Bjørklund, 1971) indicate that the preservation of siliceous shells is rather closely correlated to their relative abundance in deep-sea sediments; see also Goll and Bjørklund (1974). The reason for this is that the original proportions of opal are usually rather low. Thus, if the initial proportion of siliceous fossils in the sediment is 20% and two-thirds are dissolved, the final proportion will be 6·6%. In contrast, calcareous oozes with more than 60% carbonate do not show such quasi-linear correlations, as is evident from equation (29.1). It may be argued that for opaline shells the supply is positively correlated with preservation because of increased burial rates (Riedel, 1959), but also, and perhaps more importantly, because of the buffering action of easily dissolved diatom and phaeodarian remains.

Kling (in Bukry et al., 1971) has suggested that opaline skeletons are more easily dissolved in the upper part of the water column, where temperatures are relatively high and silica concentrations are low. This is in agreement with

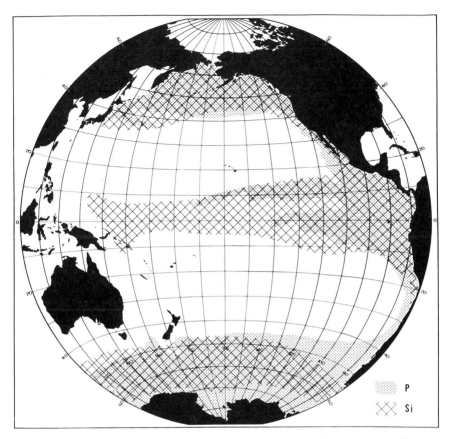

FIG. 29.25. Distribution of sediments rich in siliceous fossils ("Si") with respect to fertile areas, as indicated by the regions where $PO_4^{3-} - P > 1$ μg-at. $l^{-1}$ at 100 m depth ("P"). From Berger (1970b).

field experiments (Fig. 29.17). He proposed that there is a silica compensation level *above* which the rate of dissolution exceeds the rate of supply. The evidence for such a level is extremely meagre. Dilution with carbonate (Riedel, 1959) as well as redeposition effects have to be taken into account since radiolarians and diatoms, like other slowly settling particles, have a tendency to avoid topographical highs and to accumulate in lows (Bramlette and Bradley, 1942; Bramlette, 1946). In addition, the postulated compensation depth would intersect the sea floor in the barren areas below the central gyres (Fig. 29.25; see also the maps by Nigrini, 1970, and Goll and Bjørklund, 1971). Thus, any such level would cut across depth contours rather drastically. Furthermore, the increased dissolution of siliceous tests parallel to the productivity decrease from east to west in equatorial sediments (Nigrini, 1967) suggests that fertility-related effects predominate over depth related effects on the deep-sea floor. However, a depth-related effect on preservation in relatively shallow areas is not excluded. See Johnson (1974) for a recent discussion on preservation patterns.

From the general correlation of the preservation of siliceous shells with their relative abundance, a dissolution scale can be established from least to most resistant; (1) silicoflagellates, (2) diatoms, (3) delicate polycystins, (4) robust polycystins (heavy spumellarians and nassellarians, especially orosphaerids), and (5) sponge spicules (see Arrhenius, 1952; Riedel and Funnell, 1964; Riedel, 1959; Berger and von Rad, 1972; Schrader, 1972c). There is also variation with species among the various groups (for radiolarians see Friend and Riedel, 1967; Moore, 1969; Petruschevskaya, 1971a; Goll and Bjørklund, 1971: for diatoms see Kolbe, 1957; Lewin, 1961; and Schrader, 1971b, 1972a, 1972c) (see also Plate 29.1–Figs. 2 and 3).

Schrader has recently carefully documented diatom dissolution in both plankton and sediment off Portugal and Morocco (1971b, 1972c). He has identified *Coscinodiscus, Actinoptychus, Asterolampra, Asteromphalus* and *Actinocyclus,* among others, as genera containing resistant species, whereas *Rhizosolenia, Chaetoceros, Thalassiothrix* are much more easily destroyed, apparently largely in transit. Dissolution in the sediment was found to be virtually complete. Despite high burial rates only a few diatoms, radiolarians and sponge spicules are preserved (Fig. 29.26). Because of the resistance of sponge spicules their relative abundance is greatly enhanced in the sediment below the uppermost layer. Certain diatoms also survive dissolution. These forms are not necessarily distinguished by size or wall thickness, and a difference in composition and/or structure may therefore be involved (Schrader, 1971c). Lewin (1961) found great differences in solubility between Recent and fossil diatoms, and suggested that differences in chemical structure and the smaller specific surface area of fossil diatoms (by a factor of 5) makes them more resistant to attack (ibid., p. 195). She also considered the effects of surface coatings, which block reactions between the opal and the surrounding water. Indications of variations in radiolarian opal mineralogy can be deduced from anomalously high refractive indices found for skeletal remains from certain deep-sea sediments; see Goll and Bjørklund (1974). Goll and Bjørklund (1971) found such well preserved forms in the Guinea Basin, which is a fertile region with a tendency for cyclonic circulation (Mazeika, 1967). High refractive indices were also found for radiolarians occurring below the productive regions of the northern gyre margin and polar front area (ibid., Fig. 1). A systematic study of refractive indices of siliceous tests should prove interesting in mapping fertility and preservation patterns.

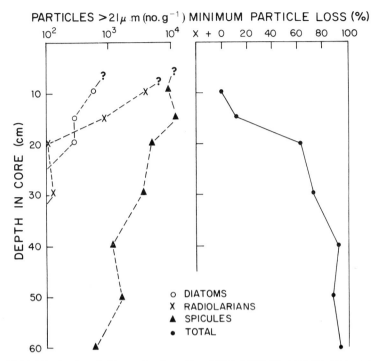

FIG. 29.26. Distribution of siliceous fossils in Core 8057B (37°41′ N; 10°5′ W; 2811 m). Profiles are, in part, influenced by redeposition processes, especially those of sponge spicules and diatoms. Data from Schrader (1972c). "Minimum particle loss' values are probably a lower limit because they are based on a 0% loss at a depth of 10 cm down the core; in fact, the loss at the surface of the sediment will be less than that a depth of 10 cm.

29.4.9. SILICA LEACHING FROM THE SEA FLOOR

Over large areas of the ocean floor radiolarians are restricted to the uppermost few centimetres of Quaternary sediments (Riedel and Funnell, 1964). It may be assumed that these sediments will ultimately be barren of radiolarians (Riedel, 1971a), and that the dissolution of the latter constitutes an input of silica to the overlying waters. Similarly, concentration gradients between interstitial waters and the overlying ocean water suggest that silica is diffusing out of the sediment (Fanning and Schink, 1969). The exact value of most interstitial water measurements is now in doubt because of a large temperature effect caused by failure to store the cores at the correct temperature (see Chapter 31 and Fanning and Pilson, 1971). However, the general conclusion that silica is leached from the sea floor is still tenable; see the comprehensive reviews by Heath (1974) and Wollast (1974). See also Hurd (1973), Fanning and Pilson (1973) and the recent reviews by Heath in Riedel and Saito (1975), and by Calvert in Hsü and Jenkyns (1974).

Most of the silica precipitated as siliceous shells is redissolved in the upper waters of the oceans, concurrently with the oxidation of organic matter and the remineralization of phosphorus. The simultaneous release of dissolved silica, phosphate, and $CO_2$ is thought to be responsible for the observed rough proportionality between the concentrations of these substances in the upper water layers (see Redfield et al., 1963). Such a correlation does not exist for silica in deep waters, suggesting that this element is supplied to such waters after oxidation processes have virtually ceased, presumably mainly from the remaining siliceous skeletons.

The rate of supply of post-oxidative silica to deep waters has been estimated as $0.2$ g-at Si $m^{-2} yr^{-1}$ which corresponds with a flux of $4.3 \times 10^{15}$ g $SiO_2$ $yr^{-1}$, most of which is assumed to be delivered from the sea floor and the remainder from the dissolution of shells in deep water (Berger, 1970b). If this flux is assumed to be correct then the input from the sea floor to ocean water exceeds that from weathering ($4.3 \times 10^{14}$ g $SiO_2 yr^{-1}$; Livingston, 1963) by a factor of between 5 and 10. Heath (1973) has made a survey of other possible sources of silica, and has concluded that they are only of secondary importance. The flux from the sea floor of $0.1$ to $0.2$ g-at Si $m^{-2} yr^{-1}$ (say $10$ g $SiO_2$ $m^{-2} yr^{-1}$) is an overall average and does not reflect regional variations. Such variations are expected to be considerable because of the large differences of flux between silica rings and silica belts on the one hand, and barren areas on the other. As a working hypothesis it can be suggested that redissolution of recently deposited siliceous skeletons in fertile areas contributes the bulk of the flux from the sea floor, and that weathering of volcanic ashes, alteration of basalt, and corrosion of Tertiary siliceous fossils and zeolites contribute relatively minor amounts. There are two kinds of measurements which can be used to test this hypothesis: (1) determination of dissolved silica gradients in interstitial waters, and (2) comparison of the sediment composition in the uppermost layer with that below this layer.

The first method, interstitial water chemistry, relies on the use of diffusion coefficients for the calculation of fluxes. Burrowing and reworking occur especially in the very areas which are thought to deliver the bulk of the flux, i.e. the fertile regions. Apparent diffusion coefficients are markedly increased by such benthic activity to an undetermined degree. Conversely, the barren areas, for which laboratory-determined physical diffusion coefficients are appropriate, may be expected to contribute little to the total silica flux. Diffusion of dissolved silica upwards from the sediment below the mixed surface layer would likewise be expected to be trivial compared with the contribution from the sediment mixed layer itself, which still contains the most easily dissolved frustules and skeletons, and which is in intensive exchange with the bottom water (see Schink et al., 1974).

The second method, comparison of solids, has not yet been applied to deep-sea sediments (see, however, Johnson, 1974). Indeed, it is questionable whether or not true surface sediments are available which could be used to compare the sediment supplied with that incorporated into the record since the common coring procedures tend to lose the uppermost layer (Soutar, 1972, personal communication). Schrader's (1972c) results from the continental slope off Portugal and Morocco show that the highest loss of silica occurs from the most recently deposited sediments (Fig. 29.26). Within a layer which has been deposited for 2000 to 4000 years dissolution is virtually complete. The depth below the sediment surface at which this has occurred reflects both the rate of dissolution and the downward mixing of fresh material. Schrader (1972c) has suggested that extensive down-mixing can be excluded because the more fragile diatoms were not found in subsurface sediments. Unfortunately, the weights of the skeletons are not given. If a weight of between 1 and 5% of the total sediment is assumed, by analogy with the work of Goll and Bjørklund (1971), the silica flux will be 1 to 5 g m$^{-2}$ yr$^{-1}$ for the sedimentation rate given.

The suggested overall average silica flux of 10 g m$^{-2}$ yr$^{-1}$ from the uppermost layer of deep-sea sediments compares with a suggested fixation rate of roughly 200 g m$^{-2}$ yr$^{-1}$ (Lisitzin, 1972) and a sedimentation rate of about 1 g m$^{-2}$ yr$^{-1}$ (which is assumed to be equal to the river influx, ignoring any other contributions which are usually assumed to be negligible; see however, Hart, 1973; Apollonio, 1973 and Section 29.3.5). The fixation–redissolution cycle in the upper waters, therefore, completely dominates the silica flux patterns in the oceans. Because of the overriding effects of fertility distribution on both the intensity of the fixation–redissolution cycle and the supply–releaching cycle at the sea floor, the fertile coastal regime is probably most important in controlling the silica budget of the ocean (cf. Heath, 1974).

29.4.10. DIFFERENCES IN PRESERVATION OF SILICA BETWEEN ATLANTIC AND INDO-PACIFIC OCEANS

There is reliable evidence that one or more factors are operative to prevent the preservation of siliceous fossils in the Atlantic, especially the North Atlantic (Haeckel, 1887; Kolbe, 1955; Nigrini, 1967; Goll and Bjørklund, 1971). Productivity differences between the North Atlantic and other ocean areas appear insufficient to explain the scarcity of siliceous sediments in this ocean. Thus, even off the shelves of Portugal and Morocco, in a coastal zone influenced by an eastern boundary current, siliceous skeletons are rare (Schrader, 1971b). Similarly, differences in the rates of burial between the various oceans cannot be used as an explanation for overall differences in preservation; such rates are generally relatively high in the North Atlantic

(Ericson *et al.*, 1961) because of high terrigenous input, and because of the relatively good preservation of carbonates even at great depths. The rarity and poor state of preservation of siliceous tests in the North Atlantic can be accounted for if it is assumed that those produced are relatively weakly silicified compared with those produced elsewhere, that remineralization is rapid in the water and/or that the interstitial waters in contact with the sediments are more aggressive than elsewhere. Arguments in favour of all these hypotheses can be put forward for the northern central Atlantic Ocean. Dissolved silicate concentrations are generally much lower here than in other oceanic areas (Fig. 29.27); they often approach zero in the euphotic zone, because organisms have extracted essentially all the silicate present. From the slopes of the silicate/phosphate plots it is clear that in the Atlantic relatively less silica will be precipitated per unit of metabolized phosphate than elsewhere (see Fig. 29.27). This condition can be met either by the weaker silicification of all shells, or by a shift in the relative proportions of the various species in favour of the more weakly silicified diatoms. Unless one of these mechanisms is operative it is difficult to explain Kolbe's (1955) observation that Atlantic diatoms dissolve very readily. A preponderance of weakly silicified tests, of course, would enhance dissolution in the water column to the detriment of deposition. Thus, accumulating radiolarians would largely lack the diatom buffer provided elsewhere (evidence for this is provided by the relatively low initial diatom to radiolaria ratios in Fig. 29.26). A direct effect of low silicate concentrations on radiolarian shell growth is also possible and any such effect could be sought for by comparing radiolarians from the eastern tropical Pacific with those from the Mediterranean.

Interstitial waters near the sediment-water interface should be more aggressive in the central Atlantic than elsewhere because of the low silicate content of the bottom water. In addition, organic matter, phosphate etc., which possibly inhibit dissolution by blocking the surface of the opal, are relatively rare in the sediments because *all* nutrient material tends to be transferred out of the North Atlantic with the deep waters.

29.4.11. CAUSES OF DISSOLUTION AND A BIODYNAMIC OCEAN MODEL*

Part of the total biogenous supply to the sea floor is redissolved; another part becomes incorporated into the sediment. The proportion remaining on the sea floor is ultimately determined by the input of dissolved calcium carbonate and dissolved silicate to the oceanic system from other systems, including the continents, the old oceanic crust and the mantle. At the present

---

* See Berger and Roth (1975) for an extension of this model to fertility control of the ocean through geological time.

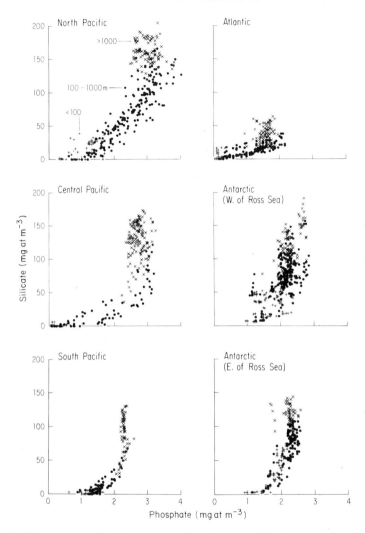

FIG. 29.27. Silicate versus phosphate concentrations in waters of various oceanic regions and various depths. Note the low values in the Atlantic (north subtropical) and the high values in the North Pacific. Differences in slopes indicate fractionation of silicate from phosphate. Silicate tends to be depleted in the euphotic zone before phosphate does. From Berger (1970b). ×, deeper than 1000 m; ● 100–1000 m; + less than 100 m.

time much of this input is essentially introduced by river run-off (Revelle and Fairbridge, 1957; Calvert, 1968; Heath, 1974), although input from submarine weathering of basalts (Hart, 1973) and from glacial activity (Apollonio, 1973; Schutz and Turekian, 1965) may not be negligible. During those periods, such as the present, when the ocean is well mixed it is highly fertile and supplies biogenic solids to the sea floor in excess of the input of the constituents from which they were derived. This "excess supply" drives the ocean towards undersaturation with respect to the substances involved. The more resistant the solids are to dissolution, the greater will be the under-saturation for a given deep water residence time (Bogoyavlenskiy, 1967). Thus, the rate of dissolution controls the concentrations of dissolved shell substances, and a feedback loop is established whereby dissolution controls supply and supply controls dissolution. The average solubility of the biogenic solids supplied is obviously only of limited interest with respect to the cycle involving the sea floor. The rate of dissolution of the most resistant portion of the excess (but not the total) supply is the limiting factor for overall under-saturation of the oceanic system with the biogenic substances. According to this "biodynamic" model of the oceanic chemical system a well mixed ocean has a low biogenic sedimentation to supply ratio, or "preservation factor". The consequence of this is that in a fertile ocean the biogenic record is, on the whole, poorly preserved. At present, the preservation factor is between 0·1 and 0·2 for both silica and calcium carbonate (Berger, 1970b). A similar factor has not been calculated for phosphate, although arguments for its existence are entirely analogous to those given above (cf. Arrhenius, 1963; Broecker, 1971b).

The possibility that the oceanic silica system is controlled by thermo-dynamic equilibria has been considered (see, for example, Sillén, 1961; MacKenzie and Garrels, 1966; Siever, 1968). Dissolved silicate concentra-tions vary greatly between different ocean basins (see, for example, Sverdrup *et al.*, 1942, p. 245). Thermodynamic equilibrium with any one unique solid phase, or phase combination, therefore, does not exist. Furthermore, old oxygen-depleted waters are rich in silicate and young well-oxygenated ones are poor in silicate. Any control by clay minerals, therefore, would have to be by the uptake of silica in the upper waters, and its release at depth. How-ever, this process is exactly the reverse of that predicted by thermodynamics and also of that found experimentally, since upper waters are much more undersaturated than deep waters. Thus, sea water–clay interactions are unlikely to control present day silica concentrations in the water. Geo-chemically important silicate–clay interactions are expected to take place during diagenesis (see Chapter 5 and Wollast, 1974). Such reactions are evident in the formation of zeolites, sepiolites and palygorskites *within* sediments. The presence of zeolites on the sea floor in areas of low sedimen-tation rates (Arrhenius, 1963) is, of course, no proof of their formation at the sediment–water interface. These zeolites may in fact be dissolving, together with any associated Tertiary radiolarians. Such fossils are widespread over large areas of the Pacific (Riedel, 1971a).

29.5. INTERPRETATION OF THE RECORD

29.5.1. BIOGEOGRAPHY AND TAPHOGEOGRAPHY

The water mass and temperature of species of shelled plankton *in the water column* (make-up of the assemblage = biocoenoses; its distribution = biogeography), and the corresponding parameters for their remains on the sea floor (taphocoenoses and taphogeography; see Fig. 29.1) form the basis for palaeoclimatical interpretations, especially of Pleistocene temperature variations. Recent surveys of the species distribution of pelagic diatoms are available (Bogorov, 1967; Hulburt, 1966; Kozlova and Mukhina, 1967; Hasle, 1969; Venrick, 1969; Kozlova, 1970). The distribution of diatoms in sediments has been described by Lohman (1942), Kolbe (1954, 1955, 1957), Kanaya and Koizumi (1966), Mukhina (1966); Kozlova and Mukhina (1967), Jousé et al. (1969, 1971) and Kozlova (1970, 1971). When biogeographic and taphogeographic distributions are compared the species lists are more or less similar. However, the proportions can be drastically changed and species that are scarce in the plankton may become dominant in the sediment, and vice versa (Murray, 1889; Kolbe, 1954; Calvert, 1966a).

The biogeography of coccolithophores has recently been studied with a view to its geological applications (McIntyre and Bé, 1967; McIntyre et al., 1970). Earlier work has been reviewed by Gaarder (1971); see also Hulburt (1966, 1967, 1970), Marshall, (1968), Hasle, (1969), Okada (1970), Okada and Honjo (1970) and Ushakova (1971). The relevant information on temperature and habitats are summarized in Table 29.6. The data from the Atlantic and the Pacific agree very well except that there is a slight tendency for warm water species to inhabit somewhat colder waters in the Atlantic. Such a shift may be due to similar nutrient levels occurring in waters having lower temperatures in the Atlantic than in the Pacific. On the sea floor differential dissolution removes delicate open-structured forms, thus enriching the assemblage with the more solid cold-adapted forms (Fig. 29.28).

The main patterns of the biogeography of planktonic foraminifera (Table 29.7) were established by Bradshaw (1959). They have been reviewed by Berger (1969a) and by Bé and Tolderlund (1971) (see also Bé and Hamlin, 1967; Bé et al., 1971; Beers and Stewart, 1970, 1971; Boltovskoy, 1969, 1971; Cifelli and Smith, 1970; Cifelli, 1971; Tolderlund and Bé, 1971; see Bé in Ramsay (in press) for a recent review). Recent studies on the taphogeography of planktonic foraminifera have been made by Belyaeva (1969), Barash (1970), Parker (1971), Parker and Berger (1971), and Imbrie and Kipp (1971).

The change in species compositions from bio- to taphogeography have been documented in some detail (Berger, 1968a, 1971b). The depth dis-

TABLE 29.6

*Habitat of coccolithophores** (T-range (°C) refers to common occurrence; sequence reflects cold to warm ranking approximately).

| Species | T-range | T-optimum | Nutrient concn. |
|---|---|---|---|
| *Emiliana huxleyi* (cold water form) | 1·5–25 ± 3 | 5–13 | high |
| *Cyclococcolithina leptopora* var. C. | 6–27 | 7–13 | medium |
| *Coccolithus pelagicus* | 7–14 | 8–9 | medium, high |
| *Gephyrocapsa caribbeanica* | 7–27 | 14–26 | n.d. |
| *Gephyrocapsa ericsonii* | 13–24 | 18–20 | medium |
| *Emiliania huxley* (warm-water form) | 7–>30 | 19–23 | high, medium |
| *Rhabdosphaera clavigera* | 18(± 2)–29 | 21 ± 2–28 | low |
| *Gephyrocapsa oceanica* | 16(± 3)–>27 | 22 ± 5–>24 | medium, low |
| *Umbillicosphaera mirabilis* | 17–25 | n.d. | n.d. |
| *Umbellosphaera tenuis* | 16–29 | 20–28 | low |
| *Cyclococcolithina fragilis* | n.d. | <22–>27 | low (Mediterranean) |
| *Discosphaera tubifera* | 18–30 | 21–25 | low |
| *Syracosphaera pulchra* | n.d. | n.d. | low |
| *Helicopontosphaera kamptneri* | 16–26 | 21–25 | n.d. |
| *Cyclolithella annula* | 20–29 | 22–28 | medium, low |
| *Cyclococcolithina leptopora* var. B. | 22–30 | 26–28 | n.d. |
| *Umbellosphaera irregularis* | 22–>30 | 27–>29 | low, medium |

* Data from McIntyre and Bé, (1967) and McIntyre *et al.*, (1970), with additional information from Hentschel (1936) and Hulburt (1966, 1967).

tribution of living planktonic foraminifera plays an important role in these changes. There are two kinds of depth stratification, primary and secondary. Secondary stratification is a submergence phenomenon; in any one region, species from adjacent colder regions tend to occur in the colder waters below the mixed layer. Primary stratification refers to the restriction of some species to surface waters and others to depths below the mixed layer (e.g. Jones, 1967). In general, spined species are shallow water forms, whereas smooth forms occur over a wide depth range. Some forms are found in deeper waters only during seasons unfavourable for reproduction; an example of such a form is *G. truncatulinoides* in the Sargasso Sea (Cifelli, 1962; Bé *et al.*, 1971) and in the Mediterranean (Glaçon *et al.*, 1971). Further, thin-shelled forms are typical of surface waters and thick shelled ones of deeper layers (see Table 29.8). Details of depth distributions of foraminifera have been described by Berger (1969a) and by Bé and Tolderlund (1971).

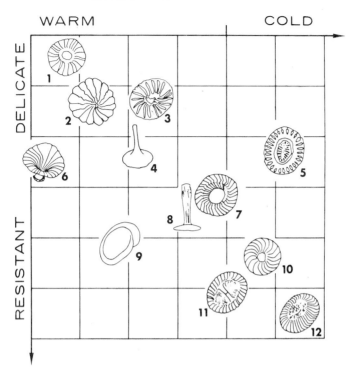

FIG. 29.28. Common modern coccoliths in a temperature dissolution diagram, showing that differential dissolution changes the temperature aspect of a coccolith assemblage. The more solidly calcified forms are concentrated in residue assemblages. Based on data in McIntyre and Bé (1967). 1. *Cyclolithella annula*; 2. *Cyclococcolithina fragilis*; 3. *Umbellosphaera tenuis*; 4. *Discosphaera tubifera*; 5. *Emiliania huxleyi*; 6. *Umbellosphaera irregularis*; 7. *Umbilicosphaera mirabilis*; 8. *Rhabdosphaera stylifera*; 9. *Helicopontosphaera kamptneri*; 10. *Cyclococcolithina leptopora*; 11. *Gephyrocapsa* sp. (includes *G. oceanica* and *G. caribbeanica*; the latter is figured, *G. oceanica* would appear to the left; 12. *Coccolithus pelagicus*. From Berger (1973b).

The particular distribution of morphologies (with open-structured, porous, spiny and globular forms being characteristic of shallow waters, and compact, dense smooth and commonly flat keeled forms being found in deeper waters) has important implications for the interpretation of sediments. Shallow water forms are generally more easily dissolved than are deep water ones, so that the sediment assemblage changes in predictable ways during dissolution (see Sections 29.2.5 and 29.4.3). Since shallow water forms are dominant in well preserved, but not in poorly preserved assemblages, there is a degree of dissolution which results in the diversity (equitability) of an assemblage reaching a maximum (Berger and Parker, 1970). The findings of Sliter (1972) suggested a similar dichotomy between shallow

TABLE 29.7.

*Characteristic habitat parameters for common planktonic foraminifera\*.*

| Species | Surface Water Temperature (°C) | | Food supply | Remarks |
|---|---|---|---|---|
| | Optimum | Common range | | |
| I *Globigerina pachyderma* | 2–7 | 0–19 | high to low | Typical cold water forms, also transition |
| *Globigerina quinqueloba* | 5–16 | 0–20 | high | |
| *Globigerina bulloides* | 8–19 | 2–22 | high to medium | |
| II *Globorotalia inflata* | 10–22 | 4–25 | low to medium | Typical for cold margin of central gyre |
| *G. truncatulinoides* | 14–24 | 8–27 | low to medium | |
| III *Globoquadrina dutertrei* | 14–25 | 10–28 | high to medium | Ubiquitous forms |
| *Globigerinita glutinata* | 14–27 | 4–30 | low | |
| *Orbulina universa* | 14–29 | 8–30 | high to low | |
| *Globigerinella spihonifera* | 17–27 | 12–30 | low to medium | |
| IV *Globorotalia tumida* | 23–30 | 20–30 | medium | Ubiquitous warm-water forms |
| *Globigerina calida* | ~25 | 18–29 | low to medium | |
| *Hastigerina pelagica* | 22–28 | 19–29 | low | |
| *Globigerinoides tenellus* | 21–29 | 19–29 | medium to low | |
| *Globigerina rubescens* | 21–29 | 19–29 | medium to low | |
| *Globigerinoides ruber* | 20–29 | 12–30 | low to medium (wide spread) | |
| V *Sphaeroidinella dehiscens* | 24–28 | 21–30 | medium to low | Typical tropical forms |

| | | | |
|---|---|---|---|
| *Pulleniatina*<br>*obliquiloculata* | 22–28 | 20–30 | medium to low |
| *Globorotalia*<br>*menardii* | 22–28 | 20–30 | medium to low |
| *Globigerinoides*<br>*conglobatus* | 22–28 | 20–30 | medium to low |
| *Globigerinoides*<br>*sacculifer* | 22–29 | 20–30 | medium to low |

* Data mostly from Bradshaw (1959) and from Bé and Tolderlund (1971). Other sources, Berger (1969a), Tolderlund and Bé (1971) and Parker and Berger (1971). Species are ranked from cold to warm-water forms.

TABLE 29.8.

*Depth distribution of planktonic foraminifera in productive warm water regions (Berger, 1969a). Note the relationship of morphology to habitat.*

| Species | Morphology |
|---|---|
| **Shallow** | |
| *Globigerinoides ruber* | Spined, globular, secondary apertures |
| *Globigerinoides sacculifer* | Spined, globular, secondary apertures |
| *Globerinoides conglobatus* | Spined, globular, secondary apertures |
| | |
| Intermediate | |
| *Globoquadrina dutertrei* | Smooth, globular, compact test |
| *Pulleniatina obliquiloculata* | Smooth, globular, thick-walled |
| | |
| Deep | |
| *Globorotalia tumida* | Smooth, flat, keeled, thick-walled |
| *Sphaeroidinella dehiscens* | Smooth, globular, thick-walled |
| *Globorotalia crassaformis* | Hirsute, flat, thick-walled |
| | |
| Variable | |
| *Globorotalia menardii* | Smooth, flat, keeled |

and deep forms extending back to the Cretaceous. From a study of the assemblages off the North American west coast, Sliter reported that hetero-helicid and hedbergellid forms are abundant in those sediments which are thought to be shallow water deposits, whereas globotruncanid forms domi-nate in the slope assemblages. Environmentally, the former correspond to modern globigerinids, the latter to globorotalids. Differential dissolution effects on these forms are analogous, with globotruncanids being enriched in partially dissolved assemblages. (Douglas, 1971).

Information on radiolarian biogeography may be found in Petrushevskaya (1966, 1971a, b, c), and in Strelkov and Reshetnyak (1971). Evidence for the importance of water mass concepts in radiolarian distributions has been summarized by Casey (1971a, b, 1973).

Much less is known about the horizontal and vertical distributions of radiolarian species than about those of foraminifera. Much of the biogeography of radiolarians has to be inferred from studies carried out on sediments (Riedel, 1958; Hays, 1965; Kruglikova, 1966, 1969; Nigrini, 1967, 1968, 1970; Petrushevskaya, 1966, 1967; Casey, 1971a, b; Sachs, 1973). Haeckel's observation on radiolarians (1887, p. CXLVIII) that, "the richest development of forms and the greatest number of species occurs between the tropics, whilst the frigid zones (both Arctic and Antarctic) exhibit great masses of individuals but relatively few genera and species", applies equally to the other shelled plankton. Renz (1973) has compared plankton distributions with sediment patterns in the Central Pacific.

The distribution of pteropods has been reviewed by McGowan (1971). Their empty shells are abundant on the sea floor in some areas, especially in warm saline seas at relatively shallow depths at which "pteropod ooze" is deposited, (see, for example, Chen, 1968; Pastouret, 1970; Herman, 1971; Sarnthein, 1971). Pteropod shells have been used for palaeoclimatic interpretations, and it would appear that differential dissolution is likely to be important in modifying their tapho-assemblages.

### 29.5.2. PALAEOTEMPERATURES, TAPHOTEMPERATURES AND THE TRANSFER PARADOX

The species found in Pleistocene deep-sea sediments are essentially the same as those found in the present day ocean, and they are assumed to have occupied similar habitats. Pleistocene temperature fluctuations, therefore, have largely been inferred from downcore variations of plankton fossils, i.e. the inferred palaeotemperatures are taphotemperatures, which may, or may not, be true palaeotemperatures depending on the changes undergone by the sediment assemblage. Extensive work has been carried out on foraminifera in this respect (see, for example, Schott, 1935; Phleger, et al., 1953; Ericson and Wollin, 1956; Parker, 1958; Ericson, et al., 1964; Beard, 1969; Kennett, 1970; Wollin et al., 1971; Imbrie and Kipp, 1971; Ruddiman, 1971). Emiliani (1955, 1966, 1970, 1972) has pioneered the application of the oxygen isotope method to studies of Pleistocene foraminifera. In addition, radiolarians, diatoms, coccoliths and silicoflagellates have received increasing attention in recent years as potential palaeotemperature indicators (radiolarians: Hays, 1965, 1967; Nigrini, 1970; diatoms: Jousé et al., 1962; Kanaya and Koizumi, 1966; Donahue, 1967, 1970; coccoliths: Cohen, 1964; McIntyre, 1967; Bartolini, 1970; Gartner, 1972; Geitzenauer, 1972; McIntyre, et al., 1972; silicoflagellates: Jendrzejewski and Zarillo, 1972). For a discussion of the variables involved in determining palaeotemperatures from oxygen isotope ratios see, for example, Epstein et al. (1953), Emiliani (1955); Craig (1965); Shackleton (1967); Dansgaard and Tauber (1969); Duplessy et al. (1970); van Donk (1970); Douglas and Savin (1971); Shackleton et al. (1973); Savin and Douglas (1973).

In general, these studies of long deep-sea cores have shown that the oceanic Pleistocene consisted of alternating climatic states which according to Phleger (1948) and Arrhenius (1952) appear to be closely correlated with glacial and interglacial periods. The number of alternations of climate is considerable, and there seems to be no well defined lower boundary to them although their amplitude decreases towards, and within, the late Tertiary (see Berggren (1972a) for a chronology of Late Pliocene–Pleistocene glaciation). The exact magnitude of the temperature fluctuations has been a matter of some contention as there are several fundamental difficulties in establishing the amplitudes of glacial–interglacial temperature cycles (see Berger and Gardner (1975) for a detailed discussion of Pleistocene palaeotemperature determination).

Firstly, amplitudes will differ according to the geographical area considered. Areas which are transgressed by a boundary between climatic regimes experience a drastic change of climate and large ranges of temperatures, whereas areas which are all in the same regime experience lesser variations. This dichotomy is evident from the study of quasi-synoptic difference maps for the central Atlantic which indicate the various ranges of temperature variations experienced by adjacent regions (Berger, 1968a). The principle of climatic transgression was suggested by Phleger et al. (1953), and is clearly illustrated by the North Atlantic polar front movements (Olausson and Jonasson, 1969; McIntyre et al., 1972; see Ruddiman and McIntyre (1973) for a detailed exposition of the frontal movements).

Secondly, in areas with pronounced seasonal variations in productivity it cannot be assumed that the sediment assemblage records the "average temperature" of the euphotic layer, even if the record is perfectly preserved and if growth at depth can be ignored. In effect, the conditions recorded will be mainly those at the time of maximum shell supply. This effect can be illustrated by the following example. Chierici et al. (1962) have studied several cores from the Adriatic Sea, four of which were collected from water depths of between 100 and 1000 m. Preservation should be nearly perfect in this area, and as a consequence the taphocoenosis (the sediment assemblage) should approximate to a thanatocoenosis, i.e. to an empty shell assemblage as delivered from the biocoenosis (see Fig. 29.1). The average composition (125 µm screen, medians) was as follows: *G. bulloides*, 48%; *G. pachyderma*, 6%; *G. quinqueloba*, 17%; *G. glutinata*, 3%; *G. conglobatus*, 8%; *G. ruber*, 8%; *G. sacculifer*, 6%; *G. scitula*, 1%; *O. universa*, 3%. The average tapho-temperature ($T_{av}$) may be calculated according to a "palaeoecological equation" (a term introduced by Imbrie and Kipp, 1971), which relates it to the most probable temperature (or other environmental parameter) and to the composition of the sediment assemblage:

$$T_{av} = \sum_{i=1}^{n} T_i p_i / \sum_{i=1}^{n} p_i \qquad (29.3)$$

where $T_i$ is the temperature at 10 m of a typical habitat of a species $i$ collected using a fine mesh net, and $p_i$ is the proportion of the species in the assemblage (Berger, 1969a). From the above data the weighted average temperature aspect of an assemblage composed of the species $1, 2, \ldots, n$ ($T_{av}$) was calculated to be 16·5 °C. The maps in Chierici *et al.* (1962) show temperatures of 24 °C in summer and 13 °C in winter for the surface waters overlying the area of the coring. Thus, according to this analysis, the winter temperature has roughly twice the weight of the summer temperature in determining the tapho-temperature. If production occurs largely in spring and autumn, these observations are internally consistent.

The above procedure may be criticized on the basis that eco-temperatures have been derived from one area (North and Central Pacific, where Bradshaw (1959) took the plankton tows on which $T_i$ in equation (29.3) is based) and applied to another (the Mediterranean). This criticism is justified and applies to all palaeoclimatic work in which patterns from one matrix (in the present) are directly transferred to another (in the past) with an entirely different climatic setting. The difference between the present and the past climatic setting is commonly established by mapping present distributions of sediment properties (faunal composition, oxygen isotope ratios, and others) onto present environmental parameters and using the resulting correlations to infer past environments from observed fossil distributions (see, for example, Phleger *et al.*, 1953; Ericson and Wollin, 1956; Ericson *et al.*, 1964; Emiliani, 1966; Berger, 1968a; Ruddiman, 1971). There is little doubt that this "transfer procedure", which is the basis for all palaeoecological reconstruction, works within certain limits, and also that it can be considerably improved by multivariate statistics using "transfer equations" (Imbrie and Kipp, 1971). These equations provide quantitative pathways from one matrix (sediments) to the other (environmental parameters). However, if the past conditions were much different from the present ones, there would be some reason to suspect that the transfer equations representing the correlations between sediment record and environment also changed through time.

Thus, the very transfer procedure that established a difference between present and past conditions provides a reason to doubt its accuracy. This "transfer-paradox" is the third fundamental difficulty in establishing amplitudes. It is at the basis of the oxygen isotope controversy, about whether variations in the $^{18}O/^{16}O$ ratios in shell materials are produced by temperature or by ice effects.

The fourth difficulty is associated with dissolution effects. Partial dissolution enriches assemblages with resistant species which may either indicate colder or warmer conditions than would the original assemblage (foraminifera: Section 29.4.3; coccoliths: see Fig. 29.28). In addition, within each species the thick-shelled, deeper-living forms are preferentially preserved, and this should considerably influence oxygen-isotope ratios in foram shells from areas where dissolution occurs (Berger, 1971b, see Savin and Douglas, 1973). Thus, for example, dissolution in the Caribbean was especially pronounced during the warm stages of the Pleistocene Period (Berger, 1971b; Broecker, 1971a). so that the temperature range of Quaternary surface waters deduced from the shell assemblage would be decreased. This effect disappears if the $^{18}O$ enrichment of ocean waters during glacials is somewhat underestimated, thus increasing the apparent palaeo-temperature range to its actual value.

29.5.3. PLEISTOCENE DISSOLUTION CYCLES

The effects of carbonate dissolution during the Pleistocene are poorly understood. The superposition of foram ooze on red clay or diatom ooze in cores from southern latitudes led Philippi (1910) to suggest that the areal extent of red clay in glacial times was greater than it is at present and that, therefore, Antarctic bottom waters had a greater distribution concomitant with an advance of the polar front toward lower latitudes. Philippi also noted that in the South Atlantic subsurface samples tend to have lower carbonate contents than do surface samples, and he stressed the importance of dilution effects arising from the increased influx of terrigenous material during glacial time.

Philippi's ideas, which he developed in part with Murray (Murray and Philippi, 1908), had a profound influence on later work carried out on Pleistocene deep-sea deposits from the *Meteor* (Wüst, 1935; Schott, 1935, 1939) and *Swedish Deep Sea Expeditions* (Arrhenius, 1952; Olausson, 1960a) to present day research (e.g. Johnson, 1972b). The concept of increased Antarctic bottom water activity during glacial periods has led Arrhenius (1952) to suggest that the dissolution of carbonate material increased during these periods in the Eastern Equatorial Pacific. However, he also correlated glacials with high rates of carbonate deposition within the carbonate cycles which he discovered. This has been supported by Emiliani's (1955) oxygen isotope studies. To reconcile high rates of carbonate deposition with simultaneously increased dissolution Arrhenius (1952, 1963) proposed greatly increased production during glacial periods. He believed that this increase resulted from an intensified trade wind system and the correspondingly intensified upwelling near the equator. The resultant increase in productivity was assumed to be large enough to far exceed the proposed enhanced rate of dissolution resulting from the more aggressive glacial bottom water.

On the basis of more recent evidence it has been suggested that carbonate dissolution may, in fact, have decreased during glacial periods in the central and eastern Pacific (Olausson, 1965), in the equatorial Pacific (Hays, et al., 1969), and in the North Pacific (Berger, 1970a). Arrhenius' data also appear consistent with pronounced dissolution during the period of low carbonate sedimentation and much less dissolution during those times when the carbonate sedimentation rate was high (Olausson, 1965, 1967, 1971; Broecker, 1971a). One obstacle to accepting this correlation was Arrhenius' belief that *Globigerina* is highly resistant to dissolution and that *Globorotalia* is highly susceptible to it, whereas the converse has been shown to be correct (Berger, 1967, 1968a). The possibility arises, therefore, that the carbonate cycles are largely a result of dissolution cycles, and the need for invoking a great variation in productivity becomes only of secondary importance.

To test these hypotheses Arrhenius' data from the Swedish Deep-Sea Expedition (SDSE) Core 59 (Equatorial Pacific) have been re-examined (Fig. 29.29). This core was selected because it shows large fluctuations in all its biogenous components. The carbonate record shows the now familiar variations between marl ($<60\%$ carbonate) and chalk ooze ($>60\%$ carbonate). The carbonate content varies between 40 and 90%, i.e. the non-carbonate content varies between 60 and 10%. If this variation is due completely to changes in productivity, and if the non-carbonate fraction accumulates at a constant rate, then the rate of accumulation of carbonate must have varied by a factor of at least 6 ($\frac{90}{10}$ divided by $\frac{60}{40}$). It appears doubtful from the evidence that such a large change in productivity could have occurred in the equatorial Pacific, which is relatively fertile even during the present low carbonate accumulation stage.

The effect of variation in dissolution rate appears to be a more promising explanation of these cycles than do variations in productivity. The fluctuations in the concentrations of entire tests of planktonic forams per gram (Fig. 29.29, second column) are in agreement with the dissolution hypothesis, which also explains the corresponding opposite variation in radiolarians. If productivity were the main controlling factor, a strong tendency for in-phase variations of forams and radiolarians would be expected. The varia-

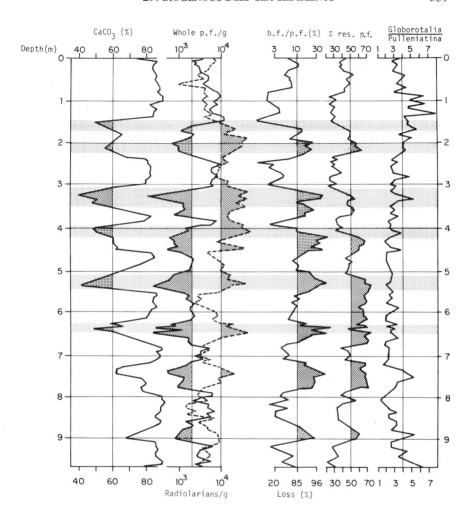

FIG. 29.29. Re-evaluation of SDSE Core 59 from the Equatorial Pacific (Arrhenius, 1952). Marly layers ( <60% carbonate) are stippled. Carbonate percentage (CaCO₃%), whole plank-tonic forams/gram (whole p.f./g, solid line), radiolarians/gram (dashed line), ratio of ben-thonic to planktonic forams (b.f./p.f.), proportion of forams dissolved (loss %, see text for calculation), sum of resistant planktonic forams (res. p.f., *Globorotalia menardii* + *G. tumida* + *Pulleniatina obliquiloculata*), and ratio between the two types of resistant species (*G. menardii-tumida/P. obliquiloculata*) are plotted versus depth in core. Shaded regions bounded by the curves identify values which could be produced by intensive dissolution. Note the common coincidence of such shaded regions with marly zones. Thus, low carbonate stages coincide with low foram numbers, high benthonic percentages, and high percentages of the resistant genera *Globorotalia* and *Pulleniatina*. The ratio of *Globorotalia* to *Pulleniatina* shows no consistent correlation with carbonate values, negating large changes in fertility or temperature regime (small ratios expected for increased temperatures and decreased fertility). From Berger (1973c).

M

tion of the b.f./p.f. ($\times 100$) ratio between benthonic and planktonic foraminifera (Fig. 29.29, 3rd column) is also in agreement with dissolution
being the overriding controlling factor. This b.f./p.f. ratio varies from less
than 3% to more than 30%. Following Arrhenius, the loss of foraminifera
was estimated by assuming that the initial b.f./p.f. ratio is invariant with time
and is altered only by dissolution, not by changing productivity. A minimum loss can be calculated from the expression $L = 1 - (B_0/B)$, where $B_0$
is the initial concentration of benthic forams expressed as a percentage of the
total forams and $B$ the final one (see Fig. 29.30, scale A). In this calculation

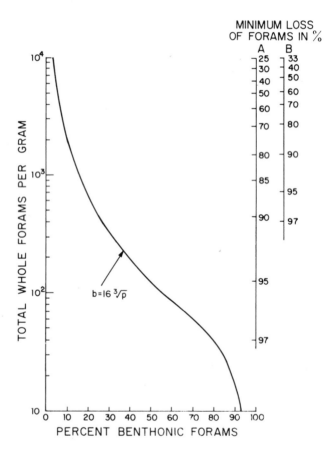

FIG. 29.30. Relationship between whole tests per gram and percent benthonic forams, and scales
showing the corresponding minimum losses calculated from the equation $L_f = 1 - B_0/B$
(scale A), and according to assumptions about differential solution of planktonic and benthonic
forams (scale B, see text). $B_0$ is the initial percentage of benthonic forams and $B$ the final one.
The empirical formula $b = 16^3\sqrt{p}$ is given by Arrhenius (1952), and refers to benthic and planktonic forams per gram of sediment. From Berger (1973c).

it is assumed that there is no dissolution of benthic forams. From the relationship $b = 16\sqrt[3]{p}$ (where $b$ and $p$ are respectively the numbers of benthic and planktonic forams per gram of sediment) given by Arrhenius (1952) it can be estimated that benthic forams will dissolve three times more slowly than will planktonic ones. A simple numerical model based on this (i.e. for the dissolution of each 9% of the total planktonic forams a total of 3% of the benthic forams will dissolve) leads to a new dissolution scale (Fig. 29.30, scale B) which is more realistic than the minimum loss scale. If the initial proportion of benthic forams is taken as 2·5% ,this calculation predicts that the initial concentration of forams in all samples was on average about 14 000 per gram. For a b.p./p.f. ratio of 0·10 (i.e. 10%) the minimum loss is 85% as indicated in column 3 of Fig. 29.29 (scale at base).

The quantitative evaluation of the relative abundance of benthic forams suggests dissolution losses ranging between 20% and 96%. The following equation

$$L = \left(1 - \frac{R_0}{R}\right)\bigg/(1 - R_0), \tag{29.4}$$

where $R_0$ is the initial non-carbonate proportion and $R$ the final one, may be used to obtain a similar range of values from the carbonate record alone. The $R_0$ values required are about 5%, and have a tendency to decrease down the core (from 6% near the top to 2·5% near the base). The low proportion of non-carbonate material near the base of the core thus appears responsible for the small amplitude of the carbonate variation, which nevertheless corresponds to considerable fluctuations in dissolution intensity as monitored by the benthic foram concentrations (compare the lower parts of columns 1 and 3 in Fig. 29.29). This effect can be illustrated by reference to the extreme case of a calcareous ooze with no non-calcareous components which would always show 100% $CaCO_3$, no matter how much dissolution had occurred.

The case for dissolution variations being the cause of the carbonate fluctuations, through high rates of dissolution at times of low carbonate accumulation, appears to be strong. Questions which remain unanswered are whether there is evidence for changes in either the temperature or fertility of the surface waters, or both. The warm water species *Globorotalia menardii, G. tumida,* and *Pulleniatina obliquiloculata* are indeed abundant during stages of low carbonate accumulation which apparently suggests a warming of the surface waters. However, these species are highly resistant to dissolution, and behave essentially in the same manner as do benthic forams (compare columns 3 and 4, Fig. 29.29).

If we assume that *G. menardii, G. tumida* and *P. obliquiloculata* are concentrated by differential dissolution to about the same degree, then the ratio of one group to the other may still yield ecologic information (column 5, Fig. 29.29). However, the ratio of the *G. menardii* + *G. tumida* complex to *P. obliquiloculata* shows no apparent relationship to the carbonate fluctuations. According to the results of Parker and Berger (1971) this ratio increases from west to east along the Pacific equatorial zone, and a high ratio may be interpreted, therefore, as indicative of increased fertility in the overlying water. According to this hypothesis there is a general trend toward increased fertility from near the base of the core to the last glacial period; this is undisturbed by the cycles contained in the interval. The results of this analysis are in agreement with Nigrini's (1971) statement that the radiolarian faunas examined in SDSE Cores 61 and 62 do not reflect cyclic fluctuations in oceanic conditions during the Pleistocene (see, however, Johnson and Knoll, 1974).

Analogous arguments with respect to dissolution, to those which have been applied to SDSE Core 59 also apply to other cores from this area which have carbonate cycles (including Cores 61 and 62), as well as to cores from the western equatorial Pacific (Olausson, 1960a) and the equatorial Indian Ocean (Olausson, 1960a; Oba, 1969; see also Olausson, 1971).

The possibility that some variations in productivity and temperature have occurred is not denied by this analysis. In view of Emiliani's (1955) oxygen isotope results, and the variations in diatom size distributions (Fig. 29.31), it is clear that other processes besides carbonate dissolution play a role, albeit a distinctly secondary one. It is also apparent that the carbonate cycles do not correspond exactly to the dissolution cycles deduced from benthonic foram proportions, but that there is a discernible phase shift.

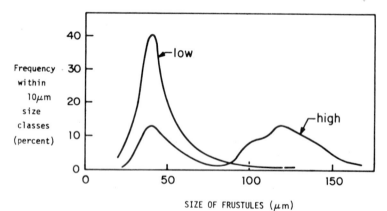

SIZE OF FRUSTULES ($\mu$m)

FIG. 29.31. Size distributions of *Coscinodiscus nodulifer* in a low carbonate accumulation stage (SDSE Core 49, Sample 40) and a high carbonate accumulation stage (SDSE Core 59, Sample 259). Adapted from Arrhenius (1952). (Compare Fig. 29.4.)

Such a shift could be produced, for example, by differential upward mixing of coarse benthonics and fine carbonate. There is no *a priori* reason why the carbonate cycles should exactly parallel the glaciation cycles as defined by amount of ice on land (see Broecker, 1971a). There is evidence from the North Atlantic, for example, for lithified-nonlithified carbonate alternations (Bartlett and Greggs, 1969) in which the lithified layers have a karst-like upper surface reminiscent of hardgrounds. The lithified layers contain warm water forams, and the nonlithified ones contain cold water species as well as redeposited shelf material. Dissolution, or non-deposition, at the end of warm periods appears to have occurred in samples from the areas and depths studied by Bartlett and Greggs (loc. cit.).

Olausson (1965, 1971) has shown that in the Atlantic the interglacials correspond to stages of high carbonate deposition, and the glacials to stages of low carbonate deposition (Fig. 29.32). This contrasts with the situation in the equatorial Pacific. Phillipi (1910) anticipated this correspondence between glacials and low carbonate content on the basis of the examination of only a few cores, and he suggested that it was due to dilution with terrigenous material. The extent to which carbonate dissolution influences the carbonate cycles in the North Atlantic is not clear, despite considerable discussion (Schott, 1935; Olausson, 1967, 1969, 1971; Berger 1968a; Broecker, 1971a, Thiede, 1971b). Broecker (1971a) has estimated the carbonate accumulation rates in a core from the North Atlantic using the time scale of Broecker and van Donk (1970). He suggested that more $CaCO_3$ was lost by dissolution during interglacials than during glacial times. The validity of this suggestion depends on whether, or not, the core analyzed is representative of a large area and a wide depth range, and whether, or not, the time scale adopted is correct.

The foraminiferal record is difficult to interpret because of the complex effects of temperature and dissolution variations on faunal composition. If only the proportion of resistant forams is considered, without regard to temperature-related changes, it would appear that dissolution at depths below the lysocline during the present interglacial is indeed greater than it was in the past (Berger, 1968a). However, when it is remembered that cooling of the surface waters increased the supply of resistant forms (*G. inflata, G. truncatulinoides, G. dutertrei*), differences in dissolution between glacials and interglacials below the lysocline may be small. Nonetheless, in either case, pronounced dissolution apparently started at shallower depths in the last glacial, that is the lysocline was several hundred metres higher.

Additional information on dissolution cycles has been obtained by a preliminary analysis of the foram data of Phleger *et al.* (1953) in conjunction with the carbonate data of Olausson (1960d). The following SDSE cores were analyzed and classified as equatorial (E), tropical (T), sub-

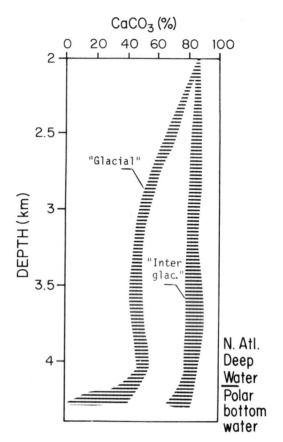

Fig. 29.32. Maximal ("interglacial") and minimal ("glacial") carbonate concentrations in cores from the Eastern North Atlantic. The approximate present boundary between the North Atlantic Deep Water and the Polar bottom water is shown. Note the correspondence between this boundary and the depth below which present carbonate concentrations decrease markedly. Adapted from Olausson (1965 and 1971).

tropical (S), mid-latitude (M), high latitude (H), shallow, and deep (S denotes above present lysocline and D denotes deep): 246 (ES), 243 (ES), 236 (ED), 234 (TS), 233 (TS), 235 (TD), 223 (SS), 280 (MS), 281 (MS), 279 (MD), 288 (HS), 292 (HS), 293 (HS), 296 (HD), 284 (HD). To avoid as much as possible the interferences of taphotemperature and taphosolution effects, ratios were calculated for temperature indices between forams of similar dissolution potential, and ratios for dissolution indices were obtained between species of similar temperature habitat. The following grouping was applied:

(A) *G. pachyderma*; (B) *G. inflata, G. truncatulinoides, G. punctulata*; (C) *G. bulloides*; (D) *G. ruber, G. sacculifer, G. conglobatus*; (E) *G. menardii, G. tumida, P. obliquiloculata, S. dehiscens.*

The following ratios apply for temperature indices: $A/(A + B)$; $B/(E + B)$; $C/(D + C)$; for dissolution indices: $E/(D + E)$; $B/(C + B)$ and similar ratios based on shortened species list were used to keep average indices near 0·5. Relationships between these indices and between percent $CaCO_3$, and percent benthic forams, were established on scatter plots by drawing horizontal and vertical median lines through the scatter and accepting 4 to 1 splits, or better, between adjacent quadrants. The results (Fig. 29.33) show that there is, in general, a positive correlation between $CaCO_3$ and increasing temperature (square 4); i.e. stages of high carbonate deposition correspond to warm temperatures, in agreement with the assessment made by Olausson (1965). However, for Core 281 (mid-latitude, shallow) the converse appears to be indicated (square 1). There is also a striking overall positive correlation between the resistant planktonic foraminifera and the benthonic forams (square 9). This would be expected if both are enriched by differential dissolution. In contrast, several shallow water cores are exceptions to this (square 12). The third correlation is the negative one between cold palaeo-

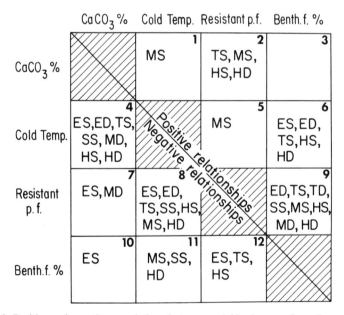

FIG. 29.33. Positive and negative correlations between variables in cores from the central and northern Atlantic. For symbols see text. Based on data from Phleger *et al.* (1953). From Berger (1973c).

temperature and resistant planktonic forams (square 8) indicating that dissolution tends to be greater during warm periods than during cold ones, with core 281 (MS) again being anomalous (square 5). From the positive correlation of resistants and benthics it is clear that a common positive correlation between decreasing temperature and percent benthonics (square 6) is due to redeposition processes and not to dissolution. A positive correlation between resistant forams and percent $CaCO_3$ (square 2) in a number of cores from the equator supports the idea that dissolution is increased during stages of high carbonate deposition, although again exceptions exist (square 7). In general, therefore, periods of high carbonate deposition, interglacials, and times of increased dissolution coincide in the area studied by Phleger *et al.* (1953), but regional differences and local complications may obscure this relationship considerably. On the whole, the present results from both the Pacific and the Atlantic, and observations for the Caribbean (Zobel cited in Berger, 1971b) support the hypothesis that dissolution rates increased in pelagic regions, at least in low latitudes, during interglacials. The reasons for this increase in dissolution in interglacial periods are by no means clear. They may include factors such as productivity variation (Arrhenius, 1952; Berger, 1968a), the total amount of carbonate available for marine sedimentation, as well as the increase in the area of deposition during interglacials when the shelves were flooded (Olausson, 1967) and when considerable calcareous deep-sea deposition took place at high latitudes (Philippi, 1910; Olausson, 1967; McIntyre *et al.*, 1972).

### 29.5.4. COCCOLITH/FORAM RATIOS AND DIFFERENTIAL REDEPOSITION

On a regional scale, redeposition processes can greatly modify the patterns produced by production–preservation relationships, as exemplified by the differences in sedimentation rates between abyssal hills and valleys (Shor, 1959), and by the lack of sediments in ridge crest areas (Ewing and Ewing, 1967; Larson, 1971; see also Davies and Laughton, 1972; Ewing and Hollister, 1972; Ruddiman, 1972). Much of the inferred redeposition is apparently differential in nature and results from the great differences in size between foraminifera and coccoliths. For example, the sediments on submarine ridges commonly contain relatively large numbers of foraminifera (Correns, 1939), whereas there is a distinct tendency for coccoliths to accumulate in basins, unless removed by dissolution (Olausson, 1961a). Suspension studies also suggest that "fines" are removed from topographic highs (Lisitzin, 1972). Coccolith/foram ratios, therefore, have to be investigated from the standpoint of differential transport. Differential dissolution may also play a part in their geographical and chronological distributions, since coccoliths are said to be more resistant to dissolution than are foraminifera (McIntyre and McIntyre, 1971; Hsü and Andrews, 1970; Hay, 1970). However, the winnowing of sediments leading to the production of lag deposits, and the wafting of fines to other places, which can produce so-called "chaff" deposits, is a well documented process (Bramlette and Bradley, 1942; Heezen and Hollister, 1964; Fox and Heezen, 1965; Emiliani and Milliman, 1966; Heezen et al., 1966; Lowrie and Heezen, 1967; Johnson and Johnson, 1970; Ewing and Hollister, 1972; Davies and Laughton, 1972; Moore et al., 1973). The overall difference between forams and coccoliths in their susceptibility to dissolution is not well documented. Certainly, there is a large overlap in dissolution behaviour (see Section 29.4.5), while settling behaviour differs by orders of magnitude.

Strong evidence for long-term local control of carbonate texture is provided by data from the JOIDES Leg 14 (Berger and von Rad, 1972). For example, when data for Neogene to Quaternary chalk and marl oozes are plotted on a sand-silt-clay diagram clusters are formed which correspond with sites of deposition, rather than with parameters such as age, rate of deposition or percentage of carbonate. Dissolution influences the clustering only when it is pronounced ($FS = 6$ to $7 \cdot 5$). Fluctuations in the abundance of siliceous fossils and variations in the lysocline indicate that fertility and preservation varied greatly throughout the Neogene. Thus, the local control of texture suggests that sedimentary processes on the ocean floor had a first order effect on coccolith/foram ratios.

Taking coccolith percentages as a measure of coccolith/foram ratios two striking general gradients are evident; one with depth (Fig. 29.34, compare topography and coccolith percentages), and the other with time (Fig. 29.35; compare Pliocene with Recent coccolith percentage values). The enrichment of the sediments of the Angola Basin with coccoliths is an example of the depth trend, because this basin is one area on the sea floor in which dissolution effects are moderate even at depths of 5000 m or more (see Fig. 29.23). Olausson (1961a) has suggested several possible mechanisms to account for the high proportions of nannofossils in this area. (1) A relative increase in the output of coccoliths in low fertility areas. This effect would be expected to be minor in this instance because of the low proportions of coccoliths in the sediments of the Mid-Atlantic Ridge underlying the subtropical gyre (see also McIntyre and McIntyre, 1971). (2) Selective removal of forams. This process could certainly contribute to the enrichment of coccoliths, but the distribution of fragments indicates that it is probably only of secondary importance. (3) Differential transport of coccoliths down the slopes by near bottom currents could lead to the enrichment of forams on ridges and to that of coccoliths in basins. This hypothesis is probably the most likely in view of the argument for local control, and because fertility and dissolution have been discounted as dominating factors (see Berger and

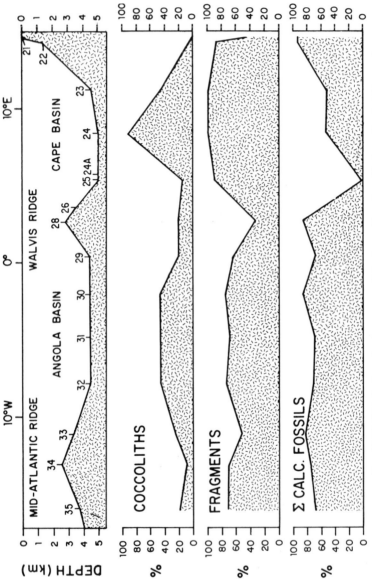

FIG. 29.34. Distribution of calcareous particles in a sediment profile from South Africa to the Mid-Atlantic Ridge. Based on data from Pratje (1939), figure adapted from Olausson (1961a). (Coccoliths expressed as a percentage of the calcareous skeletal remains. Fragments shown as a percentage of the total foraminiferal remains (whole tests + fragments). Calcareous fossils expressed as a percentage of the total sediment. Percentages are based on microscopic counts, i.e. they refer to the area covered on a slide and not to weight or volume).

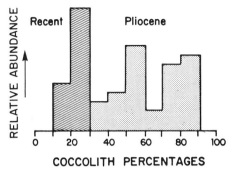

FIG. 29.35. Frequency of estimated coccolith percentages from Recent oozes (Mid-Atlantic Ridge data from McIntyre and McIntyre, 1971), and from Pliocene oozes (Sites 135, 136, 139, 140, 141, JOIDES Leg 14; estimates by W. E. Benson, W. H. Berger, P. R. Supko, and U. von Rad; see Berger and von Rad, 1972). Note the absence of overlap along the percentage scale.

von Rad, 1972). The data of McIntyre and McIntyre (1971, Figs. 16.1, 16.2) would seem to be in agreement with the differential transport mechanism advocated here since the highest concentrations of coccoliths ($>0.3$ g of coccoliths per g of sediment) in the Atlantic occur in the Northwest Atlantic Basin, and there seems to be little correlation of coccolith abundances with the productivity contours drawn by these authors.

The great change in coccolith/foram ratios which occurred towards the end of the Tertiary has been pointed out by Bramlette (1958) and also by Heath (1969). Seen against the background of Neogene deep-sea sedimentation, the coccolith proportions in Quaternary sediments, especially those of the Mid-Atlantic Ridge (Fig. 29.35) are low, and this requires an explanation. Increased fertility with the onset of the Quaternary may have delivered relatively more foraminifera than coccoliths to the sea floor, not only through a change in the standing stock ratio, although this in fact was not very pronounced (Fig. 29.11), but also through a change in the transfer mode of coccoliths brought about for example, by an increase in the destructive predation of nannoplankton. Increased fertility would also have increased the supply of excess carbonate to the sea floor which increased dissolution effects, and thus altered the ratio in favour of the more resistant forms. If coccoliths are indeed more resistant than forams, as suggested by various authors, this effect would increase the coccolith/foram ratio rather than decrease it. Here again, the most efficient mechanism influencing the ratio appears to be size sorting by bottom currents (cf. Bramlette et al., 1959; Heath, 1969) including weak, distal turbidity currents. Fertility patterns have an effect on this mechanism in so far as benthonic organisms tend to resuspend fines so that they can be carried to an environment in which there is less water movement. In general, the benthic biomass appears to be negatively correlated with depth and positively correlated with surface productivity (Rowe, 1971) which means that most resuspension takes place in relatively shallow, fertile regions (cf. Laughton, 1963). Resuspension by organisms would be more or less independent of the purely physical relationships between cohesiveness and texture of sediment, current strength, and bottom roughness. However these factors cannot be excluded as possible controls of differential transport. As coccoliths are more susceptible to resuspension and long-distance transport than are forams, they have a greater chance than the latter of reaching depths at which they may dissolve. Thus, at times when redeposition processes on the ocean floor are most active because of high fertility (which engenders benthic activity and resuspension) and strong bottom circulation (which provides a mechanism for transport), the foraminiferal content of calcareous oozes is likely to be high because of the removal of coccoliths to greater depths at which they may dissolve.

### 29.5.5. EXPONENTIAL MIXING

Reworking of older sediments into younger ones is a common phenomenon on the sea floor, as is shown by the redistribution of volcanic ash and by the mixing of micro-fossils of various ages (Bramlette and Bradley, 1942; Riedel, 1963). Two main processes contribute to this reworking, i.e. vertical mixing and horizontal redeposition (Bramlette and Bradley, 1942). Vertical mixing is commonly ascribed to the activities of burrowing organisms (Arrhenius, 1952). Horizontal redeposition is thought to be caused by bottom currents, the effects of which are often visible in bottom photographs (Heezen and Hollister, 1964). Vertical mixing has been studied by Arrhenius (1952, 1963) who found that mean mixing depths are about 5 cm, although single burrows may occasionally penetrate 20 cm or more (see also: Goldberg and Koide, 1962; Clarke, 1968; Griggs et al., 1969; Piper and Marshall, 1969; Donahue, 1971).

Berger and Heath (1968) have derived an equation to describe the vertical distribution of particles sedimented and mixed vertically under idealized conditions. The results from this agree well with field data, under conditions in which the frequencies of discrete burrows are low (Fig. 29.36).

This vertical mixing model can be generalized to include all mixing in which the probability of the relocation of a particle (as measured by sediment thickness) decreases exponentially with the time after its deposition. This generalized model includes both vertical and horizontal redeposition, but considers only upward mixing. The concentration $P_D$ at a vertical distance $D$ from an original concentration $P_O$ is given by

$$P_D = P_O \cdot \exp(-D/m') \tag{29.5}$$

INCREASING DEPTH
IN CORE

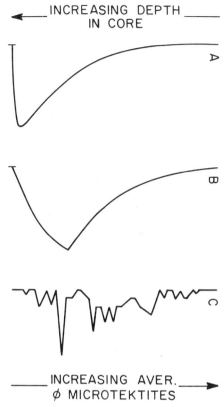

INCREASING AVER.
ϕ MICROTEKTITES

FIG. 29.36. Typical abundance patterns of microtektites in deep sea cores, with (A) little, (B) intermediate and (C) extensive burrowing. Redrawn from Glass (1969). The shape of curve B agrees well with distributions predicted from theory after making simplifying assumptions (constant rates of sedimentation and tracer supply between appearance and disappearance, homogeneous mixing; Berger and Heath, 1968).

where $m'$ is the virtual vertical mixing length. This length is the mixed layer thickness that would produce the observed distribution if vertical mixing were its only cause. The equation yields the virtual mixing length or "mixing criterion"

$$m' = D/(\ln P_O - \ln P_D) \qquad (29.6)$$

$P_O$ and $P_D$ can be measured anywhere along an exponential curve where $D$ is the distance between the two points. For non-exponential mixing, a variable value of $m'$ is obtained. The magnitude and the distribution of $m'$ should facilitate discussion of the mixing processes involved; i.e. simple vertical mixing, discrete burrowing, horizontal redistribution, together with other factors such as gradual extinction of the species concerned.

Information is available to test this model. Glass (1969), Table 2 has described the vertical distribution of microtectites in eight deep-sea cores. The average and median length of the microtectite zone below peak abundance is 10 cm, giving a maximum value for $m'$ if ideal vertical mixing is assumed. Lengths above peak abundance range from 25 cm to more than 90 cm, with an average of 45 cm. The median value is 35 cm. A single microtektite represents between 1% and 6% of the total population of particles counted (Glass, 1969). Equation (29.5) suggests that mixing through one mixed layer thickness will lead to a decrease in abundance of the particle concerned by a factor of $1/e$. After three mixed layer thicknesses its concentration will be $0.37^3$, or about 5% of the original proportion. Thus, the length of 35 cm would appear to represent about 3 mixing lengths, that is $m' = 12$ cm. The average mixing length based on these data would, therefore, seem to be of the order of 10 cm, indicating that reworking occurs mainly by vertical mixing.

McIntyre et al. (1967) studied a floral and faunal boundary in seven deep-sea cores; this boundary was identified by Ericson et al. (1963, 1964) as that separating the Pliocene and Pleistocene. The contacts found might be uncomformable (Bandy and Wade, 1967). If sedimentation was indeed discontinuous across this boundary it may considerably affect the interpretation of the mixed zone, which was estimated by Ericson et al. (1964) to be about 15 cm in width compared to the 30 cm or more estimated by McIntyre et al. (1967). McIntyre et al. (1967) found discoasters within the sediments above the boundary, and postulated that "if they become extinct at the boundary, then organic burrowers must have mixed them vertically and/or the dead tests were transported from another area where older beds were being eroded". They concluded that their observations "confirm that the discoaster distribution is the result of vertical mixing of an extinct group".

The hypothesis of McIntyre et al. can be tested by means of the mixing criterion (Fig. 29.37). The quasi-exponential coccolith distributions are reduced to straight lines having a slope $m'$ on a semilog plot, according to equation (29.6). The sediment less than 60 cm above the boundary in cores V15–164 and V16–21 contains discoasters which have a distribution indicating a vertical mixing depth $m'$ of $\sim 19$ cm. This result supports their suggestion that the distributions are the result of vertical mixing. The value of 19 cm is higher than that of 12 cm for the much coarser microtektites (see above), but appears reasonable. All other values of $m'$ are distinctly greater than 10–20 cm (i.e. 50, 70, 180 cm) and seem to be too high to be consistent with the hypothesis of vertical mixing. It is probable that they reflect a horizontal influx which was decreasing in an exponential manner. This suggestion is in

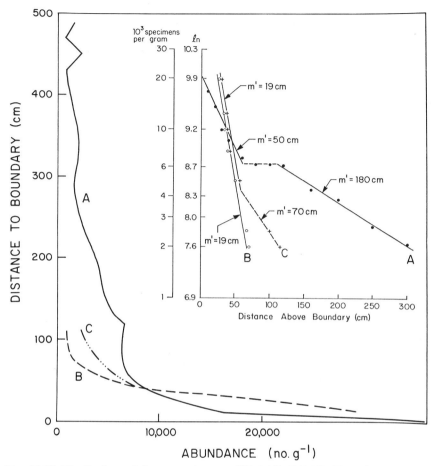

FIG. 29.37. Distributions of discoasters in cores V12-5 (A), V15-164 (B), and V16-21 (C), according to McIntyre *et al.* (1967), and analysis for exponential mixing (inset). $m'$ is the virtual mixing length, or "mixing criterion"; $m' = D(\ln P_0 - \ln P_D)$.

agreement with the statement that "the long length of the discoaster curves in V12–5 also could be the result of influx from another source area in addition to vertical mixing" (McIntyre *et al.*, 1967).

Ruddiman and Glover (1972) have investigated the distribution of ash shards in cores from the North Atlantic in order to examine vertical mixing. The upper two ash zones and the exponential decrease upwards of the ash content of the sediment had been previously described by Bramlette and Bradley (1942). The average total thickness of the ash zones described by Ruddiman and Glover is ~40 cm, and the ash distribution is skewed with

14 cm below maximum abundance and 26 cm above. Analysis of individual curves (see above) yields $m' \sim 6 \pm 2.6$. This low average value supports the contention of Ruddiman and Glover (1972) that vertical mixing, rather than horizontal redistribution effects, was responsible for the distribution of the shards. This calculated depth also agrees with their visual estimate of 5 cm, which was based on mixing across colour contacts. However, the variation in thickness of the ash zone is considerable, and this may be attributed to inhomogenous mixing and discrete burrowing.

Reworking processes appear to be ubiquitous. Upward mixing and downward mixing are not symmetrical, and as a result of this a deposited layer tends to yield a date older than it should do. Different size grades may have different apparent mixing lengths, so that their ages may disagree (see Olsson and Eriksson, 1965). Another result of reworking is that it makes sharp transitions less well defined, thus inhibiting the interpretation of the Pleistocene record.

### 29.5.6. BIOCLASTIC TURBIDITES, ALLOLYTIC LIMESTONES

Deep-sea sands with shallow water characteristics were described by Murray and Philippi (1908) and by Philippi (1910), and following the ideas expressed by Daly (1936) were subsequently recognized to be indications of discrete redeposition events. Thus, Bramlette and Bradley (1942) invoked turbidity currents to explain sandy layers in North Atlantic deep-sea cores. Work on depth zonation and displacement of benthic forams, as well as laboratory experiments on transport by high density currents (Kuenen and Migliorini, 1950; Natland and Kuenen, 1951; Phleger, 1951a), led to a general acceptance of the turbidity current hypothesis for the origin of deep sea sands (Ericson et al., 1952, 1961). The study of cable break sequences (Heezen and Ewing, 1952) lent strong support to the concept of catastrophic redepositional events. This concept is now widely believed to be of first order importance in understanding the processes of deep-sea sedimentation in areas which have an influx of continental material (see Shepard, 1963; Menard, 1964; Horn, et al., 1972).

Bioclastic turbidites were described by Meischner (1964) in his paper on allodapic limestones in which he provides a simple explanation for the alternation of limestone beds with marls or clay; a phenomenon which is commonly found in the geological record. Deposits that are thought to be bioclastic turbidites are common on the present deep-sea floor (Rusnak and Nesteroff, 1964; Connolly and Ewing, 1967; Davies, 1968; Connolly, 1969; Van Andel and Komar, 1969; Eade and van der Linden, 1970; Bornhold and Pilkey, 1971; Griggs and Fowler, 1971; Davies and Jones, 1971), and in the ancient deep-sea deposits sampled during the JOIDES program

(Beall and Fischer, 1969; Heath and Moberly, 1971b; Bernoulli 1972; Berger and von Rad, 1972). Ideally this theory should be supported by evidence showing: differential settling from suspension; a source for the material of the extraneous layer which is at a depth shallower than that of the site of deposition; and demonstration that the directions of the main depositing current and the palaeoslope were similar, thus indicating that gravity was the driving force of the current. Of these factors, evidence is commonly found for differential settling and downslope displacement of bioclastics derived from shelf and slope deposits (Phleger *et al.*, 1953; Rusnak and Nesteroff, 1964; Connolly, 1969).

In pelagic turbidites, differential settling leads to both size grading and species grading (see, for example, Berger and von Rad, 1972). The planktonic forams *G. hexagona, G. iota, G. glutinata, O. universa, G. digitata* and *C. nitida* have been identified as "stragglers" which tend to be concentrated toward the top of calcareous turbidites, or within chaff deposits (Berger and Piper, 1972). Downslope redeposition leads to intercalation of calcareous sediments with clay or, above the compensation depth, of relatively well preserved with relatively poorly preserved assemblages. In such alternating sediment layers, the ones containing the more dissolved calcareous assemblages may be regarded as "authilytic" (dissolved *in situ*), and the layers with the relatively well preserved assemblages as "allolytic" (dissolved elsewhere). For purely pelagic turbidites it may be exceedingly difficult to distinguish between the effects of dissolution cycles and those of redeposition on fluctuating degrees of preservation in a given sediment section.

In the central Atlantic, complex dissolution patterns are produced by factors such as: the overall lowering of the CCD and foram dissolution zones since the Miocene; the rather poor preservation of Neogene forams in areas adjacent to the continents; and the displacement of better preserved assemblages into areas in which the indigenous faunas have undergone greater dissolution (especially in the Ceara Abyssal Plain; see Fig. 29.38). In some instances, the preservation patterns of siliceous fossils also appear to be related to discrete redeposition events. For example, the state of preservation of siliceous fossils in sediments from Site 138 of JOIDES Leg 14 varies considerably (Berger and von Rad, 1972). The abundance of the associated quartz (Plate 29.1–Fig. 6) suggests that these fluctuations may have been produced by the redeposition of continental siliceous sediments rich in terrigenous material from near-continental areas.

Both epigenetic and diagenetic processes affect allodapic and indigenous layers in different ways with the result that further variations are produced in the lithology, such as the degree of cementation. Facies association, composition and grading in certain Cretaceous deep-sea limestones obtained on JOIDES Leg 14 suggest emplacement of calcareous turbidites into a poorly

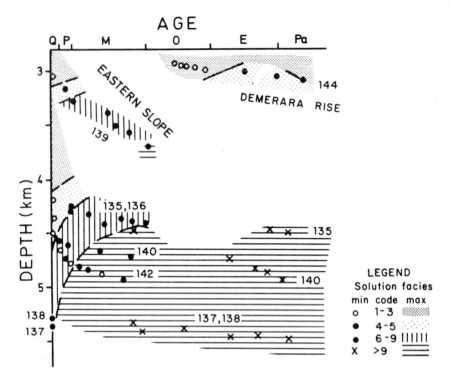

FIG. 29.38. Foraminiferal solution facies in a depth-age diagram, JOIDES Leg 14. Depth = water depth + drilling depth. 3 digit numbers refer to sites. Code numbers in the legend on the figure refer to state of dissolution: 1 to 3, least dissolved; 6 to 9, most dissolved, X, completely dissolved. Symbols refer to least dissolved sample in any one core at any one age; patterns refer to most dissolved sample in any one core and age. For each age the distribution of patterns shows the shallowest depths to which any one solution facies reaches, the distribution of symbols indicates the greatest depths. Note the considerable (> 1000 m) apparent displacement of well preserved samples into indigenous non-calcareous facies at Site 142 (intercalation of authilytic and allolytic facies). Adapted from Berger and von Rad (1972).

oxygenated basin (Berger and von Rad, 1972). Shallow parts of this basin were probably more oxygenated than deeper ones, and turbidity currents could have brought down the oxygen necessary for the development of an infauna (burrowing fauna) on top of the turbidite layers. Because of their general downward displacement all such calcareous turbidites are likely to cut across dissolution facies zones, and therefore allodapic limestones in the deep-sea would be generally expected to be derived from allolytic calcareous ones, a circumstance which should greatly enhance their susceptibilities to diagenesis and lithification (cf. Ewing *et al.* 1969).

   Where siliceous shells of radiolarians and diatoms participate in the turbidite redeposition they tend to settle out slowly (see Table 29.3) and to concentrate at the top of the sequence, together with nannofossils and organic material. Dissolution of carbonate may then concentrate the silica (and any carbonaceous matter) on top of the calcareous layer to produce ultimately a carbonate-silica couplet (Berger and von Rad, 1972). Any dolomite present may also be concentrated by differential dissolution of calcite by analogy with the selective removal of aragonite and high magnesian calcite at shallower depths (Chave et al., 1962; Friedman, 1965). The dolomite-pelite cycles described by Berger and von Rad (1972), may have originated in such a fashion.

If dissolution of carbonate proceeds far enough, and if the original layered structure is preserved, then bedded cherts could be formed from a sequence of allolytic bio-turbidites. Any remaining calcite would be present at the base of individual layers of such radiolarites (cf. Bernoulli, 1964; Beall and Fischer, 1969). The role of redeposition and/or dissolution in the formation of certain limestones and radiolarites of the Alps and other young mountain chains has been considered by Hollman (1964), Garrison and Fischer (1969) and Scholle (1971a,b). Bernoulli (1972) has discussed the similarities, and differences, between these sediments and those found in the North Atlantic by deep-sea drilling. For a treatment of the diagenetic processes in deep sea carbonates see Fischer and Garrison (1967), Bathurst (1971), Wise and Hsü (1971), Wise and Kelts (1972) and Adelseck et al. (1973). For cherts, see Ernst and Calvert (1969), Calvert (1971), Garrison (1971), Heath and Moberly (1971a), Jones and Segnit (1971), von der Borch et al. (1971), Berger and von Rad (1972) and Lancelot (1973); see also Calvert, von Rad and Rösch in Hsü and Jenkyns (1974), and Kastner and Keene (1975).

### 29.5.7. CIRCULATION REVERSALS AND MEDITERRANEAN SEDIMENTATION

Major biofacies patterns appear to reflect oceanic fractionation by estuarine and anti-estuarine deep-sea circulation (see Section 29.4.1). Circulation reversals, therefore, should lead to a predictable succession of biofacies. The Mediterranean Pleistocene record supports this hypothesis. The modern Mediterranean Sea is a prime example of anti-estuarine circulation. At present evaporation by far exceeds precipitation. There is a deep water outflow of about $2 \times 10^6 \, \text{m}^3 \, \text{s}^{-1}$ (2 sverdrups) over the Gibraltar sill, and this outflow is largely replaced with surface water from the Atlantic Ocean. Consequently, the Mediterranean Sea is filled with water having the characteristics of ocean surface water, viz. a high degree of $CaCO_3$ saturation and low silica and nutrient concentrations. These "shallow water" characteristics result in a strong predominance of coccolithophores in the phytoplankton (Olausson, 1961b), a depressed pteropod compensation depth (pteropods have been reported down to about 3500 m in places; Olausson, 1960c), and in a relatively low production and a rapid dissolution of siliceous shells. Not surprisingly, therefore "the Mediterranean cores are generally characterized by an absence of diatoms" (Olausson, 1961b).

Diatoms and radiolarians are, however, found in the sapropelitic layers in cores from the eastern Mediterranean (Olausson, 1960c, 1961b; Dumitrica, 1973). During times of increased precipitation and runoff, the deep circulation may have reversed, so that "the undercurrent in the Straits of Gibraltar will bring Atlantic water into the Mediterranean Sea, and conditions will be similar to those existing in the Black Sea" (Kullenberg, 1952; see also Huang et al., 1972; Nesteroff, 1973). This estuarine circulation would tend to prevent the outflow of nutrients (and silica) from the Mediterranean, since the surface waters moving into the Atlantic would be largely stripped of the nutrient elements by biological activity. The deep water, therefore, would eventually be highly enriched with nutrients, greatly enhancing both the production and preservation of siliceous, organic (and phosphatic) sediment. The co-occurrence of diatoms and sapropelitic muds would be a necessary consequence of the circulation reversal. In the present interpretation, salinity changes would be incidental as a factor influencing sedimentation, although a lowered salinity would certainly be expected for the period of surface water outflow from the Mediterranean (Olausson, 1961b). Any influx of nutrient salts *via* rivers would be incidental in this model, since the enrichment effect is a matter of deep-sea circulation, although the final steady state is attained somewhat later without this supply than with it.

The foraminiferal data (Olausson, 1961b) show that the sapropelitic mud is characterized by *Globigerina bulloides, Globoquadrina dutertrei* ("*G. eggeri*") and *G. quinqueloba,* an association typical for the subtropical upwelling regions off California. Extreme *G. dutertrei* values appear to be the result of selective destruction prior to burial of *G. bulloides* in the thin oxidized layer on top of the sapropelitic sediment (Berger and Soutar, 1970). Below this layer dissolution would have ceased quickly because of the increased alkalinity caused by sulfate reduction. Thus, a relatively large proportion of forams was preserved. The absence of benthic forams suggested that in some instances the oxidized layer was apparently entirely missing. This may then have resulted in a true death assemblage (thanatocoenosis) free from dissolution effects (see Olausson, 1961b). Detailed information on foraminiferal assemblages in Eastern Mediterranean cores has been given by Parker (1958). For a general background to the sedimentary geology of the Mediterranean see Blanc (1968), and for evidence on the changes in circulation and sedimentation through geological time see Ryan et al. (1973).

The Red Sea, another basin with deep-water outflow, also contains indications of a possible circulation reversal during recent times. Goll (1969) has described radiolarian sediments which are apparently restricted to the time interval 12 000–9000 years BP, which corresponds to a period of deglaciation. This occurrence of siliceous sediments suggests that circulation was of an estuarine type during this period. The fact that radiolarians are extremely rare in present Red Sea sediments (Olausson, 1960c; Goll, 1969) is in keeping with the concepts that high salinity and low nutrient concentrations inhibit production, and that the relatively high temperature and low silicate values of the water favour rapid dissolution of shells.

### 29.5.8. STRATIGRAPHY AND DATING

Historically, the development of the study of shelled micro- and nanno-plankton has stemmed from their use in stratigraphy. The accuracy of relative age-dating by microfossils (i.e. $\pm 1$ million years) greatly exceeds that attainable by methods based on radioactive decay (i.e. $\pm 5$ million years for the Late Cretaceous, and somewhat better for the Tertiary). The absolute ages assigned to stratigraphic units (Fig. 29.39) have been, and are being, therefore, continuously revised. Relative age determinations of fossiliferous pelagic sediments are of course not affected by such revisions; see Berger and Roth (1975) for a summary of recent work on deep-sea stratigraphy.

The stratigraphical application of planktonic foraminifera was pioneered by Bolli and his co-workers (see Loeblich et al., 1957), that of radiolarians by Riedel (1952, 1957), and that of coccolithophores by Bramlette (Bramlette and Riedel, 1954; Bramlette and Sullivan, 1961). Hanna (1932) has demonstrated the stratigraphical usefulness of diatoms in a series of sediments from the west coast of the United States. Some of the most recent Tertiary stratigraphical literature has been surveyed in the "Initial Reports of the Deep Sea Drilling Project"; this covers radiolarians (standard reference: Riedel and Sanfilippo, 1971; see also Hays, 1970; Kling, 1971b; Moore, 1971; Goll, 1972; Riedel and Sanfilippo, 1973) and foraminifera (see esp. Bolli, 1970; Brönnimann and Resig, 1971; Jenkins and Orr, 1972; Poag, 1972; Berggren, 1972b). Additional useful references to the Tertiary biostratigraphy of planktonic foraminifera are found in Parker (1967), Blow (1969) and Postuma (1971); see also Berggren and van Couvering (1974). Bukry (1971) has given a summary of coccolith stratigraphy based on work by Bramlette and Wilcoxon (1967); Hay et al. (1967); Gartner (1969); Bukry and Bramlette (1970); Roth (1970) and additional studies by Bukry and others especially on JOIDES cores (see Fig. 29.40a,b; see also Martini, 1971a). The stratigraphy of diatoms is treated in the comprehensive paper by Schrader in the JOIDES Leg 18 report (see Fig. 29.41). Useful recent references for diatom zonation have been collected by Burckle (1972), Donahue (1970), Kanaya (1957; 1959), Kanaya and Koizumi (1970), Koizumi (1968) and Wornardt (1967). For data on silicoflagellates the reader is referred to Mandra (1968),

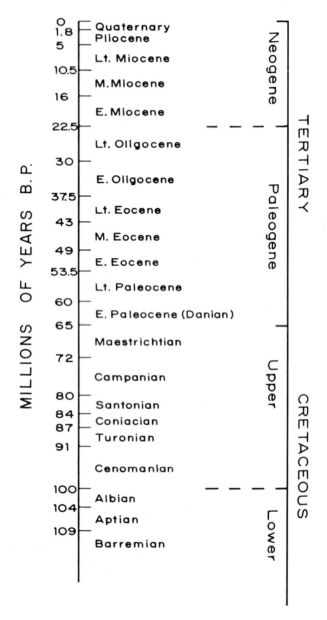

FIG. 29.39. Time scale used in JOIDES Leg 14 report. Tertiary after Berggren (1972b); Cretaceous, estimated from various sources. Cretaceous age assignments are imprecise (compare van Hinte, 1972). From Berger and von Rad (1972).

Martini (1971b) and Ling (1972), and for acheomonads to Hajos (1968) and Tynan (1971). The stratigraphical applications of fish teeth have been described by Helms and Riedel (1971) and by Belyaev and Glikman (1970).

The various concepts and difficulties associated with stratigraphical dating, correlation and evolution have been reviewed by Riedel (1973) and

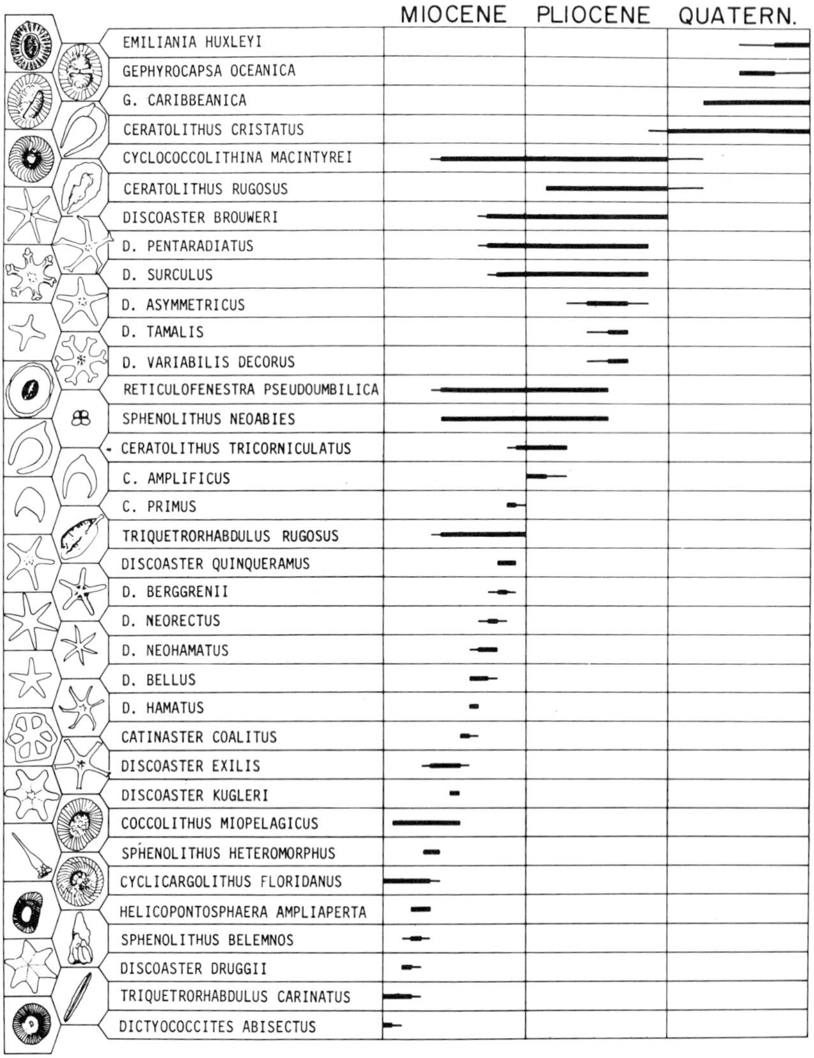

FIG. 29.40a. Neogene–Quaternary coccolith zonation. Ranges from Bukry (1971). Illustrations redrawn from various authors (drawings by Judy Lachmund Clinton).

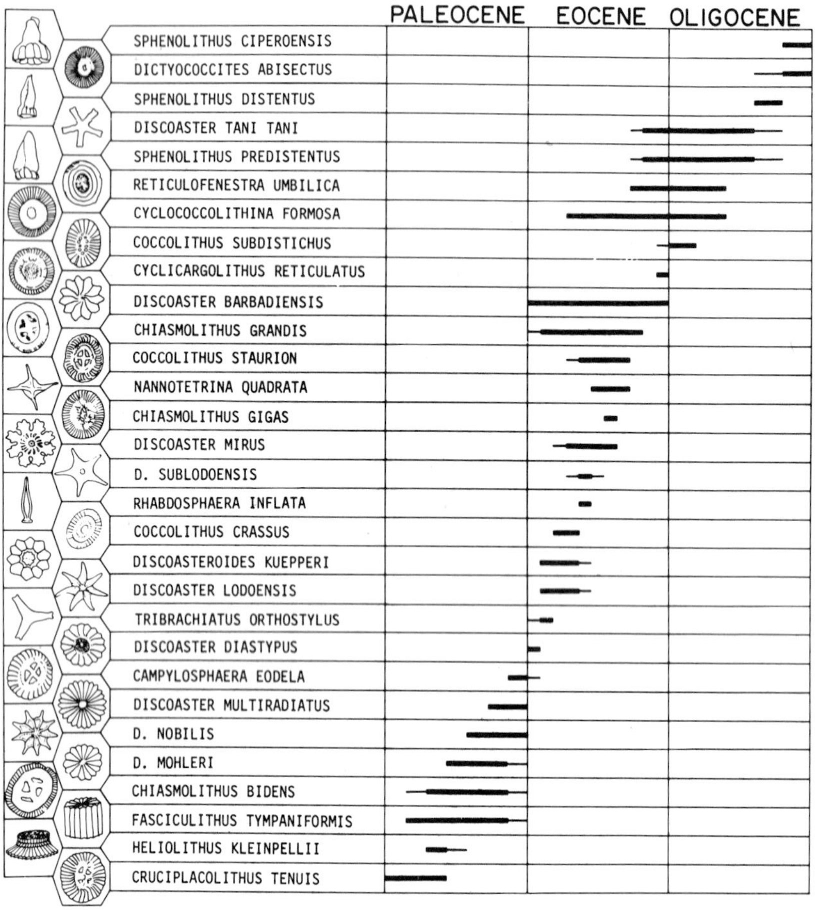

| | PALEOCENE | EOCENE | OLIGOCENE |
|---|---|---|---|
| SPHENOLITHUS CIPEROENSIS | | | ▬ |
| DICTYOCOCCITES ABISECTUS | | | ▬ |
| SPHENOLITHUS DISTENTUS | | | ▬ |
| DISCOASTER TANI TANI | | ▬▬▬▬▬ | |
| SPHENOLITHUS PREDISTENTUS | | ▬▬▬▬▬ | |
| RETICULOFENESTRA UMBILICA | | ▬▬▬ | |
| CYCLOCOCCOLITHINA FORMOSA | | ▬▬▬▬ | |
| COCCOLITHUS SUBDISTICHUS | | ▬ | |
| CYCLICARGOLITHUS RETICULATUS | | ▪ | |
| DISCOASTER BARBADIENSIS | | ▬▬▬▬ | |
| CHIASMOLITHUS GRANDIS | | ▬▬▬ | |
| COCCOLITHUS STAURION | | ▬ | |
| NANNOTETRINA QUADRATA | | ▬ | |
| CHIASMOLITHUS GIGAS | | ▪ | |
| DISCOASTER MIRUS | | ▬ | |
| D. SUBLODOENSIS | | ▬ | |
| RHABDOSPHAERA INFLATA | | ▪ | |
| COCCOLITHUS CRASSUS | | ▬ | |
| DISCOASTEROIDES KUEPPERI | | ▬ | |
| DISCOASTER LODOENSIS | | ▬ | |
| TRIBRACHIATUS ORTHOSTYLUS | | ▪ | |
| DISCOASTER DIASTYPUS | ▪ | | |
| CAMPYLOSPHAERA EODELA | ▬ | | |
| DISCOASTER MULTIRADIATUS | ▬▬ | | |
| D. NOBILIS | ▬▬ | | |
| D. MOHLERI | ▬▬ | | |
| CHIASMOLITHUS BIDENS | ▬▬▬ | | |
| FASCICULITHUS TYMPANIFORMIS | ▬▬▬ | | |
| HELIOLITHUS KLEINPELLII | ▬ | | |
| CRUCIPLACOLITHUS TENUIS | ▬ | | |

FIG. 29.40b. Palaeogene coccolith zonation. Ranges and illustrations as in Fig. 29a.

by Bramlette and Riedel (1971). Jenkins and Orr (1971, 1972) have emphasized the effects of differential dissolution on foram stratigraphy. Similar considerations also apply to the stratigraphy of the other fossils. The various methods of dating by radionuclides and their range of applicability have been reviewed by Goldberg and Bruland (1973).

An important new stratigraphical tool has been developed from the discovery that reversals of the earth's magnetic field are recorded in deep-sea sediments (Harrison and Funnell, 1964). Developments in this field have been reviewed by Opdyke (1972) and by Watkins (1972); see also Opdyke *et al.* (1974) and Theyer and Hammond (1974). Data on sedimentation rate

|  |  | CRETACEOUS | PALEOCENE | EOCENE | OLIGOCENE | MIOCENE | PLIOCENE | PLEISTOCENE | RECENT |
|---|---|---|---|---|---|---|---|---|---|
| 1 | PSEUDOEUNOTIA DOLIOLUS |  |  |  |  |  |  | ——— | — |
| 2 | RHIZOSOLENIA CURVIROSTRIS |  |  |  |  |  |  | —— |  |
| 3 | DENTICULA SEMINAE |  |  |  |  |  | ——— |  |  |
| 4 | LITHODESMIUM CORNIGERUM |  |  |  |  |  | —— |  |  |
| 5 | NITZSCHIA FOSSILIS |  |  |  |  |  | —— |  |  |
| 6 | RHIZOSOLENIA BARBOI |  |  |  |  | · | ——— |  |  |
| 7 | DENTICULA HUSTEDTII |  |  |  |  | ——— |  |  |  |
| 8 | DENTICULA NICOBARICA |  |  |  |  | — |  |  |  |
| 9 | COSCINODISCUS LEWISIANUS |  |  |  |  | — |  |  |  |
| 10 | CRASPEDODISCUS COSCINODISCUS |  |  |  |  | — |  |  |  |
| 11 | ACTINOCYCLUS INGENS |  |  |  | —— |  |  |  |  |
| 12 | CRASPEDODISCUS EXCAVATUS |  |  |  | —— |  |  |  |  |
| 13 | COSCINODISCUS OBLONGUS |  |  | — |  |  |  |  |  |
| 14 | PYXILLA SPP. |  | ---— |  |  |  |  |  |  |
| 15 | TRINACRIA REGINA |  | ——— |  |  |  |  |  |  |
| 16 | GLORIOPTYCHUS CALLIDUS |  |  |  |  |  |  |  |  |

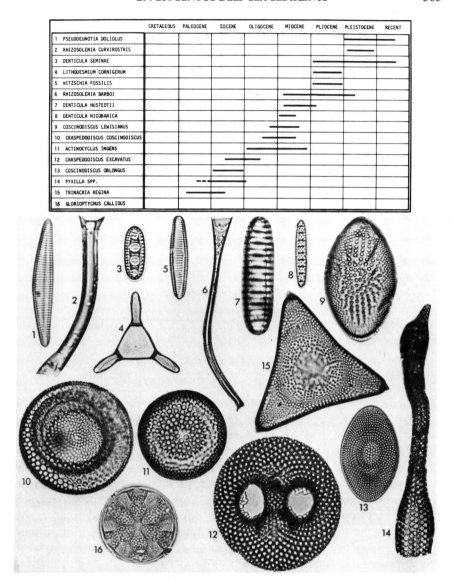

FIG. 29.41. Ranges of some stratigraphically useful diatoms. Data and photographs by H. J. Schrader (Kiel). Photograph of *Lithodesmium cornigerum* from Hanna (1930). Range of *Nitzschia fossilis* adjusted from the data of Koizumi (1972).

distributions and variations are being greatly refined by this method (see, for example, Hays *et al.*, 1969). These refinements make it obvious that many small scale breaks occur in the deep-sea record, such as those caused by periods of anomalously low sedimentation rate, nondeposition, or erosion (see Opdyke and Glass, 1969; Kobayashi et al., 1971). It can be shown that the effect of the conditions leading to such breaks is to enhance the co-occurrence of faunal extinction and magnetic reversal noted by Harrison and Funnell (1964), and in subsequent studies. A causal relationship, emphasized by Hays (1971), among others, may also exist in places. However, the significance of such a relationship would be difficult to assess, since most species experience a large number of reversals before becoming extinct.

Another new method with great potential for dating deep-sea sediments deposited over the last several million years has been introduced (Bada et al., 1970; Wehmiller and Hare, 1971; Bada and Schroeder, 1972). It is based on the racemization of amino acids, that is on the fact that L-forms of amino acids, which completely dominate living matter, are converted to an equilibrium mixture of L- and D-forms after death. Investigations have concentrated on the L-isoleucine vs. D-alloisoleucine reaction, because of the ease of measurement. Rates of racemization depend on temperature and on chemical interactions between reactants and environment. The reliability of this new tool, therefore, hinges on establishing the appropriate rate constants, and on the confidence with which the physicochemical history of a given sediment can be reconstructed.

### 29.5.9. PLATE STRATIGRAPHY: THE EASTERN TROPICAL PACIFIC

The current revolution in geological thinking associated with the concepts of sea-floor spreading (Hess, 1962; Vine and Matthews, 1963; Wilson, 1965), and of plate tectonics (McKenzie and Parker, 1967; Morgan, 1968; LePichon, 1968; Isacks et al., 1968; Menard, 1969; Sclater et al., 1971) has a profound implication for the interpretation of ancient deep sea sediments. Sediments are deposited on moving subsiding plates, so that they change both their geographical location and depth during geological time as was initially proposed by Hess (1962) and as has been confirmed by Tertiary fossil distributions on the sea floor (Burckle et al., 1967; Riedel, 1967; Funnell and Smith, 1968). Corresponding subsurface distributions on ridge flanks have become known through the JOIDES project (see, for example, Peterson et al., 1970). The results of the JOIDES studies immediately suggested that the present distributions of facies boundaries are, in part, a result of sea floor subsidence and horizontal plate motions. The effects of such motions on facies patterns have been discussed by Hsü and Andrews (1970, Leg 3), Benson et al., (1970, Leg 4), Fischer et al., (1970, Leg 6), Hays et al. (1972, Leg 9), Berger and von Rad (1972, Leg 14), van Andel and Heath (1973, Leg 16) and Heezen et al. (1973, Leg 20).

The concepts necessary for the understanding of sediment bodies on a moving plate ("plate stratigraphy") have been sketched by Menard (1972), and the appropriate operations have been outlined in some detail for Cretaceous and Cenozoic sediment patterns of the Atlantic (Berger, 1972; Berger and von Rad, 1972) and for the Cenozoic of the Central Pacific (Winterer, 1973; Berger, 1973a; see also Berger and Winterer in Hsü and Jenkyns, 1974 for exposition of the concept). The fundamental assumption on which plate stratigraphy is based is that the present-day age-depth relationship of the sediments on the mid-ocean ridge flanks, shown by Sclater et al. (1971), is valid for the past. This assumption of an age-depth constancy allows the depth of deposition of a dated sediment sample overlying a dated basement to be established.

The eastern equatorial Pacific is the classic area of pelagic biosedimentation as it was here that the major features of carbonate and silica deposition were first established in detail (Arrhenius, 1952, 1963). Much additional information is now available for this area on sediment thickness (Shor, 1959; Ewing et al., 1968), and on microfossil distributions (Riedel, 1963; Riedel and Funnell, 1964; Nigrini, 1968; Hays et al., 1969; Riedel, 1971a; Parker and Berger, 1971). Russian work for the same region has been summarized by Bezrukov (1969, 1970) and Lisitzin (1971a, b, 1972).

In Fig. 29.42 the main facies trends are projected onto a latitude-depth plane which indicates the kind of sediment which is encountered between 100° W and 150° W in the tropical Pacific. The sedimentary lysocline (i.e. the facies boundary between well preserved and poorly preserved calcareous assemblages) appears to be diffuse in this region and its position is not well known north of 10° S. The CCD is fairly well defined, and its depression to a greater depth below the high fertility zone is very obvious. Its rise to 3 km at about 10° N produces the east-west trending carbonate-clay boundary on the ocean floor proposed by Arrhenius (1952) and confirmed by Bramlette (1961). In part, this rise may be due to erosion by bottom currents along the Clipperton Fracture Zone (Riedel, 1963; Johnson, 1972a). The belt of siliceous ooze marking the equatorial high fertility zone extends somewhat beyond the carbonate belt, especially in the north (Arrhenius, 1952). Below the CCD siliceous ooze accumulates as its preservation is insensitive to depth in abyssal areas.

The JOIDES project has extended our knowledge of past Pacific equatorial facies. It has permitted the questions originally raised by Arrhenius (1963) and by Riedel (1963, 1967), regarding changes in the productivity and dissolution patterns along the Equatorial Divergence and in the location of the Divergence itself relative to the present equator, to be explained in terms of polar wandering or plate motions. An excellent summary of the available data and the processes concerned has been given by van Andel and Heath

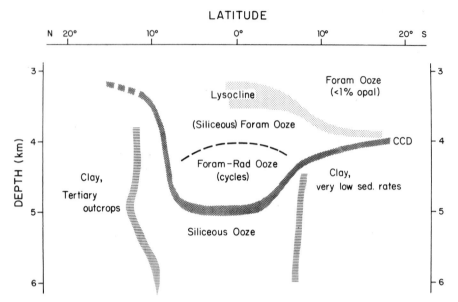

FIG. 29.42. Sketch of facies distributions in the eastern tropical Pacific, between 100° W and 150° W, in a latitude-depth matrix. Data mainly from Arrhenius (1952), Riedel and Funnell (1964), Hays *et al.* (1969), Riedel (1971), Parker and Berger (1971) and JOIDES results (in Berger, 1973a). The lysocline position is tentative.

(1973) in their synthesis of sedimentation in the Central Equatorial Pacific west of the East Pacific Rise. Winterer (1973) has constructed a model of sedimentation in the equatorial Pacific in which sedimentation patterns and plate motions are integrated. This model identifies the important factors which must be considered in plate stratigraphy in addition to subsidence and horizontal motion of the plate receiving the deposits; these are pro- ductivity and dissolution variations. The model predicts sediment thickness distributions which are in remarkably good agreement with those found by Ewing *et al.* (1968).

Matrices of the type presented in Fig. 29.42 can be reconstructed for specified geological periods of the Cenozoic by backtracking sedimentary sequences from individual drilling sites along their probable subsidence path. The post-Eocene matrices for individual periods are similar to each other (Berger, 1973a), but show a characteristic northward shift with age (Fig. 29.43). On the basis of maximum sedimentation rate distributions this shift has been estimated by Winterer (1973) to have occurred at a rate of 0·23° per million years ($\sim 2\cdot5$ cm yr$^{-1}$) for the Cenozoic. This rate agrees well with that necessary to obtain optimum congruency of matrices, and to align the deepest palaeodepths of the CCD of the various periods. Using this vector a

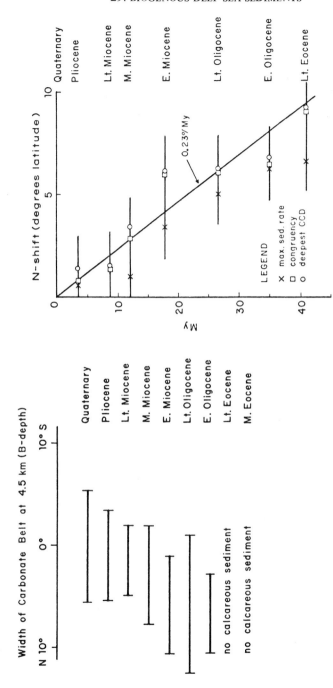

FIG. 29.43. Sedimentary record on the Pacific Plate, in the eastern tropical region. Left: width of carbonate belt at 4·5 km backtracking depth. Right: amount of northward movement as inferred from shifts in maximum sedimentation rate, optimum congruency of facies patterns, and maximum depth of CCD (from Berger, 1973a). Bars indicate estimated range of uncertainty.

palaeolatitude-palaeodepth matrix, incorporating the post Eocene facies patterns in a generalized fashion, has been constructed, (Fig. 29.44). The difference between this general post-Eocene matrix and the present one (Fig. 29.42) are not very pronounced. This suggests that biogenic accumulation in the Eastern Equatorial Pacific has been quite constant over the last 35 million years. The deviations of the ancient facies patterns from the present day ones found by Riedel (1963) and also during the JOIDES project can therefore largely be explained by plate motions.

The path of a sedimentary sequence from a drilling site in the palaeolatitude-palaeodepth matrix obtained by back-tracking has the shape of a subsidence curve, with palaeolatitude corresponding to age (Fig. 29.45). Various properties of biogenous sediments of different ages can be viewed within this framework. For example, Moore (1971) has noted that there is a progressive northward shift of good silica preservation with age, "similar to the progression of the "axis" of sediment accumulation"; thus implying that maximum preservation of radiolarians delineates the equatorial high productivity belt. The present analysis (Fig. 29.45) agrees exactly with this implication. (van Andel et al., 1975 have given a comprehensive account of the plate stratigraphy of the equatorial Pacific).

### 29.5.10. CENOZOIC CCD FLUCTUATIONS

Several authors (Revelle, in Bramlette, 1958; Arrhenius, 1963; Heath, 1969; Hay, 1970, Ramsay, 1972), have suggested that the CCD has fluctuated. Such fluctuations are to be expected because compensation depths differ between ocean basins, and because the oceanographic setting producing these differences must have changed through time in response to tectonic and climatic changes. The chief difficulty in reconstructing CCD fluctuations, i.e. the separation of tectonic movements from depth variations of the facies boundary, can be provisionally removed by adopting age-depth constancy (Berger, 1972; Winterer, 1973). The resulting CCD palaeodepth fluctuations (Fig. 29.46) have interesting properties.

The minimum depth of the Atlantic curve occurs in the Middle Miocene, indicating an environment unfavourable for preservation of carbonate. Significantly, siliceous fossils tend to be relatively well preserved and fish debris (phosphate) is abundant (Berger and von Rad, 1972), suggesting an estuarine deep-sea circulation pattern.

In general, the shallow CCD is part of a Miocene sedimentation pattern reminiscent of the present day North Pacific facies regime. The maximum depth of the CCD characterizes the Quaternary and is typical of the present Atlantic facies regime which is a result of anti-estuarine deep-sea circulation. The factors responsible for the drastic fall of the CCD in the Late Miocene

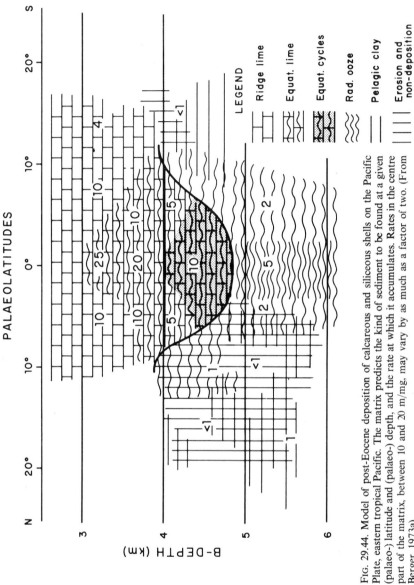

FIG. 29.44. Model of post-Eocene deposition of calcareous and siliceous shells on the Pacific Plate, eastern tropical Pacific. The matrix predicts the kind of sediment to be found at a given (palaeo-) latitude and (palaeo-) depth, and the rate at which it accumulates. Rates in the centre part of the matrix, between 10 and 20 m/mg, may vary by as much as a factor of two. (From Berger, 1973a).

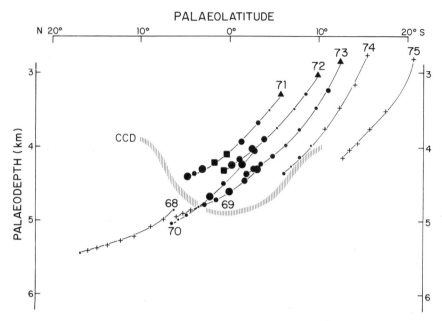

FIG. 29.45. Preservation patterns of radiolarians in a palaeolatitude–palaeodepth matrix, eastern tropical Pacific. Numbers refer to drill sites. Preservation estimates by Moore (1971); grading from well preserved (squares and large circles) to poorly preserved (small circles). No radiolarians = crosses; chert = triangles. Optimum preservation occurs at the palaeo-equator.

are poorly understood, but are thought to involve changes in bottom water circulation due to climatic variations (e.g. the influx of Antarctic Bottom Water and the onset of the North Atlantic Deep Water production), and to tectonic events (e.g. the closure of the Panama Isthmus; see Woodring, 1966). Closure of this isthmus prevented large scale transport of Atlantic surface waters into the Pacific, thus making the North Atlantic a receptacle for water with "shallow" characteristics (i.e. water which is well oxygenated, nutrient-poor, and close to saturation with respect to calcite). This kind of hydrography will lower the CCD (see Section 29.4.1).

The minimum depth of the Equatorial Pacific curve occurs in the Eocene, and again, the minimum is associated with a great abundance of siliceous fossils. The maximum depth was apparently reached during the Late Oligocene. As the largest ocean basin, the Pacific would be expected to be influenced to a lesser extent than the Atlantic by regional controlling factors. World-wide changes in sedimentation patterns are therefore likely to be reflected in the Pacific CCD curve, and discussion of the CCD fluctuations will be greatly facilitated when more is learned about the carbonate budgets of the Indian and the Antarctic Oceans.

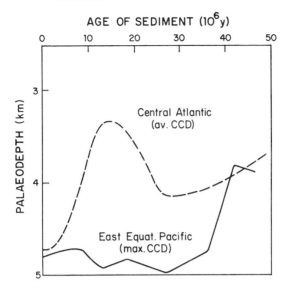

FIG. 29.46. Sketch of CCD fluctuations in the eastern equatorial Pacific, west of East Pacific Rise (maximum CCD) and in the central Atlantic (average CCD) based on backtracking of JOIDES drill sites. Curves from Berger and von Rad (1972) (Atlantic) and Berger (1973a) (Pacific).

## NOTE ADDED IN PROOF

Since this chapter was completed in 1973, much new information on biogenous deep-sea sediments has become available, largely as a result of the intensive study of the sample material which is constantly being provided by the Deep-Sea Drilling Project. The results and references of most of this work are readily accessible in the Initial Reports of the Project. Several recent references have been added in the text.

In addition, useful summary articles on recent work in this general field are available in the following:

Drooger, C. W. (ed.). (1973). "Messinian Events in the Mediterranean". Koninklijke Nederlandse Akademie van Wetenschappen, Amsterdam.

Fraser, R. (ed.). (1973). "Oceanography of the South Pacific 1972". New Zealand National Commission for UNESCO, Wellington.

Smith, L. A. and Hardenbol, J. (eds.). (1973). "Proceedings of Symposium on Calcareous Nannofossils". Society of Economic Paleontologists and Mineralogists, Houston, Texas.

Symposium about to be published:

CLIMAP. (1976). "Studies in Quaternary Oceanography." Geological Society of America, Memoir Series.

## Acknowledgements

In am greatly indebted to friends and colleagues for advice, discussion and preprints of their work. For advice, I sincerely thank P. H. Roth (coccoliths), H. J. Schrader and E. L. Venrick (diatoms), H. Poelchau (silicoflagellates), F. L. Parker (foraminifera), and W. R. Riedel (radiolarians). Errors remain my own, of course. Suggestions by G. Arrhenius, K. Banse, T. Gieskes, G. R. Heath, and discussions with C. Adelseck, R. G. Douglas, E. D. Goldberg, D. C. Hurd, T. Imbrie, T. Johnson, F. B. Phleger, E. L. Winterer and J. Yount were very much appreciated. R. Batiza assisted with the analysis of Pleistocene dissolution cycles in the Atlantic. Judy Lachmund Clinton and Sharon Weldy drafted the illustrations.

Supported by the National Science Foundation, Oceanography Section, Grants GB-21259, GA 36697, GA 35451 and GA 36697.

## References

Adelseck, C. G., Geehan, G. W. and Roth, P. H. (1973). *Bull. Geol. Soc. Amer.* **84**, 2755.

Adelseck, C. G. and Berger, W. H. (1975). *In* "Dissolution of Deep Sea Carbonates" (Sliter, W. V., Bé, A. W. H. and Berger, W. H., eds.). *Cushman Found. Foram., Res., Spec. Publ.* **13**.

Antia, N. J., McAllister, C. D., Parsons, T. R., Stephens, K. and Strickland, J. D. H. (1963). *Limnol. Oceanogr.* **8**, 166.

Apollonio, S. (1973). *Science, N.Y.* **180**, 491.

Arrhenius, G. (1952). *Rep. Swed. Deep-Sea Exped.* **5**, 1.

Arrhenius, G. (1963). *In* "The Sea" (Hill, M. N., ed.), Vol. 3, p. 655. Interscience, New York.

Bada, J. L. and Schroeder, R. A. (1972). *Earth Planet. Sci. Lett.* **15**, 1.

Bada, J. L., Luyendyk, B. P. and Maynard, J. B. (1970). *Science, N.Y.* **170**, 730.

Bandy, O. L. (1966). *Micropaleontology,* **12**, 79.

Bandy, O. L. and Wade, M. E. (1967). *Prog. Oceanogr.* **4**, 51.

Barash, M. S. (1970). "Planktonnie foraminiferi v osadkakh severnoy Atlantiki". Izdat, Nauka, Moscow.

Barker, R. W. (1960). *Soc. Econ. Paleontol. Mineral., Spec. Publ.* **9**. 1.

Bartlett, G. A. and Greggs, R. G. (1969). *Science, N.Y.* **166**, 740.

Bartolini, C. (1970). *Micropaleontology,* **16**, 129.

Bathurst, R. G. C. (1971). "Carbonate Sediments and their Diagenesis", 620pp. Elsevier, Amsterdam.

Bé, A. W. H. (1965). *Micropaleontology,* **11**, 81.

Bé, A. W. H. (1967). "Fiches d'identification du zooplankton," Sheet 108, *Cons. Permanent Int. Explor. Mer.*

Bé, A. W. H. (1968). *Science, N.Y.* **161**, 881.

Bé, A. W. H. (1969). *Antarct. Map Folio Ser., Amer. Geogr. Soc.* **11**, 9.

Bé, A. W. H. and Ericson, D. B. (1963). *Ann. N.Y. Acad. Sci.* **109**, 65.

Bé, A. W. H. and Hamlin, W. H. (1967). *Micropaleontology,* **13**, 87.

Bé, A. W. H. and Hemleben, C. (1970). *N. Jb. Geol. Palaeontol. Abh.* **134**, 221.

Bé, A. W. H. and Lott, L. (1964). *Science, N.Y.* **145**, 823.

Bé, A. W. H. and Tolderlund, D. S. (1971). *In* "The Micropalaeontology of Oceans", (Funnell, B. M. and Riedel, W. R., eds.), p. 105. Cambridge University Press, London.

Bé, A. W. H., Vilks, G. and Lott, L. (1971). *Micropaleontology*, **17**, 31.

Beall, A. O., and Fischer, A. G. (1969). *Init. Rept. Deep-Sea Drilling Proj.* **1**, 521.

Beard, J. H. (1969). *Trans. Gulf Coast Ass. Geol. Soc.* **19**, 535.

Beers, J. R. and Stewart, G. L. (1967). *J. Fish. Res. Bd. Can.* **24**, 2053.

Beers, J. R. and Stewart, G. L. (1970). *Bull. Scripps Inst. Oceanogr.* **17**, 67.

Beers, J. R. and Stewart, G. L. (1971). *Deep-Sea Res.* **18**, 861.

Belyaev, G. M. and Glikman, L. S. (1970). *Tr. Inst. Okeanol. SSSR*, **88**, 252.

Belyaeva, N. V. (1964). *Tr. Inst. Okeanol. SSSR*. **68**, 12.

Belyaeva, N. V. (1969). *In* "Microflora and Microfauna in Recent Sediments of the Ocean" (Bezrukov, P. L., ed.), p. 73, Nauka, Moscow.

Belyaeva, N. V. (1970). *Okeanologiya*, **10**, 1016.

Benson, W. E., Gerard, R. D. and Hay, W. W. (1970). *Init. Rep. Deep-Sea Drilling Proj.* **4**, 659.

Ben-Yaakov, S. and Kaplan, I. R. (1969). *Limnol. Oceanogr.* **15**, 874.

Ben-Yaakov, S. and Kaplan, I. R. (1971). *J. Geophys. Res.* **76**, 722.

Berger, W. H. (1967). *Science, N.Y.* **156**, 383.

Berger, W. H. (1968a). *Deep-Sea Res.* **15**, 31.

Berger, W. H. (1968b). *Science, N.Y.* **159**, 1237.

Berger, W. H. (1969a). *Deep-Sea Res.*, **16**, 1.

Berger, W. H. (1969b). *J. Paleontol.* **43**, 1369.

Berger, W. H. (1970a). *Mar. Geol.* **8**, 111.

Berger, W. H. (1970b). *Geol. Soc. Amer. Bull.* **81**, 1385.

Berger, W. H. (1970c). *Limnol. Oceanogr.* **15**, 183.

Berger, W. H. (1971a). *J. Foram. Res.* **1**, 95.

Berger, W. H. (1971b). *Mar. Geol.* **11**, 325.

Berger, W. H. (1972). *Nature, Lond.* **236**, 392.

Berger, W. H. (1973a). *Bull. Geol. Soc. Amer.* **84**, 1941.

Berger, W. H. (1973b). *Deep-Sea Res.* 917.

Berger, W. H. (1973c). *J. Foram. Res.* 187.

Berger, W. H. (1974). *In* "The Geology of Continental Margins", (Burk, C. A. and Drake, C. L., eds.), p. 213. Springer-Verlag, New York.

Berger, W. H. and Heath, G. R. (1968). *J. Mar. Res.*, **26**, 134.

Berger, W. H. and Parker, F. L. (1970). *Science, N.Y.* **168**, 1345.

Berger, W. H. and Piper, D. J. W. (1972). *Limnol. Oceanogr.* **17**, 275.

Berger, W. H. and Soutar, A. (1967). *Science, N.Y.* **156**, 1495.

Berger, W. H. and Soutar, A. (1970). *Bull. Geol. Soc. Amer.* **81**, 275.

Berger, W. H. and von Rad, U. (1972). *Init. Rep. Deep-Sea Drilling Proj.* **14**, 787.

Berger, W. H. and Gardner, J. V. (1975). *J. Foram. Res.* **5**, 102.

Berger, W. H. and Roth, P. H. (1975). *Rev. Geophys. Space Phys.* **13**(3). 561.

Berggren, W. A. (1972a). *Init. Rep. Deep-Sea Drilling Proj.* **12**, 953.

Berggren, W. A. (1972b). *Init. Rep. Deep-Sea Drilling Proj.* **12**, 965.

Berggren, W. A. and van Couvering, J. (1974). *Palaeogeo., Palaeochim. Palaeoecol.* **16**, 1.

Bernard, F. (1963). *Pelagos, Bull. Inst. Oceanogr. Alger.* **1**, 5.

Bernard, F. (1969). *Oceanogr. Mar. Biol. Ann. Rev.* **7**, 205.

Berner, R. A., Scott, M. R. and Thomlinson, C. (1970). *Limnol. Oceanogr.* **15**, 544.

Bernoulli, D. (1964). *Beitrage z. Geol. Karte der Schweiz, Neue Folge, Lieferung,* **118**, 1.

Bernoulli, D. (1972). *Init. Rep. Deep-Sea Drilling Proj.* **11**, 801.

Bezrukov, P. L. (ed.). (1969). "Microflora and Microfauna in the Recent Sediments of the Pacific Ocean. *The Pacific Ocean,* **8**, 1. Akad. Nauk, Moskow.

Bezrukov, P. L. (ed.). (1970). "Sedimentation in the Pacific Ocean". *The Pacific Ocean,* **6**, 2 pts. Izdat, Nauk, Moskow.

Blackman, A. and Somayajulu, B. L. K. (1966). *Science, N.Y.* **154**, 886.

Blanc, F., Blanc-Vernet, L. and Le Campion, J. (1972). *Tethys.* **4**. 761.

Blanc, J. J. (1968). *Oceanogr. Mar. Biol. Ann. Rev.* **6**, 377.

Blankley, W. F. (1971). Auxotrophic and Heterotrophic Growth and Calcification in Coccolithophorids. Ph.D. Thesis, 186pp, Univ. of Calif., San Diego.

Blasco, D. (1971). *Inv. Pesq.* **35**, 61.

Blow, W. H. (1969). *Proc. 1st Int. Conf. Planktonic Microfossils, Geneva, 1967,* **1**, 199. Brill, Leiden.

Bogorov, V. G. (1967). "The Pacific Ocean. Biology of the Pacific Ocean," Part I— Plankton, Vol. 7, **7**, p. 1.

Bogoyavlenskiy, A. N. (1967). *Int. Geol. Rev.* **9**, 133.

Bolli, H. M. (1970). *Init. Rep. Deep-Sea Drilling Proj.* **4**, 577.

Bolli, H. M. (1971). *In* "The Micropaleontology of Oceans", (Funnell, B. M. and Riedel, W. R., eds.), p. 639. Cambridge University Press, London.

Boltovskoy, E. (1965). "Los Foraminiferos Recientes". Eudeba, Buenos Aires.

Boltovskoy, E. (1969). *Micropaleontology,* **15**, 237.

Boltovskoy, E. (1971). *Micropaleontology,* **17**, 53.

Bornhold, B. D. and Pilkey, O. H. (1971). *Geol. Soc. Amer. Bull.* **82**, 1341.

Bradley, W. H. (1965). *Science, N.Y.* **150**, 1423.

Bradshaw, J. S. (1959). *Contrib. Cushman Found. Foram. Res.* **10**, 25.

Bradshaw, J. S. (1961). *Contr. Cushman Found. Foram. Res.* **12**, 87.

Brady, H. B. (1884). *Rep. Sci. Results Voyage H.M.S. "Challenger",* **9**, 1.

Bramlette, M. N. (1946). *U.S. Geol. Survey Prof. Paper,* **212**, 1.

Bramlette, M. N. (1958). *Geol. Soc. Amer. Bull.* **69**, 121.

Bramlette, M. N. (1961). *In* "Oceanography" (Sears, M., ed.), *Amer. Assoc. Adv. Sci. Publ.* **67**, 345.

Bramlette, M. N. and Bradley, W. H. (1942). *U.S. Geol. Survey, Prof. Paper* **196-A**, 1.

Bramlette, M. N. and Riedel, W. R. (1954). *J. Paleontol.* **28**, 385.

Bramlette, M. N. and Riedel, W. R. (1971). *In* "The Micropalaeontology of Oceans" (Funnell, B. M. and Riedel, W. R., eds.), p. 665, Cambridge University Press, London.

Bramlette, M. N. and Sullivan, F. R. (1961). *Micropaleontology,* **7**, 129.

Bramlette, M. N. and Wilcoxon, J. A. (1967). *Tulane Studies Geol.* **5**, 93.

Bramlette, M. N., Faughn, J. L. and Hurley, R. J. (1959). *Geol. Soc. Amer. Bull.,* **70**, 1549.

Brandt, K. (1885). "Fauna and Flora des Golfes von Neapel". *Zool. Stat. Neapel, Monogr.* **13**, 1.

Broecker, W. S. (1971a). *In* "The late Cenozoic Glacial Ages" (Turekian, K. K., ed.), p. 239. Yale University Press.

Broecker, W. S. (1971b). *Quaternary Res.* **1**, 188.

Broecker, W. S. and van Donk, J. (1970). *Rev. Geophys. Space Phys.* **8**, 169.

Brönniman, P. and Renz, H. H. (eds.). (1969). *Proc. 1st Int. Conf. Planktonic Microfossils, Geneva, 1967*, 2 Vols. Brill, Leiden.
Brönniman, P. and Resig, J. (1971). *Init. Rep. Deep-Sea Drilling Proj.* 7, 1235.
Bukry, D. (1971). *San Diego Soc. Nat. Hist. Trans.* 16, 303.
Bukry, D. (1973). *Init. Rep. Deep-Sea Drilling Proj.* 16, 653.
Bukry, D. and Bramlette, M. N. (1970). *Init. Rep. Deep-Sea Drilling Proj.* 3, 589.
Bukry, D., Douglas, R. G., Kling, S. A. and Krasheninnikov, V. (1971). *Init. Rep. Deep-Sea Drilling Proj.* 6, 1253.
Burckle, L. H. (1972). *Nova Hedwigia, Beih.* 39, 217.
Burckle, L. H., Ewing, J., Saito, T. and Leyden, R. (1967). *Science, N.Y.* 157, 537.
Busby, W. F. and Lewin, J. (1967). *J. Phycol.* 3, 127.
Cachon, J. and Cachon, M. (1971). *Arch. Protistenk.* 113, 80.
Calvert, S. E. (1966a). *J. Geol.* 74, 546.
Calvert, S. E. (1966b). *Geol. Soc. Amer. Bull.* 77, 569.
Calvert, S. E. (1968). *Nature, Lond.* 219, 919.
Calvert, S. E. (1971). *Nature, Phys. Sci.* 234, 133.
Campbell, A. S. (1954). *In* "Treatise on Invertebrate Paleontology, Part D, Protista 3" (Moore, R. C., ed.), p. D-11.
Casey, R. E. (1971a). *In* "The Micropalaeontology of Oceans" (Funnell, B. M. and Riedel, W. R., eds.), p. 151. Cambridge University Press, London.
Casey, R. E. (1971b). *In* "The Micropalaeontology of Oceans" (Funnell, B. M. and Riedel, W. R., eds.), p. 331. Cambridge University Press, London.
Casey, R. E. (1976). *In* "Oceanic Micropaleontology" (Ramsay, A. T. S., ed.). (In press).
Casey, R. E., Partridge, T. M. and Sloan, J. R. (1971). *Proc. 2nd Plankt. Conf., Rome 1970*, p. 159. Edizioni, Technoscienza, 2 Vols., Rome.
Castracane, A. F. (1886). *Rep. Voyage H.M.S. "Challenger", Botany*, 2, 1.
Chave, K. E. (1965). *Science, N.Y.* 148, 1723.
Chave, K. E. and Suess, E. (1967). *Trans. N.Y. Acad. Sci. Ser. II*, 29, 991.
Chave, K. E. and Suess, E. (1970). *Limnol. Oceanogr.* 15, 633.
Chave, K. E., Deffeyes, K. S., Weyl, P. K., Garrels, R. M. and Thompson, M. E. (1962). *Science, N.Y.* 137, 33.
Chen, C. (1968). *Nature, Lond.* 219, 1145.
Chierici, M. A., Busi, M. T. and Cita, M. B. (1962). *Rev. Micropaleontol.* 5, 123.
Christiansen, B. O. (1965). *Pubbl. Staz. Zool. Napoli*, 34, 197.
Cifelli, R. (1962). *J. Mar. Res.* 20, 201.
Cifelli, R. (1971). *J. Foram. Res.* 1, 170.
Cifelli, R. and Sachs, K. N. (1966). *Deep-Sea Res.* 13, 751.
Cifelli, R. and Smith, R. K. (1970). *Smithson. Contrib. Paleobiol.* 4, 1.
Cita, M. B. (1971). *Rev. Micropaléontol.* 14, 17.
Clarke, R. H. (1968). *Deep-Sea Res.* 15, 397.
Cohen, C. L. D. (1964). *Micropaleontology*, 10, 231.
Conolly, J. R. (1969). *N.Z. J. Geol. Geophys.* 12, 210.
Conolly, J. R. and Ewing, M. (1967). *J. Sediment Petrology* 37, 44.
Conover, R. J. and Lalli, C. M. (1972). *J. Exp. Mar. Biol. Ecol.* 9, 279.
Correns, C. W. (1939). *In* "Recent Marine Sediments" (Trask, P. D., ed.), p. 373. Amer. Assoc. Petr. Geol., Tulsa, Okla.
Costlow, J. D. (ed.). (1971). "Fertility of the Sea". 2 Vols. Gordon Breach, New York.
Craig, H. (1965). *In* "Stable Isotopes in Oceanographic Studies and Paleotemperatures" (Tongiorgi, E., ed.), Vol. 2, p. 1. Consig. Naz. Richerche Labor. Geol. Nucleare, Pisa.

Craig, H., Chung, Y. and Fiadeiro, M. (1972). *Earth Planet. Sci. Lett.*
Cupp, E. E. (1943). *Bull. Scripps Inst. Ocean.* **5**, 1.
Daly, R. A. (1936). *Amer. J. Sci.* **31**, 401.
Dansgaard, W. and Tauber, H. (1969). *Science, N.Y.* **166**, 499.
David, A. and Stommel, H. (1970). *Cah. Oceanogr.* **22**, 343.
Davies, D. K. (1968). *J. Sediment Petrology* **38**, 1100.
Davies, T. A. and Jones, E. J. W. (1971). *Deep-Sea Res.* **18**, 619.
Davies, T. A. and Laughton, A. S. (1972). *Init. Rep. Deep-Sea Drilling Proj.* **12**, 905.
Deflandre, G. and Fert, C. (1954). *Ann. Paléontol.* **40**, 115.
Donahue, J. (1971). *Mar. Geol.* **11**, M1.
Donahue, J. D. (1967). *Prog. Oceanogr.* **4**, 133.
Donahue, J. D. (1970). *Geol. Soc. Amer., Memoir,* **126**, 121.
Douglas, R. G. (1971). *Init. Rep. Deep-Sea Drilling Proj.* **6**, 1027.
Douglas, R. G. and Savin, S. M. (1971). *Init. Rep. Deep-Sea Drilling Proj.* **6**, 1123.
Drebes, G. (1972). *Nova Hedwigia, Beih.* **39**, 95.
Dumitrica, P. (1973). *Init. Rep. Deep-Sea Drilling Proj.* **13**, 829.
Duplessy, J.-C. (1972). "La géochimie des isotopes stables du carbone dans la mer".
  Centre d'Études Nucléaires de Saclay, CEA-N-1565, 198pp.
Duplessy, J.-C., Lalou, C. and Vinot, A. C. (1970). *Science, N.Y.* **168**, 250.
Eade, J. V. and van der Linden, W. J. M. (1970). *N.Z. J. Geol. Geophys.* **13**, 228.
Edmond, J. M. (1974). *Deep-Sea Res.* **21**, 455.
Edmond, J. M. and Gieskes, J. M. (1970). *Geochim. Cosmochim. Acta,* **34**, 1261.
Edmond, J. M., Chung, Y. and Sclater, J. G. (1971). *J. Geophys. Res.* **76**, 8089.
Ehrenberg, C. G. (1838). *Abhandl. König. Preuss. Akad. Wiss. Jahrg.* **59**, (Berlin).
Ehrenberg, C. G. (1856). "Mikrogeologie. Das Erden und Felsen schaffende Wirken
  des unsichtbar kleinen selbständigen Lebens auf der Erde". Voss, Leipzig.
Ellis, B. F. and Messina, A. (1940). "Catalogue of Foraminifera. American Museum
  of Natural History", New York .(Continuing Series).
Emiliani, C. (1954). *Amer. J. Sci.* **252**, 149.
Emiliani, C. (1955). *J. Geol.* **63**, 538.
Emiliani, C. (1966). *J. Geol.* **74**, 109.
Emiliani, C. (1970). *Science, N.Y.* **168**, 822.
Emiliani, C. (1971). *Science, N.Y.* **173**, 1122.
Emiliani, C. (1972). *Science, N.Y.* **178**, 398.
Emiliani, C. and Milliman, J. D. (1966). *Earth Sci. Rev.* **1**, 105.
Emiliani, C. and Shackleton, N. (1974). *Science, N.Y.* **183**, 511.
Eppley, R. W. (1970). *Bull. Scripps Inst. Oceanogr.* **17**, 43.
Eppley, R. W. (1972). *Fish. Bull.* **70**, xxx.
Epstein, S., Buchsbaum, R., Lowenstam, H. and Urey, H. C. (1953). *Bull. Geol. Soc. Amer.* **64**, 1315.
Ericson, D. B. and Wollin, G. (1956). *Deep-Sea Res.* **3**, 104.
Ericson, D. B. and Wollin, G. (1962). *Sci. Amer.* **207**, 96.
Ericson, D. B., Ewing, M. and Heezen, B. C. (1952). *Bull. Amer. Assoc. Petrol. Geol.* **36**, 489.
Ericson, D. B. Ewing, M., Wollin, G. and Heezen, B. C. (1961). *Bull. Geol. Soc. Amer.* **72**, 193.
Ericson, D. B., Ewing M. and Wollin, G. (1963). *Science, N.Y.* **139**, 727.
Ericson, D. B., Ewing, M. and Wollin, G. (1964). *Science, N.Y.* **146**, 723.
Ernst, W. G. and Calvert, S. E. 1969. *Amer. J. Sci.* **267-A**, 114.
Ewing, J. and Ewing, M. (1967). *Science, N.Y.* **156**, 1591.

Ewing, J. I. and Hollister, C. H. (1972). *Init. Rep. Deep-Sea Drilling Proj.* **11**, 951.

Ewing, J., Ewing, M., Aitken, T. and Ludwig, W. J. (1968). *Amer. Geophys. Union Geophys. Monogr.* **12**, 147.

Ewing, M., Worzel, J. L. and Burk, C. A. (1969). *Init. Rep. Deep-Sea Drilling Proj.* **1**, 624.

Fanning, K. A. and Pilson, M. E. Q. (1971). *Science, N.Y.* **173**, 1228.

Fanning, K. A. and Pilson, M. E. Q. (1973). *Geochim. Cosmochim. Acta,* **37**, 2405.

Fanning, K. A. and Schink, D. R. (1969). *Limnol. Oceanogr.* **14**, 59.

Farinacci, A. (1969). "Catalogue of Calcareous Nannofossils." Edizioni Tecno-scienza, Vol. 1. Rome.

Farinacci, A. (ed.) (1971). *Proc. 2nd Plankt. Conf., Rome 1970.* Edizioni Tecno-scienza, 2 Vols., Rome.

Fischer, A. G., Heezen, B. C., Boyce, R. E., Bukry, D., Douglas, R. G., Garrison, R. E., Kling, S. A., Krasheninnikov, V. V., Lisitzin, A. P. and Pimm, A. C. (1970). *Science, N.Y.* **168**, 1210.

Fischer, A. G. and Garrison R. E. (1967). *J. Geol.* **75**, 488.

Foreman, H. P. and Riedel, W. R. (1972). "Catalogue of Polycystine Radiolaria." The American Museum of Natural History, New York.

Fournier, R. O. (1968). *Limnol. Oceanogr.* **13**, 693–697.

Fowler, S. W. and Small, L. F. (1972). *Limnol. Oceanogr.* **17**, 293.

Fox, P. J. and Heezen, B. C. (1965). *Science, N.Y.* **149**, 1367.

Freudenthal, H. D., Lee, J. J. and Kossoy, V. (1964). *J. Protozool.* **11** (Suppl.), 12.

Friedman, G. M. (1965). *Geol. Soc. Amer. Bull.* **76**, 1191.

Friend, J. K. and Riedel, W. R. (1967). *Micropaleontology,* **13**, 217.

Funnell, B. M. and Riedel, W. R., eds. (1971). "The Micropalaeontology of Oceans." Cambridge University Press, London.

Funnell, B. M. and Smith, A. G. (1968). *Nature, Lond.* **219**, 1328.

Gaarder, K. R. (1971). *In* "The Micropalaeontology of Oceans" (Funnell, B. M. and Riedel, W. R., eds.), p. 97. Cambridge University Press, London.

Garrison, R. E. (1971). *Init. Rep. Deep-Sea Drilling Proj.* **6**, 1212.

Garrison, R. E. and Fischer, A. G. (1969). *Soc. Econ. Paleontol. Mineral., Spec. Publ.* **14**, 20.

Gartner, S. (1969). *Gulf Coast Assoc. Geol. Soc. Trans.* **19**, 585.

Gartner, S. (1972). *Paleogeogr. Palaeoclimatol. Palaeoecol.* **12**, 169.

Gartner, S. and Bukry, D. (1969). *J. Paleontol.* **43**, 1213.

Geitzenauer, K. R. (1972). *Deep-Sea Res.* **19**, 45.

Gemeinhardt, K. (1931). *Deut. Südpolar Exped., 1901–1903, Zoologie,* **20**, 221.

Gemeinhardt, K. (1934). *Wiss. Ergeb. Deut. Atl. Exped. "Meteor", 1925–1927,* **12**(1), 274.

Gieskes, J. M. (1970). *"Meteor" Forsch.-Ergebn. Reihe A,* **8**, 12.

Glaçon, G., Grazzini, C. V. and Sigal, M. J. (1971). *Proc. 2nd Plankt. Conf., Rome 1970,* p. 555.

Glass, B. P. (1969). *Earth Planet. Sci. Lett.* **6**, 409.

Glezer, Z. I. (1966). *In* "Cryptogamic Plants of the USSSR", Vol. 7, p. 1, Akad. Nauk. SSSR.

Goldberg, E. D. and Bruland, K. (1974). *In* "The Sea", Vol. 5 (Goldberg, E. D., ed.), Interscience, New York. p. 451.

Goldberg, E. D. and Koide (1962). *Geochim. Cosmochim. Acta,* **26**, 417.

Goldberg, E. D., Broecker, W. S., Gross, M. G. and Turekian, K. K., 1971. *In* "Radioactivity in the Marine Environment", p. 137, Acad. Sci., Washington, D.C.

Goll, R. M. (1969). *In* "Hot Brines and Recent Heavy Metal Deposits in the Red Sea" (Degens, E. T. and Ross, D. A., eds.), p. 306. Springer, New York.

Goll, R. M. (1972). *Init. Rep. Deep-Sea Drilling Proj.* **9**, 947.

Goll, R. M. and Bjørklund, K. R. (1971). *Micropaleontology*, **17**, 434.

Goll, R. M. and Bjørklund, K. R. (1974). *Micropaleontology* **20**, 38.

Grell, K. G. (1967). *In* "Research in Protozoology" (Chen, T. T., ed.), Vol. 2, p. 147, Pergamon Press, Oxford.

Griggs, G. B. and Fowler, G. A. (1971). *Deep-Sea Res.* **18**, 645.

Griggs, G. B., Carey, A. G. and Kulm, L. D. (1969). *Deep-Sea Res.* **16**, 157.

Haeckel, E. (1887). *Rep. Voyage H.M.S.* "Challenger", *Zoology*, **18**, 1.

Haecker, V. (1907). *Arch. Protistenk.* **10**, 114.

Hajos, M. (1968). *Geologica Hungarica, Ser. Paleontol.* **37**, 31.

Halldall, P. and Markali, J. (1955). *Avh. Norske Vidensk, Akad., Ser. I Mat.-Nat. Kl.* **7**, 1.

Hamilton, E. L. (1957). *Micropaleontology*, **3**, 69.

Hanna, G. D. (1930). *J. Paleontol.* **4**, 189.

Hanna, G. D. (1932). *Proc. Calif. Acad. Sci. 4th Ser.* **20**(6), 161.

Hardy, A. C. (1955). *Deep-Sea Res.* **3** (Supplement), 7.

Harrison, C. G. A. and Funnell, B. M. (1964). *Nature, Lond.* **204**, 566.

Hart, R. A. (1973). *Nature, Lond.* **243** (5402), 76.

Harvey, H. W. (1957). "The Chemistry and Fertility of Sea Water", 240 pp. Cambridge University Press, London.

Hasle, G. R. (1959). *Deep-Sea Res.* **6**, 38.

Hasle, G. R. (1960). *Nytt. Mag. Botan.* **8**, 77.

Hasle, G. R. (1969). *Hvalrådets Skrifter*, **52**, 1.

Hawley, J. and Pytkowicz, R. M. (1969). *Geochim. Cosmochim. Acta*, **33**, 1557.

Hay, W. W. (1970). *Init. Rep. Deep-Sea Drilling Proj.* **4**, 669.

Hay, W. W. (ed.) (1974)." Studies in Paleo-oceanography." *Soc. Econ. Paleont. Min., Spec. Publ.* **20**.

Hay, W. W., Mohler, H. P., Roth, P. H., Schmidt, R. A. and Boudreaux, J. E. (1967). *Trans. Gulf Coast Assoc. Geol. Soc.* **17**, 428.

Hays, J. D. (1965). *Amer. Geophys. Union, Antarctic Res. Series,* **5**, 125.

Hays, J. D. (1967). *Prog. Oceanogr.* **4**, 117.

Hays, J. D. (1970). *Mem. Geol. Soc. Amer.* **126**, 185.

Hays, J. D. (1971). *Bull. Geol. Soc. Amer.* **82**, 243.

Hays, J. D., Saito, T., Opdyke, N. D. and Burckle, L. H. (1969). *Bull. Geol. Soc. Amer.* **80**, 1481.

Hays, J. D., Cook, H., Jenkins, G., Orr, W., Goll, R., Cook, F., Milow, D. and Fuller, J. (1972). *Init. Rep. Deep-Sea Drilling Proj.* **9**, 909.

Heath, G. R. (1969). *Bull. Geol. Soc. Amer.* **80**, 689.

Heath, G. R. (1973). *In* "Studies in Paleo-oceanography" (Hay, W. W., ed.). Soc. Econ. Paleontol. Mineral., Spec. Publ. **20**, 77.

Heath, G. R. and Culberson, C. (1970). *Bull. Geol. Soc. Amer.* **81**, 3157.

Heath, G. R. and Moberly, R. (1971a). *Init. Rep. Deep-Sea Drilling Proj.* **7**, 991.

Heath, G. G. and Moberly, R. (1971b). *Init. Rep. Deep-Sea Drilling Proj.* **7**, 1009.

Hecht, A. D. and Savin, S. M. (1972). *J. Foram. Res.* **2**, 55.

Heezen, B. C. and Ewing, M. (1952). *Amer. J. Sci.* **250**, 849.

Heezen, B. C. and Hollister, C. D. (1964). *Mar. Geol.* **1**, 141.

Heezen, B. C., Hollister, C. D. and Ruddiman, W. F. (1966). *Science, N.Y.* **152**, 502.

Heezen, B. C., MacGregor, I. D., Foreman, H. P., Forristal, G., Hekel, H., Hesse, R., Hoskins, R. H., Jones, E. J. W., Kaneps, A., Krasheninnikov, V. A., Okada, H. and Ruef, M. H. (1973). *Nature, Lond.* **241**, 25.

Helms, P. B. and Riedel, W. R. (1971). *Init. Rep. Deep-Sea Drilling Proj.* **7**, 1709.

Hendey, N. I. (1964). "An Introductory Account of the smaller Algae of British Coastal Waters. V. Bacillariophyceae." G. B. Min. Ag. Fish Food, Fish. Invert., Ser. 4, 5.

Hentschel, E. (1936). *Wiss. Ergeb. Deutsche Atl. Exped. "Meteor", 1925–1927,* **11**, 1.

Herman, Y. (1971). *In* "The Micropalaeontology of Oceans" (Funnell, B. M. and Riedel, W. R., eds.). p. 463. Cambridge University Press, London.

Hess, H. H. (1962). *In* "Petrological Studies: A volume in Honor of A. F. Buddington (Engel, A. E. J., *et al.*, eds.), p. 599. Geol. Soc. Amer., New York.

Hofker, J. (1959). *Contrib. Cushman Found. Foram. Res.* **10**, 1.

Hofker, J. (1968). "Studies of Foraminifera", Vol. 1. Naturhist. Genotschap Limburg, Goffin, Maastricht.

Hollande, A. and Enjumet, M. (1960). *Arch. Mus. Nat. Hist. Nat., Paris, Sér.* 7, **7**, 7.

Hollman, R. (1964). *Neues Jb. Geol. Paläont. Abh.* **119**, 22.

Holmes, R. W. (1967). *Amer. J. Bot.* **54**, 163.

Holmes, R. W. and Reimann, B. E. F. (1966). *Phycologia,* **5**, 233.

Honigberg, B. M., Balamuth, W., Bovee, E. C., Corliss, J. O., Gojdics, M., Hall, R. P., Kudo, R. R., Levine, N. D., Loeblich, A. R., Jr., Weiser, J. and Wenrich, D. H. (1964). *J. Protozool.* **11**, 7.

Horn, D. R., Ewing, J. I. and Ewing, M. (1972). *Sedimentology,* **18**, 247.

Hsü, K. J. and Andrews, J. E. (1970). *Init. Rep. Deep-Sea Drilling Proj.* **3**, 445 and 464.

Hsü, K. J. and Jenkins (eds.) (1974). "Pelagic Sediments on Land and under the Sea". *Spec. Publ. Int. Ass. Sedimentol.* **1**.

Huang, T.-C., Stanley, D. J. and Stuckenrath, R. (1972). *J. Mar. Techn. Soc.* **6**(4), 25.

Huang, W. H. and Vogler, D. L. (1972). *Nature, Phys. Sci.* **235**, 157.

Hulburt, E. M. (1962). *Limnol. Oceanogr.* **7**, 307.

Hulburt, E. M. (1963). *J. Mar. Res.* **21**, 81.

Hulburt, E. M. (1964). *Bull. Mar. Sci., Gulf Caribb.* **14**, 33.

Hulburt, E. M. (1966). *J. Mar. Res.* **24**, 67.

Hulburt, E. M. (1967). *Deep-Sea Res.* **14**, 685.

Hulburt, E. M. (1970). *Ecology,* **51**, 475.

Hulburt, E. M. and Corwin, N. (1969). *J. Mar. Res.* **27**, 55.

Hulburt, E. M. and Mackenzie, R. S. (1971). *Bull. Mar. Sci.* **21**, 603.

Hunkins, K., Bé, A. W. H., Opdyke, N. D. and Mathieu, G. (1971). *In* "Late Cenozoic Glacial Ages" (Turekian, K. K., ed.), p. 215. Yale University Press, New Haven, Conn.

Hurd, D. C. (1972). *Earth Planet. Sci. Lett.* **15**, 411.

Hurd, D. C. (1973). *Geochim. Cosmochim. Acta,* **37**, 2257.

Hustedt, F. (1930). *In* "Krytogamen-Flora" (Rabenhorst, L., ed.), Vol. 7, Pt. I, p. 1. Leipzig.

Hustedt, F. (1959). *In* "Krytogamen-Flora" (Rabenhorst, L. ed.), Vol. 7, Pt. II, p. 1. Leipzig.

Hunstedt, F. (1961). "Kieselalgen (Diatomeen)." Franckh. Stuttgart.

Imbrie, J. and Kipp, N. G. (1971). *In* "Late Cenozoic Glacial Ages" (Turekian, K. K., ed.), p. 71, Yale University Press, New Haven, Conn.

Imbrie, J., van Donk, J. and Kipp, N. G. (1973). *Quaternary Res.* **3**, 10.
Isacks, B., Oliver, J. and Sykes, L. R. (1968). *J. Geophys. Res.* **73**, 5855.
Ivanenkov, V. N., Sapozhnikov, V. V., Chernyakova, A. M. and Gusarova, A. N. (1972). *Okeanologiia,* **12**, 243.
Jendrzejewski, J. P. and Zarillo, G. A. (1972). *Deep-Sea Res.* **19**, 327.
Jenkins, D. G. and Orr, W. N. (1971). *Rev. Españ. Micropal.* **3**, 301.
Jenkins, D. G. and Orr, W. N. (1972). *Init. Rep. Deep-Sea Drilling Proj.* **9**, 1059.
Jepps, M. W. (1956). "The Protozoa, Sarcodina." Oliver and Boyd, Edinburgh.
Johnson, D. A. (1972a). *Deep-Sea Res.* **19**, 253.
Johnson, D. A. (1972b). *Bull. Geol. Soc. Amer.* **83**, 3121.
Johnson, D. A. and Johnson, T. C. (1970). *Deep-Sea Res.* **17**, 157.
Johnson, D. A. and Knoll, A. H. (1974). *Quaternary Res.* **4**, 206.
Johnson, T. C. (1974). *Deep-Sea. Res.* **21**, 851.
Jones, J. B. and Segnit, E. R. (1971). *J. Geol. Soc. Aust.* **18**, 57.
Jones, J. I. (1967). *Micropaleontology,* **13**, 489.
Jousé, A. P., Koroleva, G. S. and Negaeva, G. A. (1962). *Tr. Inst. Okeanol. Akad. Nauk, SSSR,* **61**, 19.
Jousé, A. P., Mukhina, V. V. and Kozlova, O. G. (1969). *In* "Mikroflora i Mikrofauna v Sovremennikh Osadkakh Tikhogo Okeana" (Bezrukov, P. L., ed.), p. 7. Nauka, Moscow.
Jousé, A. P., Kozlova, O. G. and Mukhina, V. V. (1971). *In* "The Micropalaeontology of Oceans" (Funnell, B. M. and Riedel, W. R., eds.), p. 263. Cambridge University Press, London.
Kamatani, A. (1971). *Mar. Biol.* **8**, 89.
Kanaya, T. (1957). *Tohoku Univ., Sci. Rep. 2nd Ser. (Geol.),* **28**, 27.
Kanaya, T. (1959). *Tohoku Univ., Sci. Rep. 2nd Ser. (Geol.),* **30**, 1.
Kanaya, T. and Koizumi, I. (1966). *Tohoku Univ., Sci. Rep. 2nd Ser. (Geol.),* **37**, 89.
Kanaya, T. and Koizumi, I. (1970). *J. Mar. Geol. Japan,* **6**, 47.
Kastner, M. and Keene, J. B. (1975). *Proc. 9th Int. Congr. Sedimentol., Nice.*
Kato, K. and Kitano, Y. (1968). *J. Oceanogr. Soc. Japan,* **24**, 147.
Kennett, J. P. (1970). *Deep-Sea Res.* **17**, 125.
Khmeleva, N. N. (1967). *Dokl. Akad. Nauk SSSR,* **172**, 1430.
King, K. and Hare, P. E. (1972). *Science, N.Y.* **175**, 1461.
Klaveness, D. and Paasche, E. (1971). *Arch. Mikrobiol.* **75**, 382.
Kling, S. A. (1971a). *Proc. 2nd Plankt. Conf., Rome 1970,* p. 663.
Kling, S. A. (1971b). *Init. Rep. Deep-Sea Drilling Proj.* **6**, 1069.
Kobayashi, K., Kitazawa, K., Kanaya, T. and Sakai, T. (1971). *Deep-Sea Res.* **18**, 1045.
Koblentz-Mishke, O. J., Vokovinsky, V. V. and Kabanova, J. G. (1970). *In* "Scientific Exploration of the South Pacific" (Wooster, W. S., ed.), p. 183. National Acad. Sci., Washington, D.C.
Koizumi, I. (1968). *Tohoku Univ., Sci. Rep. 2nd Ser. (Geol.),* **40**, 171.
Koizumi, I. (1972). *Trans. Proc. Palaeontol. Soc. Japan, N.S.* **86**, 340.
Kolbe, R. W. (1954). *Rep. Swed. Deep-Sea Exped.* **6**, 1.
Kolbe, R. W. (1955). *Rep. Swed. Deep-Sea Exped.* **7**, 151.
Kolbe, R. W. (1957). *Rep. Swed. Deep-Sea Exped.* **9**, 1.
Kozlova, O. G. (1970). *In* "Osadkoobrazovanye v Tikhon Okeane" (Bezrukov, P. L., ed.), Vol. 1, p. 127. Nauka, Moskow.
Kozlova, O. G. (1971). *In* "The Micropalaeontology of Oceans" (Funnell, B. M. and Riedel, W. R., eds.), p. 271. Cambridge University Press, London
Kozlova, O. G. and Mukhina, V. V. (1967). *Int. Geol. Rev.,* **9**, 1322.

Krauskopf, K. B. (1959). *Soc. Econ. Paleontol. Mineral., Spec. Publ.* **7**, 4.
Kruglikova, S. B. (1966). *In* "Geokhimiya Kremnezema" (Strakhov, N. M., ed.), p. 246. Izdat, Nauka, Moscow.
Kruglikova, S. B. (1969). *In* "Microflora and Microfauna in Recent Sediments of the Pacific Ocean" (Bezrukov, P. L., ed.), p. 48. Nauka, Moscow.
Kuenen, P. H. (1950). "Marine Geology." John Wiley, New York.
Kuenen, P. H. and Migliorini, C. I. (1950). *J. Geol.* **58**, 91.
Kullenberg, B. (1952). *K. Vet. Vitt.-Samh. Handl. Ser. B*, **6**, 3.
Lalli, C. M. (1970). *J. Exp. Mar. Biol. Ecol.* **4**, 101.
Lancelot, Y. (1973). *Init. Rep. Deep-Sea Drilling Proj.* **17**, p. 377.
Larson, R. L. (1971). *Bull. Geol. Soc. Amer.* **82**, 823.
Laughton, A. S. (1963). *In* "The Sea", Vol. 3 (Hill, M. N., ed.), p. 437. Interscience, New York.
Lawrence, D. R. (1968). *Bull. Geol. Soc. Amer.* **79**, 1315.
LeCalvez, J. (1936). *Ann. Protistol.* **5**, 125.
Lee, J. J. and Zucker, W. (1969). *J. Protozool.* **16**, 71.
Lee, J. J., Freudenthal, H. D., Kossoy, V. and Bé, A. (1965). *J. Protozool.* **12**, 531.
Lee, J. J., McEnery, M., Pierce, S., Freudenthal, H. D. and Muller, W. A. (1966). *J. Protozool.* **13**, 659.
LePichon, X. (1968). *J. Geophys. Res.* **73**, 3661.
Lewin, J. C. (1961). *Geochim. Cosmochim. Acta*, **21**, 182.
Lewin, J. C. (1962a). *In* "Physiology and Biochemistry of Algae" (Lewin, R. A., ed.), p. 445, Academic Press, New York and London.
Lewin, J. C. (1962b). *In* "Physiology and Biochemistry of Algae" (Lewin, R. A., ed.), p. 457, Academic Press, New York and London.
Lewin, J. C. and Guillard, R. R. L. (1963). *Ann. Rev. Microbiol.* **17**, 373.
Li, Y. H., Takahashi, T. and Broecker, W. S. (1969). *J. Geophys. Res.* **74**, 5507.
Ling, H. Y. (1972). *Bull. Amer. Paleontol.* **62**, 135.
Lipps, J. H. (1966). *J. Paleontol.* **40**, 1257.
Lipps, J. H. (1970). *Proc. N. Amer. Paleontol. Convention, Sept., 1969, Part G*, p. 965.
Lipps, J. H. and Erskian, M. G. (1969). *J. Protozool.* **16**, 422.
Lisitzin, A. P. (1967). *Int. Geol. Rev.* **9**, 631.
Lisitzin, A. P. (1971a). *In* "The Micropalaeontology of Oceans (Funnell, B. M. and Riedel, W. R., eds.), p. 173. Cambridge University Press, London.
Lisitzin, A. P. (1971b). *In* "The Micropalaeontology of Oceans" (Funnell, B. M. and Riedel, W. R., eds.), p. 197. Cambridge University Press, London.
Lisitzin, A. P. (1972). *Soc. Econ. Paleontol. Mineralog., Spec. Publ.* **17**, 1.
Livingstone, D. A. (1963). *U.S. Geol. Survey Prof. Paper*, 440-G.
Loeblich, A. R. and Tappan, H. (1964). *In* "Treatise on Invertebrate Paleontology" (Moore, R. C., ed.), Pt. C, Protista 2, 2 Vols. Geol. Soc. Amer. and Univ. Kansas Press.
Loeblich, A. R. and Tappan, H. (1966). *Phycologia,* **5**. 81.
Loeblich, A. R. and Tappan, H. (1968). *J. Palaeontol.* **42**, 584.
Loeblich, A. R. and Tappan, H. (1969). *J. Palaeontol.* **43**, 568.
Loeblich, A. R. and Tappan, H. (1970a). *J. Palaeontol.* **44**, 558.
Loeblich, A. R. and Tappan, H. (1970b). *Phycologia*, **9**, 157.
Loeblich, A. R. and Tappan, H. (1971). *Phycologia*, **10**, 315.
Loeblich, A. R., Tappan, H., Beckmann, J. P., Bolli, H. M., Gallitelli, E. M. and Troelsen, J. C. (1957). *U.S. Nat. Museum Bull.* **215**, 1.

Loeblich, A. R., III, Loeblich, L. A., Tappan, H. and Loeblich, A. R., Jr. (1968). *Geol. Soc. Amer. Mem.* **106**, 1.
Lohman, K. E. (1942). *U.S. Geol. Survey, Prof. Pap.* **196**, 55.
Longhurst, A. R., Lorenzen, C. J. and Thomas, W. H. (1967). *Ecology,* **48** 190.
Lowrie, A. and Heezen, B. C. (1967). *Science,* **157**, 1552.
Lyakhin, Y. U. (1968). *Oceanology,* **8**, 44.
Lynts, G. W. (1971). *Micropaleontology,* **17**, 152.
MacKenzie, F. T. and Garrels, R. M. (1966). *Amer. J. Sci.* **264**, 507.
Malone, T. C. (1971). *Limnol. Oceanogr.* **16**, 633.
Mandra, Y. T. (1968). *Calif. Acad. Sci. Proc., Ser. 4,* **36**, 231.
Mandra, Y. T. (1969). *Antarctic J. U.S.* **4**, 172.
Manton, I. and Leedale, G. F. (1969). *J. Mar. Biol. Ass. U.K.* **49**, 1.
Marshall, H. G. (1968). *Limnol. Oceanogr.* **13**, 370.
Martini, E. (1971a). *Proc. 2nd Plankt. Conf., Rome 1970*, p. 85.
Martini, E. (1971b). *Init. Rep. Deep-Sea Drilling Proj.* **7**, Pt. 2, 1965.
Mazeika, P. A. (1967). *Limnol. Oceanogr.* **12**, 537.
McGowan, J. A. (1971). *In* "The Micropalaeontology of Oceans" (Funnell, B. M. and Riedel, W. R., eds.), p. 3. Cambridge University Press, London.
McIntyre, A. (1967). *Science, N.Y.* **158**, 1314.
McIntyre, A. and Bé, A. W. H. (1967). *Deep-Sea Res.* **14**, 561
McIntyre, A. and McIntyre, R. (1971). *In* "The Micropalaeontology of Oceans" (Funnell, B. M. and Riedel, W. R., eds.), p. 253, Cambridge University Press, London.
McIntyre, A., Bé, A. W. H. and Preikstas, R. (1967). *Progr. Oceanogr.* **4**, 3.
McIntyre, A., Bé, A. W. H. and Roche, M. B. (1970). *Trans. N.Y. Acad. Sci. Ser. II,* **32**, 720.
McIntyre, A., Ruddiman, W. F. and Jantzen, R. (1972). *Deep-Sea Res.* **19**, 61.
McKenzie, D. and Parker, R. L. (1967). *Nature, Lond.* **216**, 1276.
Meischner, K. D. (1964). *In* "Turbidites" (Bouma, A. H. and Brouwer, A., eds.), p. 156. Elsevier, Amsterdam.
Menard, H. W. (1964). "Marine Geology of the Pacific." McGraw-Hill, New York.
Menard, H. W. (1969). *Earth Planet. Sci. Lett.* **6**, 275.
Menard, H. W. (1972). *In* "The Nature of the Solid Earth" (Robertson, E. C., ed.), p. 440. McGraw-Hill, New York.
Millero, F. J. and Berner, R. A. (1972). *Geochim. Cosmochim. Acta,* **36**, 92.
Moore, T. C. (1969). *Bull. Geol. Soc. Amer.* **80**, 2103.
Moore, T. C. (1971). *Init. Rep. Deep-Sea Drilling Proj.* **8**, 727.
Moore, T. C., Heath, G. R. and Kowsman, R. O. (1973). *J. Geol. Soc.* **81**, 458.
Morgan, W. J. (1968). *J. Geophys. Res.* **73**, 1959.
Morse, J. W. and Berner, R. A. (1972). *Amer. J. Sci.* **272**, 840.
Mukhina, V. V. (1966). *Okeanologiia,* **6**, 807.
Mullin, M. M. (1966). *In* "Some Contemporary Studies in Marine Science" (Barnes, H., ed.), p. 545. Allen & Unwin, London.
Murray, J. (1889). *Scot. Geogr. Mag., Aug.* p. 23.
Murray, J. (1897). *Nat. Sci.* **11**, 17.
Murray, J. and Irvine, R. (1891). *Proc. Roy. Soc. Edin.* **18**, 229.
Murray, J. and Philippi, E. (1908). *Wissensch. Ergebn. Deut. Tiefsee. Exped. "Valdivia",* **10**, 77.

Murray, J. and Renard, A. F. (1891). Report on Deep-sea Deposits, H.M.S. *Challenger*, 1873–1876. Reprinted 1965 by Johnson Reprint, London.

Myers, E. H. (1943). *Stanford Univ. Publ., Biol. Sci.* **9**, 1.

Natland, M. and Kuenen, P. H. (1951). *Soc. Econ. Paleontol. Mineral., Spec. Publ.* **2**, 76.

Nesteroff, W. D. (1973). *Init. Rep. Deep-sea Drilling Proj.* **13**, 713.

Nigrini, C. (1967). *Bull. Scripps Inst. Oceanogr.* **11**, 1.

Nigrini, C. (1968). *Micropaleontology*, **14**, 51.

Nigrini, C. (1970). *Geol. Soc. Amer., Memoir* **126**, 139.

Nigrini, C. (1971). *In* "The Micropalaeontology of Oceans" (Funnell, B. M. and Riedel, W. R., eds.), p. 443. Cambridge University Press, London.

Nipkow, F. (1927). *Rev. Hydrol.* **4**, 71.

Nival, P. (1965). *Cah. Biol. Mar.* **6**, 67.

Oba, T. (1969). *Sci. Rep. Tohoku Univ. 2nd. Ser. (Geol.)*, **41**, 129.

Okada, H. (1970). *J. Geol. Soc. Japan*, **76**, 537.

Okada, H. and Honjo, S. (1970). *Pacific Geol.* **2**, 11.

Olausson, E. (1960a). *Rep. Swed. Deep-Sea Exped. 1947–1948*, **6**, 161.

Olausson, E. (1960b). *Rep. Swed. Deep-Sea Exped. 1947–1948*, **9**, 53.

Olausson, E. (1960c). *Rep. Swed. Deep-Sea Exped. 1947–1948*, **8**, 287.

Olausson, E. (1960d). *Rep. Swed. Deep-Sea Exped. 1947–1948*, **7**, 227.

Olausson, E. (1961a). *Medd. Oceanogr. Inst.* **Göteborg** 29 *(K. Vet. O. Vitterh. Samh. Handl. F. 6 Ser. B)*, **8**, 1.

Olausson, E. (1961b). *Rep. Swed. Deep-Sea Exped. 1947–1948*, **8**, 337.

Olausson, E. (1965). *Prog. Oceanogr.* **3**, 221.

Olausson, E. (1967). *Prog. Oceanogr.* **4**, 245.

Olausson, E. (1969). *Geol. Mijnbouw*, **48**, 349.

Olausson, E. (1971). *In* "The Micropalaeontology of Oceans" (Funnell, B. M. and Riedel, W. R., eds.), p. 375. Cambridge University Press, London.

Olausson, E. and Jonasson, U. C. (1969). *Geol. Fören. Stockholm Förhandl.* **91**, 185.

Olausson, E., Bilal Ul Haq, U. Z., Karlsson, G. B. and Olsson, I. U. (1971). *Geol. Foren. Stockholm Forhandl.* **93**, 51.

Olsson, R. K. (1971). *Trans. Gulf Coast Assoc. Geol. Soc.* **21**, 419.

Olsson, R. K. (1973). *J. Paleontology*, **47**, 327.

Olsson, I. U. and Eriksson, K. G. (1965). *Prog. Oceanogr.* **3**, 253.

Opdyke, N. D. (1972). *Rev. Geophys. Space Phys.* **10**, 213.

Opdyke, N. D. and Glass, B. P. (1969). *Deep-Sea Res.* **16**, 249.

Opdyke, N. D., Burckle, L. H. and Todd, A. (1974). *Earth. Planet. Sci. Lett.* **22**, 300.

Paasche, E. (1962). *Nature, Lond.* **193**, 1094.

Paasche, E. (1968a). *Ann. Rev. Microbiol.* **22**, 71.

Paasche, E. (1968b). *Limnol. Oceanogr.* **13**, 178.

Paasche, E. (1969). *Arch. Mikrobiol.* **67**, 199.

Parke, M. and Adams, I. (1960). *J. Mar. Biol. Ass. U.K.* **39**, 263.

Parker, F. L. (1958). *Rep. Swed. Deep-Sea Exped. 1947–1948*, **8**, 119.

Parker, F. L. (1962). *Micropaleontology*, **8**, 219.

Parker, F. L. (1967). *Bull. Amer. Paleontol.* **52**, 115.

Parker, F. L. (1971). *In* "The Micropalaeontology of Oceans" (Funnell, B. M. and Riedel, W. R., eds.), p. 289. Cambridge University Press, London.

Parker, F. L. and Berger, W. H. (1971). *Deep-Sea Res.* **18**, 73.

Pastouret, L. (1970). *Tethys*, **2**, 227.

Perch-Nielsen, K. (1971). *Proc. 2nd Plankt. Conf., Rome 1970*, pp. 939, 1348.
Peterson, M. N. A. (1966). *Science, N.Y.* **154**, 1542.
Peterson, M. N. A., Edgar, N. T., von der Borch, C., Cita, M. B., Gartner, S., Goll, R. and Nigrini, C. (1970). *Init. Rep. Deep-Sea Drilling Proj.* **2**, 1.
Petrushevskaya, M. G. (1966). *In* "Geokhimiya Kremnezema" (Strakhov, N. M., ed.), p. 219. Nauka, Moskow.
Petrushevskaya, M. G. (1967). *Issled. Fauny Morei, 4 (Rez. Biol. Issl. Sov. Antarkt. Exped. 1955–58)*, **3**, 5.
Petrushevskaya, M. G. (1971a). *In* "The Micropalaeontology of Oceans" (Funnell, B. M. and Riedel, W. R., eds.), p. 309, Cambridge University Press, London.
Petrushevskaya, M. G. (1971b). *In* "The Micropalaeontology of Oceans" (Funnell, B. M. and Riedel, W. R., eds.), p. 319. Cambridge University Press, London.
Petrushevskaya, M. G. (1971c). *In* "Radiolarri Mirovogo Okeana, *Issled Fauny Morei*", **9**, 5 (Bykhovskii, B. E., ed.), Izdat, Nauka, Leningrad.
Petrushevskaya, M. G. (1971d). *Proc. 2nd Plankt. Conf., Rome 1970*, p. 981.
Pflaumann, U. (1972). *"Meteor" Forsch. Ergebn.* C, **7**, 4.
Philippi, E. (1910). *Deut. Südpolar Expedition, 1901–1903*, **2**, 411.
Phleger, F. B. (1948). *Medd. Oceanogr. Inst. Göteborg*, **16**, 3.
Phleger, F. B. (1951a). *Soc. Econ. Paleontol. Mineral. Spec. Publ.* **2**, 66.
Phleger, F. B. (1951b). *Geol. Soc. Amer. Memoir* **46**, 1.
Phleger, F. B., Parker, F. L. and Peirson, J. F. (1953). *Rep. Swed. Deep-Sea Exped.* **7**, 1.
Piper, D. J. W. and Marshall, N. F. (1969). *J. Sed. Pet.* **39**, 601.
Poag, C. W. (1972). *Init. Rep. Deep-Sea Drilling Proj.* **11**, 483.
Poelchau, H. S. (1974). Ph.D. Thesis, University of California, San Diego, 165 pp. (available from Microfilms, Ann Arbor).
Pomeroy, L. R. and Johannes, R. E. (1966). *Deep-Sea Res.* **13**, 971.
Pomeroy, L. R. and Johannes, R. E. (1968). *Deep-Sea Res.* **15**, 381.
Postuma, J. A. (1971). "Manual of Planktonic Foraminifera." Elsevier, Amsterdam.
Prakash, A. (1971). *In* "Fertility of the Sea", Vol. 2 (Costlow, J. D., ed.), p. 351. Gordon & Breach, New York.
Pratje, O. (1939). *Wiss. Ergeb. Deutsch. Atlant. Exped. "Meteor", 1925–1927*, **3**, 57.
Pratje, O. (1951). *Deut. Hydrogr. Z.* **4**, 1.
Pytkowicz, R. M. (1968). *Oceanogr. Mar. Biol. Ann. Rev.* **6**, 83.
Ramsay, A. T. S. (1972). *Nature, Lond.* **236**, 67.
Ramsay, A. T. S. (ed.) (1975). "Oceanic Micropaleontology", (in press).
Raymont, J. E. (1963). "Plankton Productivity in the Oceans." Macmillan, New York.
Redfield, A. C., Ketchum, B. H. and Richards, F. A. (1963). *In* "The Sea", Vol. 2 (Hill, M. N., ed.), p. 26. Interscience, New York.
Reid, J. L. (1962). *Limnol. Oceanogr.* **7**, 287.
Reimann, B. E. F., Lewin, J. C. and Volcani, B. E. (1966). *J. Phycol.* **2**, 74.
Reiss, Z. and Luz, B. (1970). *Rev. Espan. Micropal.* **2**, 85.
Renz, G. W. (1973). "The Distribution and Ecology of Radiolaria in the Central Pacific—Plankton and Surface Sediments." Ph.D. Thesis, 251 pp. University of California, San Diego.
Revelle, R. R. (1944). *Carnegie Inst. of Washington Publ.* **556**, 1.
Revelle, R. R. and Fairbridge, R. W. (1957). *Geol. Soc. Amer. Mem.* **67**, 239.
Rhumbler, L. (1911). *Plankt. Exped. Humboldt-Stift (1909)*, **3** 1.
Riedel, W. R. (1952). *Medd. Oceanogr. Inst. Göteborg, Ser. B*, **6**, 1.

Riedel, W. R. (1957). *Rep. Swed. Deep-Sea Exped.* **6**, p. 61.
Riedel, W. R. (1958). *Rep B.A.N.Z. Antarct. Res. Exped., Ser. B*, **6**, (10), 217.
Riedel, W. R. (1959). *Soc. Econ. Paleontol. Mineral., Spec., Publ.* **7**, 80.
Riedel, W. R. (1963). *In* "The Sea", Vol. 3 (Hill, M. N., ed.), p. 866. Interscience, New York.
Riedel, W. R. (1967). *Science, N.Y.* **157**, 540.
Riedel, W. R. (1971a). *In* "The Micropalaeontology of Oceans" (Funnell, B. M. and Riedel, W. R., eds.), p. 567, Cambridge University Press, London.
Riedel, W. R. (1971b). *In* "The Micropalaeontology of Oceans" (Funnell, B. M. and Riedel, W. R., eds.), p. 649, Cambridge University Press, London.
Riedel, W. R. (1973). *Ann. Rev. Earth Planet. Sci.* **1**, 241.
Riedel, W. R. and Funnell, B. M. (1964). *Quart. Geol. Soc. Lond.* **120**, 305.
Riedel, W. R. and Sanfilippo, A. (1971). *Init. Rep. Deep-Sea Drilling Proj.* **7**, 1529.
Riedel, W. R. and Sanfilippo, A. (1976). *In* "Oceanic Micropalaeontology" (Ramsay, A. T. S., ed.), Academic Press, New York and London.
Riedel, W. R. and Saito, T. (eds.), (1975), "Marine Plankton and Sediment". Micropaleontology Press, New York.
Ronov, A. B. and Vernadskiy, V. I. (1968). *Dokl. Akad. Nauk SSSR*, **179** (3), 701.
Roth, P. H. (1970). *Eclog. Geol. Helv.* **63**, 799.
Roth, P. H., Mullin, M. M. and Berger, W. H. (1975). *Geol. Soc. Amer., Bull.* **86**, 1079.
Rowe, G. T. (1971). *In* "Fertility" (Costlow, J. D., ed.), p. 441. Gordon & Breach, New York.
Röttger, R. (1972). *Verh. Deut. Zool. Ges.* **65**, Jahresvers., p. 42.
Röttger, R. and Berger, W. H. (1972). *Mar. Biol.* **15**, 89.
Ruddiman, W. F. (1971). *Geol. Soc. Amer. Bull.* **82**, 283.
Ruddiman, W. F. (1972). *Geol. Soc. Amer. Bull.* **83**, 2039.
Ruddiman, W. F. and Glover, L. K. (1972). *Geol. Soc. Amer. Bull.* **83**, 2817.
Ruddiman, W. F. and Heezen, B. C. (1967). *Deep-Sea Res.* **14**, 801.
**Ruddiman, W. F. and McIntyre, A.** (1973). *Quaternary Res.* **3**, 117.
Rusnak, G. A. and Nesteroff, W. D. (1964). *In* "Papers in Marine Geology" (Miller, R. L., ed.), p. 488. Macmillan Co., New York.
Ryan, W. B. F., Hsü, K. J., Cita, M. B., Dumitrica, P., Lort, J. M., Maync, W., Nesteroff, W. D., Pautot, G., Stradner, H. and Wezel, F. C. (1973). *Init. Rep. Deep-Sea Drilling Proj.* **13**, 1201.
Ryther, J. H. (1963). *In* "The Sea" (Hill, M. N., ed.), Vol. 2, p. 347. Interscience, New York.
Ryther, J. H. (1969). *Science, N.Y.* **166**, 72.
Sachs, H. M. (1973). *Quaternary Res.* **3**, 89.
Saidova, Kh. M. (1968). *Dokl. Akad. Nauk, SSSR*, **182**, (2), 453.
Sarnthein, M. (1971). *"Meteor" Forsch. Ergebn. C*, **5**, 1.
Savin, S. M., and Douglas, R. G. (1973). *Geol. Soc. Amer. Bull.* **84**, 2327.
Schink, D. R., Fanning, K. A. and Pilson, M. E. (1974). *J. Geophys. Res.* **79**, 2243.
Scholle, P. A. (1971a). *Geol. Soc. Amer. Bull.* **82**, 629.
Scholle, P. A. (1971b). *J. Sediment Petrology*, **41**, 233.
Schott, W. (1935). *Dtsch. Atlant. Exped. "Meteor", 1925–1927*, **3**, 43.
Schott, W. (1939). *In* "Recent Marine Sediments" (Trask, P. D., ed.), p, 396. Amer. Assoc. Petr. Geol., Tulsa, Oklahoma.
Schrader, H. J. (1971a). *Science, N.Y.* **174**, 55.
Schrader, H. J. (1971b). *Proc. 2nd Plankt. Conf., Rome 1970*, p. 1139.
Schrader, H. J. (1971c). *Proc. 2nd Plankt. Conf., Rome 1970*, p. 1149.

Schrader, H. J. (1972a). *Nova Hedwigia, Beih.* **39**, 191.

Schrader, H. J. (1972b). *"Meteor" Forsch. Ergebn.* **D**, **10**, 58.

Schrader, H. J. (1972c). *"Meteor" Forsch. Ergebn.* **C**, **8**, 10.

Schutz, D. E. and Turekian, K. K. (1965). *Geochim. Cosmochim. Acta* **29**, 259.

Sclater, J. G., Anderson, R. N. and Bell, M. L. (1971). *J. Geophys. Res.* **76**, 7888.

Scott, G. H. (1966). *N.Z.J. Geol. Geophys.* **9**, 513.

Scott, G. H. (1967). *N.Z.J. Geol. Geophys.* **10**, 55.

Shackleton, N. J. (1967). *Nature, Lond.* **215**, 15.

Shackleton, N. J. Wiseman, J. D. H. and Buckley, H. A. (1973). *Nature, Lond.* **242**, 177.

Shackleton, N. J, and Opdyke, N. D. (1973). *Quaternary Res.* **3**, 39.

Seibold, E. (1970). *Geol. Rundschau*, **60**, 73.

Semina, G. I. (1963). *Tr. Inst. Okeanol. Akad. Nauk, SSSR*, **71**, 5.

Sheldon, R. W., Prakash, A. Sutcliffe, W. H. (1972). *Limnol. Oceanogr.* **17**, 327.

Shepard, F. P. (1963). "Submarine Geology, 2nd Edition." Harper and Row, New York.

Shor, G. G. (1959). *Deep-Sea Res.* **5**, 283.

Siever, R. (1968). *Earth Planet. Sci. Lett.* **5**, 106.

Sillén, L. G. (1961). *In* "Oceanography" (Sears, M., ed.), p. 549. *Amer. Assoc. Adv. Sci., Publ.* **67**.

Sliter, W. V. (1965). *J. Protozool.* **12**, 210.

Sliter, W. V. (1970). *Contr. Cushman Found. Foramin. Res.* **21**, 87.

Sliter, W. V. (1972). *Paleogeogr., Paleoclimatol., Paleocol.* **12**, 15.

Sliter, W. V., Bé, A. W. H. and Berger, W. H. (eds.) (1975). "Dissolution of Deep Sea Carbonates." *Cushman Found. Foram. Res., Spec. Publ.* **13**.

Smayda, T. J. (1969). *Limnol. Oceanogr.* **14**, 621.

Smayda, T. J. (1970). *Oceanogr. Mar. Biol. Ann. Rev.* **8**, 353.

Smayda, T. J. (1971). *Mar. Geol.* **11**, 105.

Smayda, T. J. and Boleyn, B. J. (1966). *Limnol. Oceanogr.* **11**, 35.

Smith, R. L. (1969). *Oceanogr. Mar. Biol., Ann. Rev.* **6**, 11.

Smith, S. V., Dygas, J. A. and Chave, K. E. (1968). *Mar. Geol.* **6**, 391.

Soutar, A. (1971). *In* 'The Micropalaeontology of Oceans" (Funnell, B. M. and Riedel, W. R., eds.), p. 223. Cambridge University Press, London.

Steemann Nielsen, E. (1963). *In* "The Sea" (Hill, M. N., ed.), Vol. 2, p. 129. Interscience, New York.

Strelkov, A. A., and Reshetnyak, V. V. (1971). *In* "Issled. Fauny Morei", Vol. 9 (17), p. 295. (Bykhovskij, B. E., ed.). Izdat, Nauk, Leningrad.

Strickland, J. D. H. (1965). *In* "Chemical Oceanography", Vol. 1 (Riley, J. P. and Skirrow, G., eds.), p. 477. Academic Press, London and New York.

Suess, E. (1970). *Geochim. Cosmochim. Acta,* **34**, 157.

Sverdrup, H. U., Johnson, M. W. and Fleming, R. H. (1942). "The Oceans—their Physics, Chemistry and General Biology." Prentice-Hall, Englewood Cliffs, New Jersey.

Tappan, H. and Loeblich, A. R. (1971a). *Geol. Soc. Amer. Spec. Paper* **127**, 247.

Tappan, H. and Loeblich, A. R. (1971b). "Evolution of the Oceanic Plankton." Presented at "Symposium on the History of the Ocean Basin", W. W. Hay, convener, Soc. Econ. Paleontol. Mineral., Houston, April, 1971. Unpublished manuscript.

Tappan, H. and Loeblich, A. R. (1973). *Earth Sci. Rev.* **9**, 207.

Theyer, F. and Hammond, S. R. (1974). *Geology*, **2**, 487.

Thiede, J. (1971a). *Deep-Sea Res.* **18**, 823.
Thiede, J. (1971b). *"Meteor" Forsch. Ergebn. Reihe* C, **7**, 15.
Tolderlund, D. S., and Bé, A. W. H. (1971). *Micropaleontology,* **17**, 297.
Towe, K. M. (1971). *Proc. 2nd Plankt. Conf., Rome 1970,* p. 1213.
Towe, K. M. and Cifelli, R. (1967). *J. Paleontol.* **41**, 742.
Tundisi, J. D. (1971). *In* "Fertility of the Sea" (Costlow, J. D., ed.), Vol. 2, p. 603. Gordon & Breach, New York.
Turekian, K. K. (1965). *In* "Chemical Oceanography" (Riley, J. P. and Skirrow, G., eds.), Vol. 2, p. 81. Academic Press, London and New York.
Turekian, K. K., ed. (1971). "The Late Cenozoic Glacial Ages." Yale University Press, New Haven, Conn.
Tynan, E. J. (1971). *Proc. 2nd Plankt. Conf., Rome 1970,* p. 1225.
Ushakova, M. G. (1971). *In* "The Micropalaeontology of Oceans" (Funnell, B. M. and Riedel, W. R., eds.), p. 245. Cambridge University Press, London.
Van Andel, T. H. and Heath, G. R. (1973). *Init. Rep. Deep-Sea Drilling Proj.* **16**, 937.
Van Andel, T. H. and Komar, P. D. (1969). *Geol. Soc. Amer. Bull.* **80**, 1163.
Van Andel, T. H., Heath, G. R. and Moore, T. C. (1975). *Geol. Soc. Amer., Memoir,* **143**, 1.
van Donk, J. (1970). "The Oxygen Isotope Record in Deep-Sea Sediments." Ph.D. Thesis, Columbia University, New York, 228 pp.
Van Falkenburg, S. D. and Norris, R. E. (1970). *J. Phycol.* **6**, 48.
van Hinte, J. E. (1972). *Proc. Koninkl. Nederl. Akad. Wetens,* 228 pp. **Ser. B, 75**, 1.
Venrick, E. L. (1969). "The distribution and Ecology of Oceanic Diatoms in the North Pacific." Ph.D. thesis, Univ. Calif., San Diego, 694 pp. Univ. Microfilms, Ann Arbor, Mich.
Venrick, E. L. (1971). *Ecology,* **52**, 614.
Venrick, E. L., McGowan, J. A. and Mantyla, A. W. (1973). *Fish. Bull.* **71**, 41.
Vine, F. J., and Matthews, P. M. (1963). *Nature, Lond.* **199**, 947.
von der Borch, C. C., Galehouse, J. and Nesterof, W. D. (1971). *Init. Rep. Deep-Sea Drilling Proj.* **8**, 819.
von Stackelberg, U. (1972). *"Meteor" Forsch. Ergebn.* C, **9**, 1.
von Stosch, H. A. (1950). *Nature, Lond.* **165**, 531.
von Stosch, H. A. (1951). *Arch. Mikrobiol.* **16**, 101.
von Stosch, H. A. (1958). *Arch. Mikrobiol.* **31**, 274.
von Stosch, H. A. and Drebes, G. (1964). *Helg. Wiss. Meeresunters.* **11**, 209.
Walsh, J. J. (1969). *Limnol. Oceanogr.* **14**, 86.
Wangersky, P. J. (1969). *Limnol. Oceanogr.* **14**, 929.
Wangersky, P. J. (1972). *Limnol. Oceanogr.* **17**, 1.
Watabe, N. and Wilbur, K. M. (1966). *Limnol. Oceanogr.* **11**, 567.
Watkins, N. D. (1972). *Geol. Soc. Amer. Bull.* **83**, 551.
Watt, W. D. (1971). *Deep-Sea Res.* **18**, 329.
Wattenberg, H. (1935). *Ann. Hydr. Marit. Meteorol.* **63**, 387.
Wehmiller, J. and Hare, P. E. (1971). *Science,* **173**, 907.
Werner, D. (1971a). *Arch. Mikrobiol.* **80**, 43.
Werner, D. (1971b). *Arch. Mikrobiol.* **80**, 115 and 134.
Wiebe, P. H. (1970) *Limnol. Oceanogr.* **15**, 205.
Wilbur, K. M. and Watabe, N. (1963). *Ann. New York Acad. Sci.* **109**, 82.
Wiles, W. W. (1967). *Prog. Oceanogr.* **4**, 153.
Wilson, J. T. (1965). *Nature, Lond.* **207**, 343.
Winterer, E. L. (1973). *Bull. Amer. Assoc. Petr. Geol.* **57**, 265.

Wise, S. W. and Hsü, K. J., 1971. *Eclog. Geol. Helv.* **64**, 273.

Wise, S. W. and Kelts, K. R., 1972. *Trans. Gulf Coast Assoc. Geol. Soc.* **22**, 177.

Wollast, R. (1974). *In* "The Sea", Vol. 5 (Goldberg, E. D., ed.), Interscience, New York 359.

Wollin, G., Ericson, D. B. and Ewing, M. (1971). *In* "The Late Cenozoic Glacial Ages" (Turekian, K. K., ed.), p. 199. Yale Univ. Press, New Haven, Conn.

Woodring, W. P. (1966). *Proc. Amer. Philos. Soc.* **110**, 425.

Wooster, W. S. (ed.) (1970). "Scientific Exploration of the South Pacific". National Acad. Sci., Washington, D.C.

Wooster, W. S. and Reid, J. L. (1963). *In* "The Sea", Vol. 3 (Hill, M. N., ed.), p. 253. Interscience, New York.

Wornardt, W. W. (1967). *Calif. Acad. Occas. Pap.* **63**, 1.

Wüst, G. (1935). *Dtsch. Atlant. Exped. "Meteor", 1925–1927*, **6**, 1.

Zeitzschel, B. (1965). *Kiel. Meeresforsch.* **21**, 55.

Zobel, B. (1968). *Geol. Jahrb.* **85**, 97.

# Subject Index

*( Numbers in bold type indicate the page on which a subject is treated most fully.)*

## A

Abyssal plains, 7
Acantharia, 283
*Actinocyclus* , 322
*Actinocyclus ingens,* 363
*Actinoptyclus,* 322
Active ridges, 153
Activities of cations in riverwater, 145
Aeolian dust, 153
Aeolian transport, 47
Aerosols, 9
Alaska, Gulf of, 41
Alkalinity, flux of, 308
Allodapic limestones, 354
Alloisoleucine, 364
Allolytic limestones, 355, 356
Aluminium oxide, solubility in water, 88
Aluminium phosphate, 204
Amazon River, 10, 11, 12, 120
Amphiboles, 130, 132
Amphiboles, weathering of, 93
Anaerobic deposits, 306
Analcime, 140, 177, 182
Anatase, 87, 93, 132, 236
Andalusite, 132
Andaman Basin, 51
Andaman Sea, 51
Andros Island, 201
Anhydrite, 202
Anorthoclase, 130
Antarctic Bottom water, 24, 44, 62
Antarctic Convergence, 294
Antarctic supply of sediment from, 11, 84
Antiestuarine circulation, 357, 358
Apatite, 140
Apatites, formation of, 202–204
Arabian Sea, chlorite in, 111, 114
   kaolinite in, 107, 110
   illite in, 118, 121
   montmorillonite in, 116
   quartz in, 128

Aragonite, 12, 140, 193
Aragonite compensation depth, 270
Aragonite oolite, 139
Aragonite precipitation of, 193, **200, 201**
   *202*
Aragonite, solubility of, 89
Archaemonads, 277
Arsenic, 190
Ash shards in sediments, 353, 354
*Asterionella,* 275
*Asterolampa,* 322
*Asteromphalus,* 322
Atacama Desert, 128
Atlantic, bottom currents in, 32–35
   carbonate sediments in, 305, 306
   CCD in, 310, 311
   chlorite in, 111, 112
   deep currents in, 32–55
   dissolution patterns in, 355
   distribution of coccoliths in, 319
   factors preventing silica deposition in,
     325, 326
   feldspars in, 128, 129
   ice conditions in, 36, 37
   ice-rafting in, 35–37
   illite in, 118, 120
Atlantic, kaolinite in, 107, 108
   lysocline in, 316, 317
   manganese nodules in, 222, 223
     228, 234, 235, 238, 242, 247, 252
   mixed layer clays in, 122, 123
   montmorillonite in, 114, 116
   near shore sediments of, 27, 28
   North, 6
   palaeo CCD in, 371
   palygorskite in, 123
   pelagic sedimentation in, 37, 38, 40
   quartz in, 124, 126, 127
Atlantic sediments of, 25–40
   map of, 26
   processes in, 27–40

389